Computational Thinking in the STEM Disciplines

Myint Swe Khine
Editor

Computational Thinking in the STEM Disciplines

Foundations and Research Highlights

Springer

Editor
Myint Swe Khine
Emirates College for Advanced Education
Abu Dhabi, United Arab Emirates

Curtin University
Bentley, Australia

ISBN 978-3-030-06683-3 ISBN 978-3-319-93566-9 (eBook)
https://doi.org/10.1007/978-3-319-93566-9

Printed on acid-free paper

This Springer imprint is published by the registered company Springer Nature Switzerland AG
The registered company address is: Gewerbestrasse 11, 6330 Cham, Switzerland

Foreword

The future is already here. It's just unevenly distributed — William Gibson

A computer terminal is not some clunky old television with a typewriter in front of it. It is an interface where the mind and body can connect with the universe and move bits of it about — Douglas Adams

In 1943, Thomas Watson, then president of International Business Machines (aka IBM) made one of the most inaccurate predictions ever. Speaking of the future of the computer business he said, "I think there is a world market for maybe five computers." Clearly, he missed the mark by a few billion in sheer numbers. But more importantly he missed the mark in seeing the dramatic shift in human life and culture that would occur. Computers are everywhere today. They drive our work, our society, and our lives in almost every way imaginable. They pervade our reality. So much so that if computers and digital technology suddenly vanished or shut down, global communication, information, and society would come to a grinding halt.

From giant supercomputing machines that can crunch quadrillions of floating-point operations per second (FLOPS), a measure of computer performance to tiny chips that are embedded in everyday objects, we exist in what is often described as the *Internet of things*—or the network of devices that connect our lives, spaces, possessions, communications, and more. More people have access to cell phones today than to safe drinking water (Casey, 2016). Computers have become integral to how we function. Our work lives are often driven by digital communications and tasks. People today buy every good or service imaginable online, from luxury items to basic necessities like food. They meet significant others online or use the Internet to connect with friends and family at near and far distances. Digital technologies are integral to how we work, think, live, and run our world. Though we can speak of the practicalities of such technology, what we are seeing is a fundamental transformation of our engagement with the world as humans.

This global shift was best captured in a talk delivered by the author Douglas Adams at a conference at Magdalene College, Cambridge, in 1998. In a wide-ranging, extempore speech, Adams covered a range of topics, in his inimitably funny yet insightful manner, including the cultural and intellectual history of the human civilization. One of the insights he shared with the audience how human technological history consists of, what he referred to as, *the four ages of sand.* Specifically, Adams seeks to describe how technological change has changed and broadened our understanding of ourselves and the world. Let us quote directly from Adams (1998), as he describes the first two ages, primarily because paraphrasing him risks losing the immediacy and power of his distinctive voice and the precise meaning in his ideas:

> *The first age of sand was the age of the telescope.* From sand we make glass, from glass we make lenses and from lenses we make telescopes. When the great early astronomers, Copernicus, Galileo, and others turned their telescopes on the heavens and discovered that the Universe was an astonishingly different place than we expected and that, far from the world being most of the Universe, with just a few little bright lights going around it, it turned out - and this took a long, long, long time to sink in - that it is just one tiny little speck going round a little nuclear fireball, which is one of millions and millions and millions that make up this particular galaxy and our galaxy is one of millions or billions that make up the Universe and that we are also faced with the possibility that there may be billions of universes, that applied a little bit of a corrective to the perspective that the Universe was ours.
>
> *The next age of sand is the microscopic one.* We put glass lenses into microscopes and started to look down at the microscopic view of the Universe. Then we began to understand that when we get down to the sub-atomic level, the solid world we live in also consists, again rather worryingly, of almost nothing and that wherever we do find something it turns out not to be actually something, but only the probability that there may be something there.

The third age of sand, according to Adams, is the discovery that sand can be used to make the silicon chip. The chip led to the digital computer and then suddenly:

> ...what opens up to us is a Universe not of fundamental particles and fundamental forces, but of the things that were missing in that picture that told us how they work; what the silicon chip revealed to us was the process. The silicon chip enables us to do mathematics tremendously fast, to model the very simple processes that are analogous to life in terms of their simplicity; iteration, looping, branching, the feedback loop which lies at the heart of everything you do on a computer.

The fourth and final age of sand according to Adams is that sand can be used to make fiber-optic cables and thus leads to a new communication platform, the Internet. This is a new form of communication (many-to-many), in contrast to previous technologies such as the telephone (which was one-to-one), or radio and TV (which were one-to-many). It is communication between people that forms the fourth age of sand.

The idea here is that each of these ages of sand changes how we, as a species, locate ourselves in relation to the world. The repercussions of the first two ages of sand are felt even today; their contribution to our knowledge of the world and our place in it is not the stuff of school textbooks. Yet we are actually now living within the next two ages (the third and fourth ages of sand), and this has significant implications for our economy, our lives, and our relationships with each other and our world. The first two ages of sand have power in human knowledge and history but are either felt indirectly or seen as knowledge removed from our everyday lives or experiences. But the third and fourth ages of sand have immediately obvious and tangible affects on our lives and experiences. Without them, our everyday acts of talking or texting with distant friends and family, buying our "stuff" online, sending work emails, or checking the immediate news would be impossible. These ages of sand pervade our experience of the world and each other. In this, they are not only ages of technological development—they become the filter or lens through which we operate and function. They drive our view and experience the world. Thus, they are woven into the fabric of our immediate lives, needs, and beliefs.

Clearly something as world-view changing as the third and fourth ages of sand require new intellectual tools to grasp, understand, control, and experience these devices and technology. It is important to remember that one of the challenges in the world of the third and fourth ages of sand is in their virtual nature. Such technologies are often *black boxes*, in that their inner workings are relatively impenetrable, unless one has explicit technical knowledge to understand them. This requires that we develop intellectual tools or ways of thinking that allow us to understand and engage with such technologies in creative, thoughtful, and appropriate ways. Along with the development of these intellectual tools, we need pedagogical tools to prepare the next generation of learners to work, live, play, engage, and design technologies appropriately, ethically, and thoughtfully. This goes beyond merely learning to program things. This means helping teachers and learners to develop a mind-set, skills, knowledge, and dispositions that allow for creative learning with technologies.

This is where computational thinking comes to the forefront. Broadly, computational thinking is thought processes for structuring and formulating problems that can be effectively carried out by computational devices, or other information processing agents. Given the breadth and depth with which technology touches our world, computational thinking cuts across disciplines and must be of interest not just to computer scientists but to every student and learner. The question of how to bring computational thinking to teachers and learners in creative and interdisciplinary ways is challenging and complex. Given the accelerating rate of technological change, with its dramatic impact on our lives and world, this is also an urgent problem, and one demanding of scholarly attention.

That is what makes this book and the range of ideas in this book so timely and important. The editor of this collection has compiled articles by top thinkers, scholars, and researchers in this emerging, growing field. The work presented in this volume touches on a range of aspects of computational thinking and helps us to better define its range, scope, and potential of learning in this new domain. We believe that this line of work can have significant impact on the future of this area, by providing ideas, guidelines, and solutions for the next generation of learners who will live, learn, and engage with these new ages of sand.

Mary Lou Fulton Teachers College
Arizona State University
Tempe, AZ, USA

Punya Mishra
Danah Henriksen

References

Adams, D. (1988). *Is there an Artificial God?* Speech at Digital Biota 2 Conference. Cambridge: Magdelene College. Retrieved from http://www.biota.org/people/douglasadams/

Casey, V. (2016, October 16). Why can people get access to mobile phones, and not safe water? Retrieved from: http://www.wateraid.org/news/blogs/2016/october/why-can-people-get-access-to-mobile-phones-and-not-safe-water

Contents

About the Authors

Charoula Angeli is Professor of Instructional Technology at the University of Cyprus. She has undergraduate and graduate studies at Indiana University-Bloomington, USA (B.S. in Computer Science, M.S. in Computer Science, and Ph.D. in Instructional Systems Technology, 1999). She also pursued a postdoctoral fellowship in LRDC (Learning Research and Development Centre) at the University of Pittsburgh, USA. Her research interests include Technological Pedagogical Content Knowledge, Computational Thinking, Instructional Design, and Complex-Problem Solving with Technology. She has published extensively in referred journals and presented her work worldwide. She also participated in several research projects funded by different agencies.

Mehmet Aydeniz is an associate professor of Science Education at the University of Tennessee, Knoxville. He received both his master and doctorate degrees in Science Education from Florida State University. His research focuses on students' engagement with scientific practices such as argumentation, engineering design and computational thinking more recently and teacher learning.

Ana Calderon is a lecturer in the Department of Computing, Cardiff Metropolitan University, Wales, United Kingdom, and editor-in-chief of the AMI journal. Completed research projects include the area of formal semantics of programming languages, relating games models to coherence space models, resulting in an in-depth analysis of composition in Game Semantics and analysing methodologies to rigorously investigate computational thinking in higher education.

Amanda Dickes is a postdoctoral research fellow on the EcoMOD project at Harvard University. Her research focuses on the design of learning environments that integrate computational modeling with other forms of scientific modeling in an attempt to understand how computation and computational modeling can become the "language" of practice in the elementary science and math classroom. Before joining Harvard, she completed her Ph.D. in the Learning Sciences at Vanderbilt University where she was a member of the *Mind, Matter and Media Lab*.

Amy Farris is an Assistant Professor in the Department of Curriculum and Instruction at Penn State University. She investigates the intersections of scientific modeling and computational thinking in elementary and middle school classrooms. Her work also addresses how learners' experience of computational modeling integrates ideas and practices both in STEM disciplines and along personal and out-of-school narratives. Farris received her Ph.D. in Learning, Teaching, and Diversity at Vanderbilt University.

Georgios Fessakis holds a BSc in Informatics, an MSc in Advanced Informatics Systems from National and Kapodistrian University of Athens/Department of Informatics and Telecommunications and a PhD in Informatics Didactics from University of the Aegean. He has many years of extensive experience as a researcher and as Computer Science teacher in secondary public schools (1999–2007). Georgios is teaching ICT, Computer Science, Mathematics Education and Learning Technology–related courses, since 2004, at the University of the Aegean, where he currently serves as associate professor. During all these years Georgios has published several articles in international and Greek journals and conferences. He also has participated and directed national and international research and development projects. His main research interests include ICT design and development for learning, Computer Science, Mathematics, Technology and Science Education, Intelligent systems and CSCL.

Francisco José García-Peñalvo received his bachelor's degree in Computing from the University of Salamanca (Spain), and the University of Valladolid (Spain), and his PhD degree from the University of Salamanca, where he is currently the director of the Research Group in Interaction and e-Learning. His main research interests focus on e-Learning, computers and education, adaptive systems, web engineering, semantic web and software reuse. He has led and participated in over 50 research and innovation projects. He has published over 200 articles in international journals and conferences. He has been a guest editor of several special issues of international journals (*Online Information Review*, *Computers in Human Behavior* and *Interactive Learning Environments*). He is the editor-in-chief of the *Education in the Knowledge Society* magazine and the *Journal of Information Technology Research.* He coordinates the Doctoral Programme in Education at the Society of Knowledge of the University of Salamanca.

Jon Good is a doctoral candidate in the Educational Psychology Educational Technology programme at Michigan State University. His research is focused on computational thinking, computer science education and creativity.

Peter Hubwieser has been teaching Mathematics, Physics and Computer Science at Bavarian Gymnasiums for 15 years. In 1995 he received his Dr.rer. nat. in Theoretical Physics at the Ludwig-Maximilians-Universität München. From 1994 to 2002 he has been delegated to the Faculty of Informatics of the Technical University of Munich on behalf of the Bavarian Ministry of Education, in charge for the implementation of new courses of studies for teacher education in Informatics and a compulsory subject of Computer Science at Bavarian secondary schools. Since June 2002 he is working as an associate professor at the Faculty of Informatics of the Technical University of Munich. From 2002 to 2015 he was additionally a visiting professor at the Alpen-Adria – University of Klagenfurt and from 2007 to 2008 at the University of Salzburg. Since 2009 he holds the position of the information officer of the TUM School of Education. In 2015 he took over the position of the scientific director of the Schülerforschungszentrum Berchtesgaden additionally to his professorship.

Kamini Jaipal-Jamani is a Professor of Education at Brock University in Canada. She obtained her MEd at Western University and her Ph.D. at UBC. Her research and writing focuses on science teaching and learning, technology integration, and teacher professional development. She has conducted research on relationships between language and science, implementation of video games and blogs in K-12 science instruction, gamification, and the development of TPACK knowledge in teacher education. Her current program of research involves an examination of the use of robotics to promote science learning and computational thinking in K-12 and teacher education contexts.

Filiz Kalelioğlu is an assistant professor of Computer Education and Instructional Technology at the Education Faculty, Baskent University. She graduated with a PhD in Computer Education and Instructional Technology from Ankara University in 2011. Her academic interest areas are e-learning, social media in education, instructional design, technology integration and computer science education. She serves as a reviewer for several journals in the field of educational technology. She has published many national and international articles and book chapters.

Myint Swe Khine is Professor of Education at the Emirates College for Advanced Education in the United Arab Emirates. He is also an Adjunct Professor at Curtin University in Australia. He received his master's degrees from the University of Southern California, Los Angeles, USA, and University of Surrey, Guildford, UK; and Doctor of Education from Curtin University, Australia. He worked in Nanyang Technological University, Singapore, and Murdoch University, Australia, before taking up position in the United Arab Emirates. He has published widely in international referred journals and edited several books. His recent publications include *Robotics in STEM Education: Redesigning the Learning Experience* (Springer, Switzerland, 2017).

Hae Young Kim holds a PhD degree in Instructional Systems and Learning Technologies and a Master degree in Educational Measurement and Statistics from Florida

State University, USA. In 2016, she joined Tallassee Community College and Department of Juvenile Justice as an instructional designer and psychometrician. She has developed e-learning courses for Florida Juvenile Officer Academy and overviewed the curriculums and certification exam items. She maintains the current certification exams and develops new test items along with subject-matter experts. Her research interests are pre-service teacher education, assessments and instructional strategies to improve student learning and performances. Currently, she is working on designing an assessment system of computational thinking in K-12 settings.

Soohwan Kim is an assistant professor at Chongshin University in Korea. He holds a PhD degree in Computer Science Education from Korea University in Korea. He taught for 15 years in elementary school. He is teaching computational thinking and programming for non-computer science major students.

From 2015, he joined SW education leading schools in Korea as a consultant and developed a textbook, some online lectures, many lesson plans and some teacher training courses for K-12 computational thinking education. He is managing teacher training course and teacher workshops about CT education. His research interests are educational equality in public education, teacher training, learner assessments and instructional strategies to improve computational thinking. Currently, he is working on designing an assessment system of computational thinking in K-12 settings.

Joyce Hwee Ling Koh is an associate professor at the Learning Sciences Academic Group of the National Institute of Education of Nanyang Technological University, Singapore. Her research interests are in the areas of design thinking, technological pedagogical content knowledge and online facilitation. She has published articles in many SSCI journals including *Computers and Education*, *Instructional Science*, and *Computers in Human Behavior*.

Vasilis Komis is currently an Associate Professor in the Department of Educational Sciences and Early Childhood Education of the University of Patras. His publications and research interests concern the teaching of computer science, the pupils' representations in the new information technologies and the representations formed during the use of computers in class room, the integration of computers in education, the conception and the development of educational software.

Johannes Krugel studied Computer Science with minor Psychology at the Free University of Berlin. From 2009 to 2016 he was teaching and research assistant at the Technical University of Munich. In 2016 he received his Dr.rer. nat. in Theoretical Computer Science at the Technical University of Munich with his dissertation on efficient algorithms and data structures for strings. For several years he was responsible for the education, support and supervision of the teaching assistants at the Faculty for Computer Science. Since 2016 he is postdoctoral researcher in the working group for Didactics in Computer Science. He designed the MOOC and coordinated the development and implementation.

Sze Yee Lye is the head of department (ICT) in Teck Whye Primary School, Singapore. She is also currently pursuing her doctorate part time at Learning Sciences Academic Group of the National Institute of Education of Nanyang Technological University, Singapore. Her research interests are in the areas of educational technology and programming. A reflective practitioner, she has presented in international conferences and has published articles in peer-reviewed journals such as *Computers in Human Behavior* and *Journal of Educational Technology & Society*.

Christiane Maday holds a B.Sc. in Construction Engineering as well as a Master of Engineering Education and worked as a student assistant in the TACCLE 3 coding project (2013–2015) at Karlsruhe Institute of Technology, Germany.

Elisavet Mavroudi holds a BSc in Informatics from the National and Kapodistrian University of Athens/Department of Informatics and Telecommunications (1991) and an MEd in "Modeling Design and Development of Educational Units" from the University of the Aegean (2015). She has been working as a Computer Science teacher in secondary public schools since 1997. She has also been working as a researcher in educational technology for almost 10 years, and she has published a couple of articles in international and Greek journals and conferences. Her main research interests include the Didactics of Informatics, ICT applications in Education and e-learning.

Greg Michaelson is professor of Computer Science at Heriot-Watt University in Edinburgh. His broad research area is in formally motivated programming language design, implementation and analysis, especially for multi-processor systems. He has enjoyed sustained funding for his research and has published widely. Dr. Michaelson has taught programming for 40 years, and has published text books on functional programming with lambda calculus and on Standard ML. Most recently, he has been exploring systematic computational thinking as a practice of problem solving.

Stavroula Prantsoudi holds a BSc degree in Informatics (Aristotle University of Thessaloniki/Department of Computer Science), an MSc degree in Information Systems (University of Macedonia) and is currently studying for an MEd degree in Interdisciplinary Education (University of the Aegean/Department of Preschool Education). She serves as a Computer Science teacher in secondary public schools (2003 - today) and has taught ICT subjects at all levels and types of Greek schools, adults training structures and vulnerable social groups. Since 2004 she has attended numerous training programmes in the areas of teaching and ICT and has implemented educational programmes in the areas of her interest.

Daniela Reimann is a researcher at Karlsruhe Institute of Technology's Institute for Vocational and General Pedagogy. Her research interest is in digital media in education and creative processes. She is currently working in the project

"Prospective further training for industry 4.0" funded by the Ministry of Economics in Baden-Württemberg, Germany.

Phil Sands is a doctoral candidate in the Educational Psychology Educational Technology program at Michigan State University. His research is focused on computer science and education.

John Sanford holds a Doctor of Engineering degree from Yale. He is Professor Emeritus of Information Systems at Philadelphia University where he pioneered and introduced a computer for student use in the 1970s and served as director of computing at the university for 10 years as the computing centre grew to a university-wide service. He introduced most of the computer programming and information systems core courses to the School of Business and has published in the area of computer education, computational thinking, fuzzy logic, and data mining. Earlier experience included systems engineering with Autonetics (later North American Rockwell) in California, and engineering management with General Electric Re-entry Systems Division in the 1960s.

Pratim Sengupta is the Research Chair of STEM Education and a professor of Learning Sciences at the University of Calgary, Canada. He is a recipient of the CAREER Award (2012) from the US National Science Foundation, and directs the Mind, Matter & Media Lab. His research programme has led to the development of ViMAP, an open-source programming language for the K-12 science and math classrooms, as well as Open Computing platforms and immersive computing experiences in public places. The work reported here was partly conducted at Vanderbilt University, where Dr. Sengupta was a faculty member and programme chair of Learning Sciences until 2015.

Phillip A. Towndrow is a senior research scientist in the Centre for Research in Pedagogy and Practice, National Institute of Education, Nanyang Technological University, Singapore. His current research and writing interests include the use of new media in multimodal communication, learning task design, classroom interactions, and teachers' professional development and learning. Phillip's work appears in various books and journals including the *Journal of Curriculum Studies*, *Teaching and Teacher Education* and the *Journal of Literacy Research*. He is currently a managing editor for the *Asia Pacific Journal of Education* (Routledge).

Michael Vallance is a professor in the Department of Media Architecture at Future University Hakodate, Japan. He has been involved in educational technology design, implementation and research for over 20 years, working closely with higher education institutes, schools and media companies in the UK, Singapore, Malaysia and Japan. His 3D virtual world design and tele-robotics research has been funded by the UK Prime Minister's Initiative (PMI2) and the Japan Advanced Institute of Science and Technology (JAIST). In 2012 he was awarded second prize in the Distributed Learning category by the United States Department of Defense for his research in virtual collaboration.

Aman Yadav is an associate professor in Educational Psychology and Educational Technology at Michigan State University. Dr. Yadav's research focuses on computational thinking, computer science education and problem-based learning. His work has been published in a number of leading journals, including *ACM Transactions on Computing Education*, *Journal of Research in Science Teaching*, *Journal of Engineering Education*, and *Communications of the ACM*.

Part I
Foundations in Computational Thinking

Chapter 1
Strategies for Developing Computational Thinking

Myint Swe Khine

1.1 Introduction

The term computational thinking came into wide use as a popular term during the 1980s that refers to a collection of computational ideas that people in computing disciplines acquire through their work in designing programs, software and computations performed by the computer hardware (Tedre and Denning 2016). It was envisioned that computational thinking will be a fundamental skill that complements to reading, writing and arithmetic for everyone and represents a universally applicable aptitude. Denning (2017) paraphrased the definition as "Computational thinking is the thought processes involved in formulating problems so that their solutions are represented as computational steps and algorithms that can be effectively carried out by an information-processing agent". Educators believed that assimilating computational thinking at a young age will assist them to enhance problem-solving skills, improve logical reasoning and advance analytical ability – key attributes to succeed in the twenty-first century (Riley and Hunt 2014).

In the age of rapid development in information and communication technologies, dramatically changing economic landscape and escalating competition, educational planners are focusing their relentless effort in equipping the young generation with real-world skills ready for the demand and challenges of the future. It is commonly believed that computational thinking will play a pivotal and dominant role in this endeavour. Wide-ranging researches on and application of computational thinking in education have emerged in the last 10 years. This book documents some of those attempts in conducting systematic, prodigious and multidisciplinary research in computational thinking and presents their findings and accomplishments. This book is divided in three parts. While the first part of the book presents foundations

M. S. Khine (✉)
Emirates College for Advanced Education, Abu Dhabi, United Arab Emirates

Curtin University, Bentley, Australia

M. S. Khine (ed.), *Computational Thinking in the STEM Disciplines*,
https://doi.org/10.1007/978-3-319-93566-9_1

in computational thinking, the second part covers computational thinking and teacher education. The last part of the book deals with computational thinking in schools and higher education.

1.2 Foundations in Computational Thinking

The first part of the book consists of chapters that discussed about the foundations of computational thinking. Kalelioglu in Chap. 2 presented a systematic literature review of studies on computational thinking. The purpose of his study was to investigate the research trends in this area by examining the databases and digital libraries using the keywords. A total of 69 publications were found between 2013 and 2017 period in the literature. The studies were conducted with the sample population ranging from kindergarten students to teachers and instructors to graduate students. It was also found that the term computational thinking has also been defined in different ways, and most works were case studies. This study has provided an insight into the characteristics of the literature on computational thinking and serves as a guide for future research in this area.

In Chap. 3, Michaelson described the teaching of programming from Papert's Logo to microworlds and Objects First. He noted that Objects First as an object-oriented programming language bridges microworlds and systematic programming. However he argued that both languages provide superficial programming skills and did not stress on problem-solving. In this chapter the author explained finding variables and expression and computational patterns with examples. The author suggested that to employ pedagogy based on strong computational thinking by teaching students how to systematically analyse the program, define the essential patterns in data and form algorithms and finally develop programs.

Pratim Sengupta and his colleagues (Chap. 4) argued the viewing coding and computational thinking as mastery to viewing them as experience. The group has conducted a series of research in K–12 classrooms and partnerships with teachers in the United States. They suggested some pedagogical guidelines for sustaining computing and computational thinking through computational modelling in the Science, Technology, Engineering and Mathematics (STEM) discipline. Among them were reframing programming and coding as "modelling" and using both visual- and text-based programming languages for curricular integration. They urged that these guidelines would help teachers and students adopt computing and computational thinking as a "language" of STEM.

In Chap. 5, Krugel and Hubwieser presented the development and delivery of a multipurpose course on computational thinking with objects-first approach in the German state of Bavaria. The chapter detailed the rationales, objectives and framework in designing the online course called Learning Object-Oriented Programming (LOOP). This course was meant for the freshmen of the university and introduces computational thinking and object-oriented concepts before the programming skill is taught. The course materials comprise of videos, quizzes and interactive exercises. The LOOP consisted of five broadly defined sections that include (i) objective-oriented modelling, (ii) algorithms, (iii) classes in programming languages, (iv) object-oriented programming and (v) associations and references. The online course is offered to any interested students on an online platform. After 5 weeks duration, the authors collected feedback from the participants with the aim of improving the course. The comments from the participants were obtained through questionnaire and online discussion. The authors reported that the feedback provided valuable information for further improvement of the course.

Sanford in Chap. 6 began with the discussion of computational thinking as complementary and additional to the reading, writing and arithmetic for early childhood education. While other attempts to introduce computational thinking and coding with software and programming languages, he noted that the value of spreadsheets for logical construction and visibility of details is overlooked. He argued that early introduction of spreadsheets as an integral part of already existing approaches to mathematics and science was necessary. Sanford demonstrated the use of spreadsheets with examples on creating functions and equations, arithmetic games, word problems, equation, simulating a problem and trigonometry that can lead to computational thinking.

1.3 Computational Thinking and Teacher Education

The chapters in the second part of the book deal with how computational thinking has been introduced in teacher education programs. In Chap. 7, Angeli and Jaipal-Jamini noted that while the importance of teaching computational thinking across K–12 curriculum has been widely recognized, teacher education programs lack the knowledge and skills to teach pre-service teachers about the subject. To address the issue, they presented a study that uses scaffolded programming scripts as one method for teaching computational thinking in teacher preparation program. These programming scripts were used in the context of educational robotics activities. The study found that significant development took place in pre-service teachers' computational thinking skills after the course despite the fact that trainees do not have any prior experience. The authors also noted that the affordances of LEGO WeDo programming language might have attributed in this process. Finally, the authors raised the issue of whether it is better to first teach students how to create abstractions and generalizations of problems before teaching them how to develop algorithms and write and debug computer code.

Sands, Yadev and Good present in-service teachers' perceptions about computational thinking in Chap. 8. This chapter reports the results from a study to examine practicing teachers' views of computational thinking and compare to the definition by the computer science education researchers. The study involved 74 elementary and secondary teachers from a Midwestern state in the United States. It was found that teachers' conceptions of computational thinking include important aspects of the computational thinking literature, but there are still several common misconceptions about the concept. The chapter discusses implications of the findings on how to engage non-computing K–12 teachers in computational thinking and develop their competencies to incorporate computational thinking within the context of their subject area. The chapter informs in-service and pre-service teacher development efforts and explains how computational thinking applies to disciplinary knowledge in school subjects.

In Chap. 9, Kim and Kim from Chongshin University in Korea shared their experience on developing a computational thinking curriculum and teacher professional development in South Korea. As a background the authors described the historical development on computing education and recent curriculum changes at elementary, middle and high school levels in the country. The goal of the curriculum is to gradually step up from information literacy to computational thinking and move towards achieving collaborative problem-solving capabilities. The authors reported that all K–12 teachers are trained to use Scratch and Entry educational programming languages. Entry is a Korean-style block programming language that allows teachers to teach online. The effectiveness of the training program was evaluated with the use of concerns-based model of teacher development. The model comprises of seven stages, and Stages of Concern Questionnaire was used to collect the data from the participating teachers. The authors concluded that the success of the computing education depends on collaboration among all stakeholders including leaders, teachers, students and parents.

1.4 Computational Thinking in Schools and Higher Education

Part III of this book covered topics on computational thinking in schools and higher education settings. Fessakis and his colleagues explored the scope and conceptualization of computational thinking at the K–12 classroom-level curriculum. In Chap. 10, they described the historical evolution of the computational thinking concept and presented the rationale of the research that involved qualitative content analysis of various computational thinking curricula and initiatives. The research questions included whether all the theoretical dimensions of computational thinking were presented in the classroom curriculum, which school subjects are utilized for the development of computational thinking in schools, which teaching and learning methods and resources are proposed for the development of computational thinking and whether any other dimensions in computational thinking is missing in the curriculum. The authors reported their findings and point out the limitation of their study. They stressed that computational thinking is important for science and technology progress and empirical studies were needed to discover how computational thinking skills in schools can be realized.

In Chap. 11, Garcia-Peñalvo and his team shared their experience with coding project that introduced computational thinking in schools. The TACCLE 3 Coding is a project that supports primary school and teachers who wish to teach computing to

4–14-year-olds. TACCLE 3 was designed to offer classroom teachers with the knowledge, skills and materials to teach in schools together with in-service training courses and other staff development events for countries in the European Union. Three objectives of the project broadly defined are to (i) equip classroom teachers, regardless of the level of confidence, with knowledge and the materials they need to teach coding effectively; (ii) develop websites that are easy to follow with innovative ideas and recourses to assist teachers to teach coding; and (iii) provide national and international in-service training courses and staff development. The chapter details the curriculum, pedagogies and activities designed for the project and different types of software and technology used to achieve the goals.

Lye and Koh presented three case studies of elementary children's engagement in computational thinking through Scratch programming in Chap. 12. The authors noted that Scratch is a programming language designed to facilitate children's engagement in computational thinking with the possibility of developing problem-solving skills. Their study that took place in a primary school in Singapore, attempts to find out the achievement made and challenges faced by the children with different programming abilities when engaging in computational thinking. The study utilized multiple-case study approach involving three cases where narratives of children's moves, utterances and behaviours during the Scratch programming were recorded and analysed. They observed that programming behaviours of three students were different, one using trial-and-error method, another using piecemeal approach and another using holistic programming. These findings confirmed that children do not approach a problem in a similar way and teachers need to scaffold different programming behaviours with differentiated form of instruction to ensure the students' progress according to their own ability.

In Chap. 13, Aydeniz provided a comprehensive view and meta-analysis of research studies conducted by computer science education community. It is a common belief that if computational thinking curriculum is introduced properly, it can enhance the students' problem-solving abilities, critical thinking, data analysis and modelling skills. These attributes are particularly important for STEM education. The chapter began with the overview of development and integration of computational thinking skills in K–12 curriculum and analysis of how these skills are integrated in science and mathematics subjects in schools. The chapter also examined how computational thinking skills can be integrated in the STEM education. The chapter looked into how computational skills have been introduced in pre-service teacher education and in-service professional development programs. The chapter concluded that challenges remain in the computer science and STEM education communities. These challenges include defining the appropriate curriculum, training teachers and equitable access for all students.

Ana Calderon from the Cardiff Metropolitan University shared the findings from her study with undergraduate students over a period of 2 years to find out whether aptitudes and preferences for STEM and humanities subjects had any impact on students' performance in computational thinking. In her chapter, she stated that the commonly agreed concepts in computational thinking includes decomposition, data representation, algorithmic thinking and abstraction and these concepts are linked together to solve complex problem. The chapter reported detail findings and differences in performance between students who studied STEM subjects such as mathematics, computing and physics and humanities subjects such as history, literature and drama. The chapter concluded with the suggestion to explore further with broader and larger sample population that consider age, educational backgrounds and careers.

The last chapter by Vallance and Towndrow presented a study that examined the learners' understanding of computational thinking concept through illustrative flow-charts. In this study undergraduate students in one university in Japan were assigned to design, build and program robots to solve a disaster scenario with the use of LEGO Mindstorms. The students reflected upon the problem-solving processes with the use of flow-chart representing metal models of computational thinking. It was found that students were able to express recognition of the computational thinking concepts on modularity, decomposition and algorithmic logic; however, they faced difficulty expressing in generalization and abstraction. From the findings, the authors suggested that more attention should be paid to task design and pedagogy in introducing computational thinking concepts.

1.5 Conclusion

The chapters in this book cover wide-ranging topics on computational thinking in various educational settings. The suggestions provided by Sengupta and his colleagues are notable. The pedagogical guidelines to sustain computational thinking through modelling in STEM education can be a practical use for promoting STEM education. The need to equip in-service and pre-service teachers with computational thinking skills has strong support from educators. Many studies have been reported in the literature that utilizes a variety of approaches. For example, one of the studies in this book describes the relationships between computational thinking concepts and object-oriented programming. Another study uses Scratch programming to engage the children with computational thinking. The chapter by Sanford discusses how spreadsheet can be used for computational thinking. It is evident from these studies that computational thinking can be engaged by many different ways.

Wang (2016) noted that understanding computing and acquiring computational thinking are two sides of the same coin. According to him, "computational thinking is the mental skill to apply fundamental concepts and reasoning, derived from modern digital computers and computer science, in all areas, including day-to-day activities". In his words "computational thinking is thinking inspired by an understanding of computers and information technologies". The foundations and research highlights described in this book relate to linking, infusing and embedding computational thinking elements to school curricula, tertiary courses and STEM-related subjects. The chapters presented best practices, critical analyses and in-depth investigations by educators and researchers in the field, highlighting the contemporary trends and issues, creative and unique approaches, innovative methods, frameworks, pedagogies and theoretical and practical aspects in computational thinking. The study on developing computational thinking is set to grow in the next decades and beyond. There is a need to collectively raise new questions that contribute to the understanding and effective implementation of the concept. It is hoped that this book will serve as a catalyst to continue the dialogue on computational thinking.

References

Denning, P. J. (2017). Computational thinking in science. *American Scientist, 105*(1), 13–17.

Riley, D. D., & Hunt, K. A. (2014). *Computational thinking for the modern problem solver*. London: CRC Press.

Tedre, M., & Denning, P. J. (2016). The long quest for computational thinking. In *Proceedings of the 16th Koli Calling Conference in Computing Education Research, Koli, Finland*, (pp. 120–129).

Wang, P. S. (2016). *From computing to computational thinking*. London: CRC Press.

Chapter 2
Characteristics of Studies Conducted on Computational Thinking: A Content Analysis

Filiz Kalelioğlu

2.1 Introduction

When the literature was examined, the first concept of computational thinking (CT) was first used by Papert in 1996. However, the particular skill of CT gained the attention of researchers following Wing's paper in 2006. According to Wing, 'Computational thinking involves solving problems, designing systems, and understanding human behaviour, by drawing on the concepts fundamental to computer science' (p. 33). Moreover, CT requires using heuristic reasoning to discover a solution beginning from planning, learning and scheduling in the presence of uncertainty (Wing 2008). For Wing, CT was seen as a kind of analytical thinking covering mathematical thinking that we might use to solve a problem. After a while, Cuny et al. (2010) proposed a new definition: 'Computational thinking is the thought processes involved in formulating problems and their solutions so that the solutions are represented in a form that can be effectively carried out by an information-processing agent' (as cited in Wing 2010, p. 1).

Guzdial (2008) defined CT as a way of thinking about computing. For Aho (2012), CT is the process of formulating problems in a way that solutions can be represented as algorithms and computational steps. Similarly, Mannila et al. (2014) defined CT as covering a set of concepts and thinking processes from computer science that help in formulating problems and their solutions. In the same line, Voogt et al. (2015) stated that CT could be seen as a thinking process required for solving problems including generalising and transferring this process to different problems in other areas.

F. Kalelioğlu (✉)
Computer Education and Instructional Technology Department, Faculty of Education, Baskent University, Ankara, Turkey
e-mail: filizk@baskent.edu.tr

M. S. Khine (ed.), *Computational Thinking in the STEM Disciplines*,
https://doi.org/10.1007/978-3-319-93566-9_2

Riley and Hunt (2014) stated that CT is a way of thinking for computer scientists and is the manner in which they reason. Google (n.d.) states that CT is a problem-solving process, such as a process that includes analysing data logically and creating solutions as ordered steps and dispositions, for example, the ability to confidently deal with complexity and open-ended problems. In more detail, Selby (2013) defined CT as a cognitive or thought process that reflects the ability to think in abstractions, decomposition, algorithm, evaluation and generalisation.

The Royal Society (2012) defined CT as, 'the process of recognising aspects of computation in the world that surrounds us, and applying tools and techniques from Computer Science to understand and reason about both natural and artificial systems and processes' (p. 29). According to the operational definition of ISTE and CSTA (2011), CT is a problem-solving process that includes the following characteristics:

- Formulating problem
- Logically organising and analysing data
- Representing data through abstractions
- Automating solutions through algorithmic thinking
- Identifying, analysing and implementing possible solutions with the goal of achieving the most efficient and effective combination of steps and resources
- Generalising and transferring solutions to a wide variety of problems

These skills are supported and enhanced by a number of dispositions or attitudes such as:

- Confidence in dealing with complexity
- Persistence in working with difficult problems
- Tolerance for ambiguity
- The ability to deal with open-ended problems
- The ability to communicate and work with others to achieve a common goal or solution

Barr and Stephenson (2011) posed a similar operational definition of CT including core CT concepts and capabilities and examples of how they might be implemented in activities through multiple disciplines. As a summary, Wing (2006) specified the following characteristics of CT as:

- Conceptualising: Thinking like a computer scientist requires thinking at many levels of abstraction and means more than being able to program a computer.
- Fundamental skill for everyone: CT is something that each person must know in order to function in this digital age.
- A way that humans think: It is a way for people to solve problems, but does not mean to think like a computer.
- Complements and combines mathematical and engineering thinking.
- Ideas: It will be the computational concepts we use to approach and solve problems, to manage our daily lives and to communicate and interact with other people.

The most important and high-level thought process in CT is the abstraction process (Wing 2008). In computing, this is necessary in order to abstract the concepts and ideas beyond the physical perspective of time and space. It is used to capture essential properties common to a set of objects while hiding irrelevant distinctions among them. Abstraction is used in defining patterns, generalising from specific instances and parameterisation. Similarly for Aho (2012), mathematical abstractions, in other words models of computation, are at the centre of computing and CT. According to Selby (2013), CT should include the idea of a thought process, the concepts of abstraction and decomposition. From analysing the findings of Kalelioğlu et al. (2016), it can be seen that they found abstraction, algorithmic thinking, problem-solving, pattern recognition and design-based thinking were the top five skills underlined by researchers, and it was obvious that the definition of CT also consists of algorithmic and design-based types of thinking.

The process of computational thinking involves many subactions and concepts. Brennan and Resnick (2012) describe a CT framework as computational concepts (sequences, loops, events, parallelism, conditionals, operators and data), computational practices (experimenting and iterating, testing and debugging, reusing and remixing, abstracting and modularising) and computational perspectives (expressing, connecting and questioning). Computing At School (CAS) states that CT involves six different concepts as logic, algorithms, decomposition, patterns, abstraction and evaluation and five approaches to working in the classroom as tinkering, creating, debugging, persevering and collaborating (CAS Barefoot 2014). Moreover, many researchers have described CT concepts and practices from different perspectives. According to the most cited papers representing the variety of perspectives and points of view, Barr and Stephenson (2011) proposed abstraction, algorithmic and procedures, automation, problem decomposition, parallel processing and simulation as a subset of core concepts and skills for CT skills. For Grover and Pea (2013), abstraction and pattern generalisation, algorithms, logic, problem decomposition, debugging, productivity and performance constraints, parallel thinking and systematic processing of information were the concepts and skills for CT skills.

When examining the studies focusing on systematic literature reviews, Benitti (2012) reviewed the literature systematically for educational benefits of robotics. She concluded that educational robotics can be used as a method for improving learning; nevertheless, she mentions other studies that do not enhance learning. Grover and Pea (2013) analysed the literature on computational thinking in the K–12 sphere and concluded that recent studies on CT focused on the definitions of computational thinking skills and tools as another issue to promote the development of CT.

In a study by Kalelioğlu et al. (2016), papers about computational thinking were analysed from between 2006 and 2014. The results revealed that the papers mainly focused on the plugged or unplugged types of activities for teaching computational thinking in the K–12 curriculum. Heintz et al. (2016) systematically analysed papers and the curriculum of ten countries in terms of computer science education in K–12 education. The countries integrated digital literacy with programming, the subject of computer science or computing. However, although computational thinking was not explicitly expressed, its ideas and some concepts were seen to be used in the curricula.

Lye and Koh (2014) reviewed nine papers on developing computational thinking through programming and concluded there was a need to conduct more research on the topic. Similarly, Moreno-León and Robles (2016) systematically analysed papers focused on the teaching of programming with Scratch and on the development of related skills within this context. In terms of developing thinking skills, seven out of nine articles concluded that students developed their problem-solving, logical thinking and creativity through learning to program. To draw more clear conclusions about learning programming through such platforms and developing thinking skills, there was seen a need for more empirical studies with larger student samples.

Approaches to teaching these skills, which are gaining popularity, are also striking. At the same time, Bundy (2007) stated that CT would gain popularity, prompting research in nearly all disciplines, both in the sciences and the humanities. On the other hand, many researchers concluded that focusing on the integration of computational thinking skills in education as inadequate (Heintz et al. 2016; Voogt et al. 2015, p. 726). Hence, the main purpose of the current study is to consider what researchers have said about CT through the following research questions:

- What are the demographic characteristics of the selected papers?
- What kind of taxonomy might characterise this entire set of literature?
- What was the purpose of the studies?
- What was the nature of the studies that have been conducted?
- What kinds of studies have been conducted?
- What are the overall results of the selected papers?

2.2 Methodology

The skill of computational thinking is important both in computer science education and in the teaching of programming skills. This skill, which has been taught since early childhood for some, continues to be taught to different age groups via a range of different approaches and methods. For this reason, this skill has been the subject of many research studies for academics working in this field. The purpose of the current study is to examine published studies in computational thinking in a systematic way and to present a history of the research and new research trends in this area. In this context, the concept of computational thinking was systematically searched in databases and digital libraries of ScienceDirect and the IEEE Xplore Digital Library.

The primary search term 'computational thinking' was entered along with the data in the abstract, title and keyword section, for publications between 2013 and 2016. The papers were then analysed according to their year of publication, target population, keywords, taxonomy type and nature of the study. The results of the inductive analysis are presented according to the research questions. After detailed document analysis, some parts of the analysed data are transformed into numerical values and illustrated through the use of graphics.

2.2.1 Demographic Characteristics of the Selected Papers

Detailed research was carried out on 65 studies that were considered worthy of review. In accordance with the American Psychological Association Publication Manual, the reviewed studies are listed in the References section, annotated with an asterisk.

2.2.1.1 Year of Publication

When the publication dates of the studies were examined, it was seen that 8 studies were published in 2013, 17 in 2014, 16 in 2015 and 24 in 2016 (see Fig. 2.1). When the distribution of publication dates is examined, it can be seen that the number of publications mostly increased in line with the year of publication.

2.2.1.2 Target Population

When the target populations of the studies were examined, 15 studies were conducted with undergraduate students, 13 with K–12 students, 7 with teachers or instructors, 6 were with high school students and 6 with middle school students (see Fig. 2.2).

The sample shows that the studies were performed with any group directly related to the topic. The studies tried to examine the effects of different subjects on different

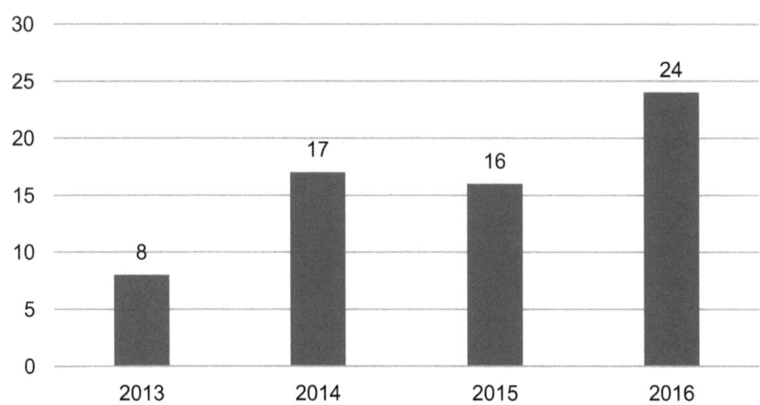

Fig. 2.1 Number of papers according to year of publication

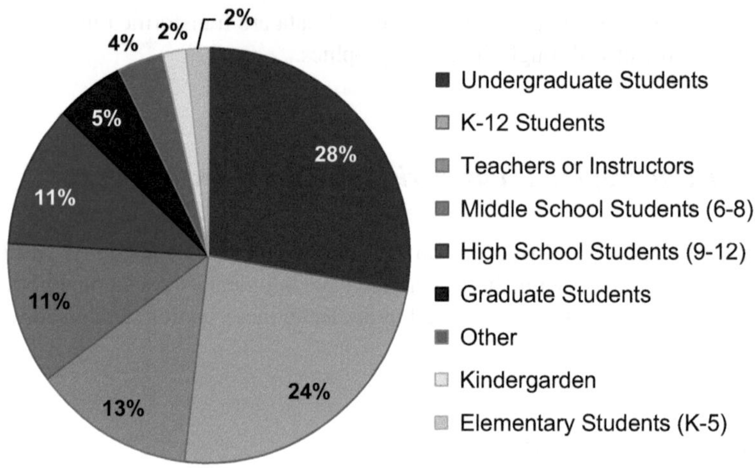

Fig. 2.2 Target populations

people. However, the studies seem to mostly target student populations. Studies conducted by university students are followed by studies with primary school students.

2.2.1.3 Keywords

When the keywords in the studies are examined, it was seen that computational thinking was the most preferred keyword ($n = 42$). In addition, programming ($n = 16$) and teaching/learning strategies ($n = 9$) are also included. Computer science ($n = 8$), educational robotics ($n = 6$), computing ($n = 5$), computer science education ($n = 5$) and Scratch ($n = 5$) were other preferred keywords (see Fig. 2.3). The chosen keywords reflect the characteristic of the sample studies. It is often seen that research focuses on CT and how to teach it. It can be said that the studies focusing on computational thinking are directly related to programming and computer science when looking at the keyword patterns related to computational thinking. It is possible to say that all the studies examined are actually about which teaching methods and techniques and which teaching tools can be used to teach this skill.

2.2.1.4 Type of Taxonomy

A fivefold taxonomy was used to classify the publications. The first category in the taxonomy was model, which includes articles that focused on discussions about CT and its scope. The second was pedagogy, which includes articles that focused on how CT can be taught: teaching methods or tools such as robots, unplugged,

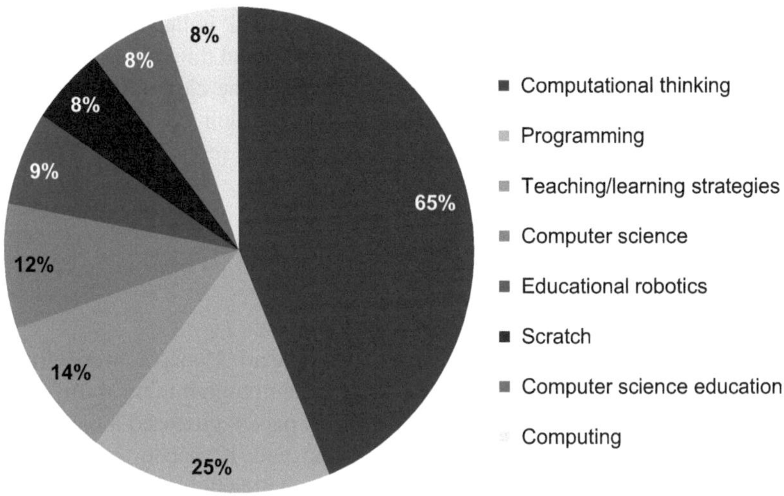

Fig. 2.3 Keywords

Fig. 2.4 Type of taxonomy

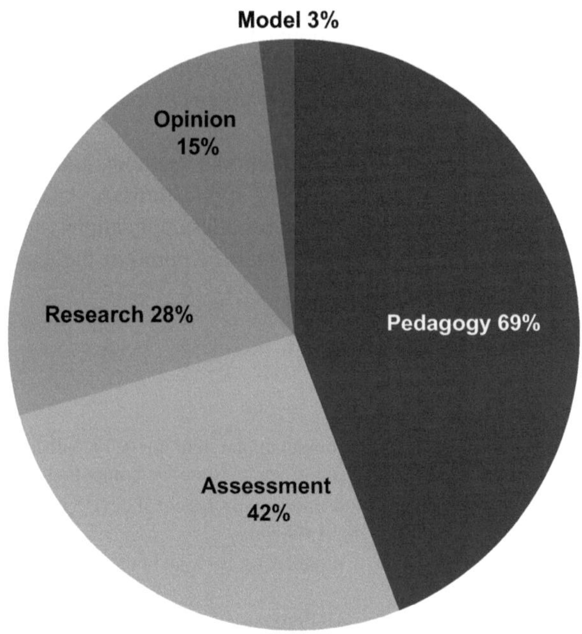

STEAM and block-based, etc. The third was assessment for articles with that focus and means on how CT can be measured and learnt. The other taxonomies were research and opinion and include articles that focused on research methodology or that share perspectives about CT (see Fig. 2.4).

It is clear that the studies primarily focus on how CT is taught ($n = 45$). The situation that emerged was similar to that of computational thinking as the keyword under the previous research question, being pedagogy the clear favourite. Second was assessment of CT publications ($n = 27$) in which the researchers were discussing or stating how to evaluate CT. Third was research for studies that measured the impact of CT ($n = 18$).

2.2.1.5 Purpose of the Papers

When the purposes of the papers were examined in detail: 19 papers were aiming to develop students' computational thinking skills or their programming ability through the application of different tools or approaches; 15 papers proposed a methodology for teaching computational thinking to students; and 12 papers had a purpose focusing on the investigation of variables on computational thinking knowledge or computer science concepts. Other purposes that were less common were six papers about assessment of computational thinking skill; six papers proposed a model for integrating computational thinking skills to curricula; five papers discussed the literature about computational thinking or programming; two papers focused on CT and its impact on K–12 science education; and two papers reported the experience and perceptions of teachers about computer science and computational thinking (see Table 2.1).

When studies on the instructional methods and tools of CT were examined in detail, most were about STEAM applications ($n = 13$) and block-based applications ($n = 13$), with publications on robotics applications ($n = 8$) in third place. Studies on unplugged practices were the least favoured of the teaching methods (see Fig. 2.5).

Table 2.1 Purpose of papers

Purpose	Number of indices
Development of students' computational thinking (CT) skills/programming in the context of educational robotics/games/mobile computing	19
Proposed a teaching method/activities/professional development for developing the computational thinking of students	15
Investigation of effects of variables on computational thinking knowledge and skills/programming	12
Assessment of computational thinking	6
Design of a computational thinking curriculum	6
Review of literature/models about CT/programming	5
Discuss CT and impacts on K–12 science education	2
Report on teachers' perceptions/experience of computer science	2

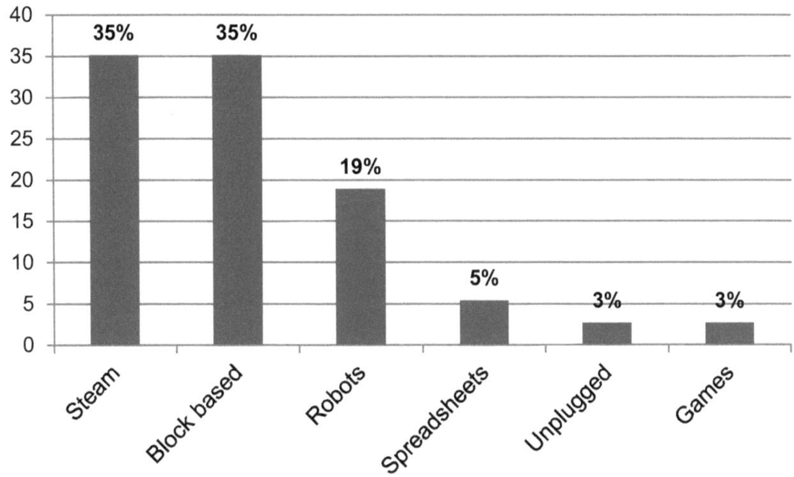

Fig. 2.5 Approaches for teaching CT

2.2.2 *Nature of the Studies Conducted*

2.2.2.1 Research Design

The method most used in the various different types of research was that of case study ($n = 13$). In addition, experimental designs ($n = 16$) were also widely preferred. Survey type ($n = 7$) and qualitative studies ($n = 7$) are the other methods found to have been used in the studies (see Fig. 2.6).

2.2.2.2 Data Collection Tools

It became clear that questionnaires ($n = 24$) were the most preferred type of data collection tool. Next were tests developed by the teacher or researcher ($n = 10$), and then interviews ($n = 8$) and observation ($n = 7$) were methods used to gather deeper information (see Fig. 2.7).

2.2.2.3 Data Analysis Method Used

A descriptive statistical analysis ($n = 26$) was the most used type of data analysis. This method is followed by other quantitative analytical methods, such as *t*-test ($n = 10$), ANOVA ($n = 4$) and ANCOVA ($n = 1$), whereas content analysis ($n = 9$) was used for qualitative data (see Fig. 2.8).

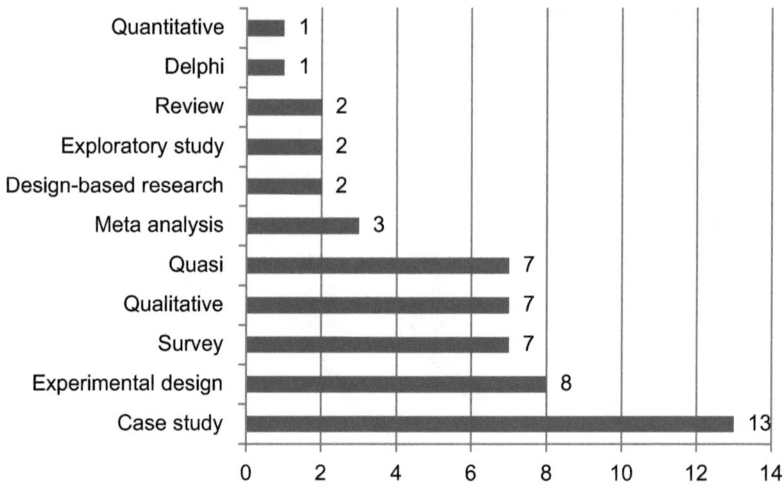

Fig. 2.6 Research design

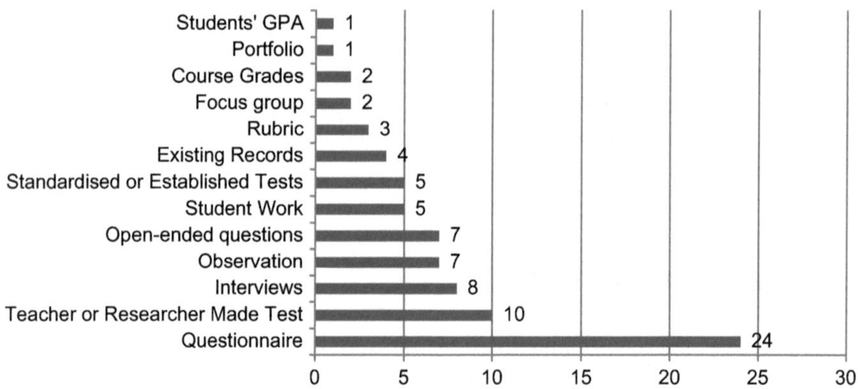

Fig. 2.7 Data collection tools

2.2.3 Summary of Results for the Selected Papers

When the results of the investigated research studies are examined, some main themes emerged as pedagogy, learning and teaching context and potential contributions. Therefore, it summarises how computational thinking is taught, how the learning environment should be, what this skill will provide and what contributions it will provide to the students.

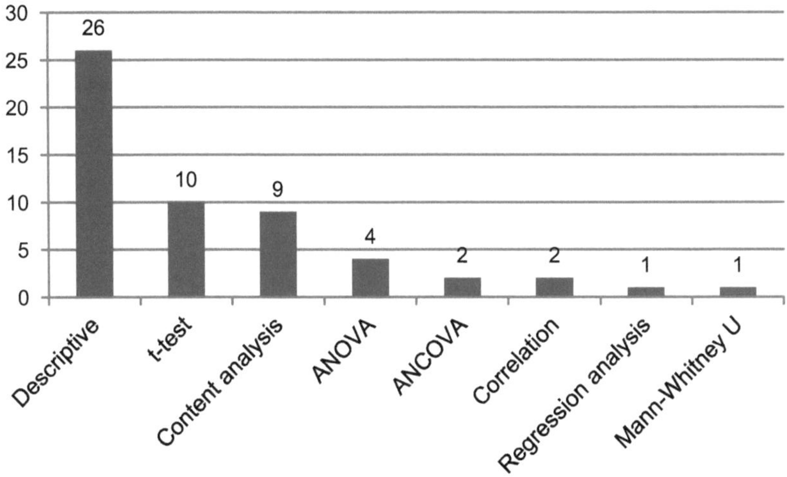

Fig. 2.8 Data analysis method used

The most common pedagogical approaches to promote the learning of CT, programming concepts, logic and computational practices are programming courses (de Araujo et al. 2016), games (Towhidnejad et al. 2014), unplugged approaches (Lee et al. 2014), visual programming languages (Bustillo and Garaizar 2014; Liu and Xu 2016; Sáez-López et al. 2016), virtual experiments (Li et al. 2015a), virtual reality and simulations (Li et al. 2016) and creative competency exercises (CCE) (Shell et al. 2014). These methods are beneficial for both the learning and teaching of computing, computational thinking and CS knowledge and skills. Moreover, learning environment should be a constructionism-based problem-solving learning environment, including scaffolding and reflection activities to support learning computational practices (Lye and Koh 2014). Moreover, pair programming and other kinds of collaborative learning environment should be designed for gender equity and socialisation (Zhong et al. 2016).

Programmes for computer science or computational thinking training impact on changing teachers perceptions about computer science (Prieto-Rodriguez and Berretta 2014) and students' learning (Rodrigues et al. 2016). Kindergartners can learn and practice many concepts and aspects of robotics, programming and computational thinking within a robotics curriculum (Bers et al. 2014). Computational thinking is an engaging way to learn science (Arraki et al. 2014; Nesiba et al. 2015; Peel et al. 2015). Moreover robotics competitions also impact on students' learning and motivation for further exploring in STEM or STEM-related fields (Eguchi 2016).

In addition to other concepts related to computational thinking, in particular the algorithms, solving problems and abstraction have emerged as the most assessed abilities (De Araujo et al. 2016). According to the experimental design results, it is found that a statistically significant correlation exists between CT and spatial ability and between reasoning ability and problem-solving ability (Moreno-León et al.

2016). Male students were better with abstract thinking and problem-solving when compared to female students. Moreover, abstract thinking improved the programming understanding of the students (Park et al. 2015). Computational activities and experiments helped to develop not only the students' exploratory skills but also their creativity (Se et al. 2015). However, there is a need for more research about supporting the CT of students with instructional media like block-based programming platforms at different education levels (Garneli 2014; Moreno-León and Robles 2016).

2.3 Discussion and Conclusion

The skill of computational thinking is important both in computer science education and in the teaching of programming skills. This skill, which has been taught since early childhood for some, continues to be taught to different age groups via different approaches and methods. For this reason, this skill has been the subject of many research studies for academics working in this field. The purpose of the current study is to examine studies in computational thinking that took place between 2013 and 2016 in a systematic way. In this context, keywords in the concept of computational thinking were systematically searched in databases and digital libraries of ScienceDirect and the IEEE Xplore Digital Library. The study described in this paper explored the literature on computational thinking and described the demographic characteristics of the literature in terms of characteristics, methods, design and data analysis.

When studies from the literature were examined in terms of their year of publication, the number of studies mostly increased in parallel with the year in which they were published. It can be said that CT will likely maintain its popularity, and more work will emerge in the coming years. As it is considered a skill necessary for programming, researchers will likely continue to study CT as a notable subject. As stated by Wing (2006), this kind of thinking will be part of the skillset of not only other scientists but of everyone else. CT should be added to every child's analytical ability besides reading, writing and arithmetic. It is thought that with the ability of computational thinking, students can solve problems in different areas (Barr et al. 2011). ISTE and CSTA (2011) stated that CT is an important skill in enhancing the success of students, preparing students for global competition and achieving success in school life and real life success. With today's digital tools, students need to be able to think cognitively in order to be able to answer the question of how tomorrow's problems can be solved. Students will need to learn and apply the new skill of computational thinking in order to benefit from the changes that come with the rapid changes in technology.

The sample reviews were performed with any group directly related to the topic. The studies tried to examine the effects of different subjects on different people. However, many countries are updating their computer science curricula to teach this skill at a young age (European Schoolnet 2015). The new computer science curriculums include pedagogical strategies to teach computer science, programming and computational thinking at different ages, with teachers and researchers seemingly able to teach this skill at a young age. Moreover, there are efforts worldwide to integrate computer science, programming and computational thinking into new curricula (e.g. Duncan 2018). As a result of this process, the researcher noted that it is expected to work with different stakeholders (e.g. teachers, students of different ages groups, etc.) in order to handle the challenges and make correct preparations and to direct learning at the student and teacher level as a natural outcome of this situation.

When the keywords in the studies were examined, it was seen that computational thinking was the most preferred keyword, programming was second and teaching and learning strategies were third. The reason that programming is the second and most used keyword is because programming is used as the context for studies focusing on computational thinking (Fletcher and Lu 2009; Hambrusch et al. 2009; Lee et al. 2011). When we look at the pattern of keywords, it was seen that studies of CT are related to the teaching strategies and technologies used for this skill. This also could be seen from the studies' themes and literature about this topic, which includes pedagogy publications, that mostly supports this finding. When studies on the instructional methods and tools of CT are examined, it was found that most were about STEAM, block-based or robotic applications. This result was comparable with that of Kalelioğu et al. (2016) who stated that the main topics covered in the papers they examined were composed of activities that promote CT in the curriculum. According to the results of their literature review of papers about computational thinking published between 2006 and 2014, while studies on integrating computational thinking to the curriculum was at the fore, pedagogical topics were second, i.e. how to teach and which tools to use. Another result, by Grover and Pea (2013), stated that 'much of the recent work on CT has focused mostly on definitional issues, and tools that foster CT development' (p. 42). On the contrary, it can be said that the studies carried out in the following years have primarily focused on the pedagogical issues.

The most used research method found in the current study was case studies. In addition to this method, experimental designs, survey type and qualitative studies were the other methods used. According to the research questions, the research methods and data collection tools may be different. In the case of teaching techniques of CT, it was observed that an in-depth study with small groups was preferred. However, Moreno-León and Robles (2016) concluded in their systematic literature review that there is a need for further studies about proving learning and developing thinking skills through programming with experimental designs and larger student samples.

The aim of the current study was to provide an insight into the characteristics of the recent literature on CT and to present the range of views found. As can be seen, the studies on CT show a pattern. It may therefore be possible to say that studies focusing on CT will increase in the years ahead. Moreover, it points towards how studies on computer science courses, whether at the university or K–12 level, should be designed and evaluated in order to remain popular. This study is limited to the databases analysed, keywords used and the analysis units obtained by the date range. It is therefore feasible to conduct further research using different databases as well as more recent publications. It is thought that the results obtained will provide researchers with further ideas for studies to be conducted on this subject.

References[1]

Aho, A. (2012). Computation and computational thinking. *The Computer Journal, 56*(7), 832–835.

*Ambrósio, A. P., Xavier, C., & Georges, F. (2014). Digital ink for cognitive assessment of computational thinking. In *2014 I.E. Frontiers in Education Conference (FIE) Proceedings* (pp. 1–7). IEEE.

*Arraki, K., Blair, K., Bürgert, T., Greenling, J., Haebe, J., Lee, G., Peel A.; Szczepanski V.; Pontelli E, Hug, S. (2014). DISSECT: An experiment in infusing computational thinking in K-12 science curricula. In *2014 I.E. Frontiers in Education Conference (FIE) Proceedings* (pp. 1–9). IEEE.

Atmatzidou, S., & Demetriadis, S. (2016). Advancing students' computational thinking skills through educational robotics: A study on age and gender relevant differences. *Robotics and Autonomous Systems, 75*(PB), 661–670.

Barr, V., & Stephenson, C. (2011). Bringing computational thinking to K-12: What is involved and what is the role of the computer science education community? *ACM Inroads, 2*(1), 48–54.

Barr, D., Harrison, J., & Conery, L. (2011). Computational thinking: A digital age skill for everyone. *Learning & Leading with Technology, 38*(6), 20–23.

*Basogain, X., Olabe, M. A., Olabe, J. C., Ramírez, R., Del Rosario, M., & Garcia, J. (2016). PC-01: Introduction to computational thinking: Educational technology in primary and secondary education. In *2016 International Symposium on Computers in Education (SIIE) Proceedings* (pp. 1–5). IEEE.

*Bean, N., Weese, J., Feldhausen, R., & Bell, R. S. (2015, October). Starting from scratch: Developing a pre-service teacher training program in computational thinking. In *2015 I.E. Frontiers in Education Conference (FIE) Proceedings* (pp. 1–8). IEEE.

Benitti, F. B. V. (2012). Exploring the educational potential of robotics in schools: A systematic review. *Computers & Education, 58*(3), 978–988.

Bers, M. U., Flannery, L., Kazakoff, E. R., & Sullivan, A. (2014). Computational thinking and tinkering: Exploration of an early childhood robotics curriculum. *Computers & Education, 72*, 145–157.

*Boechler, P., Artym, C., Dejong, E., Carbonaro, M., & Stroulia, E. (2014). Computational thinking, code complexity, and prior experience in a videogame-building assignment. In *2014 I.E. 14th International Conference on Advanced Learning Technologies (ICALT) Proceedings* (pp. 396–398). IEEE.

*Brackmann, C., Barone, D., Casali, A., Boucinha, R., & Muñoz-Hernandez, S. (2016). Computational thinking: Panorama of the Americas. In *2016 International Symposium on Computers in Education (SIIE) Proceedings* (pp. 1–6). IEEE.

Brennan, K., & Resnick, M. (2012). *New frameworks for studying and assessing the development of computational thinking*. Paper presented at the Annual American Educational Research Association meeting, Vancouver, BC, Canada.

[1]Note: References marked with an asterisk indicate studies included within inductive analysis, as per APA Publication Manual (http://blog.apastyle.org/apastyle/2012/08/alert-change-in-apa-style-on-meta-analysis-references.html)

Buckingham, L., & Hogan, J. M. (2014). Computational science for undergraduate biologists via QUT. *Bio Excel Procedia Computer Science, 29*, 1403–1412.
Bundy, A. (2007). Computational thinking is pervasive. *Journal of Scientific and Practical Computing., 1*(2), 67–69.
*Burgett, T., Folk, R., Fulton, J., Peel, A., Pontelli, E., & Szczepanski, V. (2015). Dissect: Analysis of pedagogical techniques to integrate computational thinking into k-12 curricula. In *2015 I.E. Frontiers in Education Conference (FIE) Proceedings* (pp. 1–9). IEEE.
*Bustillo, J., & Garaizar, P. (2014). Scratching the surface of digital literacy... but we need to go deeper. In *2014 I.E. Frontiers in Education Conference (FIE) Proceedings* (pp. 1–4). IEEE.
*Byrne, J. R., Fisher, L., & Tangney, B. (2015). Empowering teachers to teach CS—Exploring a social constructivist approach for CS CPD, using the Bridge21 model. In *2015 I.E. Frontiers in Education Conference (FIE) Proceedings* (pp. 1–9). IEEE.
CAS Barefoot. (2014). *Computational thinking*. CAS Barefoot. Retrieved from http://barefootcas.org.uk/barefoot-primary-computing-resources/concepts/computational-thinking/
*Chuang, H. C., Hu, C. F., Wu, C. C., & Lin, Y. T. (2015). Computational thinking curriculum for K-12 education – a Delphi survey. In *2015 International Conference on Learning and Teaching in Computing and Engineering (LaTiCE) Proceedings* (pp. 213–214). IEEE.
*Cross, J., Hamner, E., Zito, L., & Nourbakhsh, I. (2016). Engineering and computational thinking talent in middle school students: A framework for defining and recognizing student affinities. In *2016 I.E. Frontiers in Education Conference (FIE) Proceedings* (pp. 1–9). IEEE.
*Dasgupta, A., & Purzer, S. (2016). No patterns in pattern recognition: A systematic literature review. In *2016 I.E. Frontiers in Education Conference (FIE) Proceedings* (pp. 1–3). IEEE.
*De Araujo, A. L. S. O., Andrade, W. L., & Guerrero, D. D. S. (2016). A systematic mapping study on assessing computational thinking abilities. In *2016 I.E. Frontiers in Education Conference (FIE) Proceedings* (pp. 1–9). IEEE.
Duncan, C. (2018). Reported development of computational thinking, through computer science and programming, and its benefits for primary school students. In *Proceedings of the 49th ACM Technical Symposium on Computer Science Education* (pp. 275–275). ACM.
Eguchi, A. (2016). RoboCupJunior for promoting STEM education, 21st century skills, and technological advancement through robotics competition. *Robotics and Autonomous Systems, 75*(PB), 692–699.
European Schoolnet. (2015). European Schoolnet: Computing our future – computer programming and coding priorities, school curricula and initiatives across Europe. Retrieved from http://fcl.eun.org/documents/10180/14689/Computing+our+future_final.pdf/746e36b1-e1a6-4bf1-8105-ea27c0d2bbe0.
Fletcher, G. H., & Lu, J. J. (2009). Education: Human computing skills: Rethinking the K-12 experience. Association for computing machinery. *Communications of the ACM – Inspiring Women in Computing, 52*(2), 23–25.
*Folk, R., Lee, G., Michalenko, A., Peel, A., & Pontelli, E. (2015). GK-12 DISSECT: Incorporating computational thinking with K-12 science without computer access. In *2015 I.E. Frontiers in Education Conference (FIE) Proceedings* (pp. 1–8). IEEE.
*García-Peñalvo, F. J. (2016). A brief introduction to TACCLE 3—coding European project. In *2016 International Symposium on Computers in Education (SIIE) Proceedings* (pp. 1–4). IEEE.
*Garneli, V. (2014). Instructional media and teaching methods for engaging children with computer programming. In *2014 I.E. 14th International Conference on Advanced Learning Technologies (ICALT) Proceedings* (pp. 768–770). IEEE.
Google. (n.d.). *Exploring computational thinking*. Retrieved from http://www.google.com/edu/computational-thinking/.
Grout, V., & Houlden, N. (2014). Taking computer science and programming into schools: The Glyndŵr/BCS Turing project. *Procedia-Social and Behavioral Sciences, 141*, 680–685.
Grover, S., & Pea, R. (2013). Computational thinking in K–12 a review of the state of the field. *Educational Researcher, 42*(1), 38–43.

Guzdial, M. (2008). Education: Paving the way for computational thinking. *Communications of the ACM – Designing games with a purpose, 51*(8), 25–27.

Hambrusch, S., Hoffmann, C., Korb, J. T., Haugan, M., & Hosking, A. L. (2009). A multidisciplinary approach towards computational thinking for science majors. *ACM SIGCSE Bulletin, 41*(1), 183–187.

*He, S., Hang, Y., & Ding, Y. (2014). Teaching method based on computational thinking a case research. In *2014 9th International Conference on Computer Science & Education (ICCSE) Proceedings* (pp. 817–820). IEEE.

*Heintz, F., Mannila, L., & Färnqvist, T. (2016). A review of models for introducing computational thinking, computer science and computing in K-12 education. In *2016 I.E. Frontiers in Education Conference (FIE) Proceedings* (pp. 1–9). IEEE.

*Hubwieser, P., & Mühling, A. (2015). Investigating the psychometric structure of Bebras contest: towards measuring computational thinking skills. In *2015 International Conference on Learning and Teaching in Computing and Engineering (LaTiCE) Proceedings* (pp. 62–69). IEEE.

Israel, M., Pearson, J. N., Tapia, T., Wherfel, Q. M., & Reese, G. (2015). Supporting all learners in school-wide computational thinking: A cross-case qualitative analysis. *Computers & Education, 82*, 263–279.

ISTE & CSTA. (2011). *Operational definition of computational thinking for K–12 Education*. Retrieved from http://www.iste.org/docs/ct-documents/computational-thinking-operational-definition-flyer.pdf.

*Jovanov, M., Stankov, E., Mihova, M., Ristov, S., & Gusev, M. (2016). Computing as a new compulsory subject in the Macedonian primary schools curriculum. In *2016 I.E. Global Engineering Education Conference (EDUCON) Proceedings* (pp. 680–685). IEEE.

*Jung, I., Choi, J., Kim, I. J., & Choi, C. (2016). Interactive learning environment for practical programming language based on web service. In *2016 15th International Conference on Information Technology Based Higher Education and Training (ITHET) Proceedings* (pp. 1–7). IEEE.

Kalelioğlu, F., Gülbahar, Y., & Kukul, V. (2016). A framework for computational thinking based on a systematic research review. *Baltic Journal of Modern Computing, 4*(3), 583–596.

*Kloos, C. D., Gil, M. C., Rodrĺguez, P., Robles, G., Tovar, E., Manjón, B. F. (2016). Designing educational material. In *2016 I.E. Global Engineering Education Conference (EDUCON) Proceedings* (pp. 1218–1220). IEEE.

*Ko, P. (2013). A longitudinal study of the effects of a high school robotics and computational thinking class on academic achievement (WIP). In *2013 I.E. Frontiers in Education Conference Proceedings* (pp. 181–183). IEEE.

*Lee, I., Martin, F., Denner, J., Coulter, B., Allan, W., Erickson, J., Malyn-Smith J, Werner, L. (2011). Computational thinking for youth in practice. *ACM Inroads, 2*(1), 32–37.

Lee, T. Y., Mauriello, M. L., Ahn, J., & Bederson, B. B. (2014). CTArcade: Computational thinking with games in school age children. *International Journal of Child-Computer Interaction, 2*(1), 26–33.

*Li, Y. (2014). Research into the computational thinking for the teaching of computer science. In *2014 I.E. Frontiers in Education Conference Proceedings* (pp. 1–7). IEEE.

*Li, Y. (2016). Teaching programming based on computational thinking. In *2016 I.E. Frontiers in Education Conference (FIE) Proceedings* (pp. 1–7). IEEE.

*Li, F., Li, D., Zheng, J., & Zhao, S. (2015a). Virtual experiments for introduction of computing: Using virtual reality technology. In *2015 I.E. Frontiers in Education Conference (FIE) Proceedings* (pp. 1–5). IEEE.

*Li, Y., Liu, Y., & Shu, P. (2015b). Teaching research and practice of blended leaning model based on computational thinking. In *2015 I.E. Frontiers in Education Conference (FIE) Proceedings* (pp. 1–8). IEEE.

*Li, F., Chen, Y., Yu, Y., Zhang, B., & Sun, M. (2016). Virtual experiment teaching and research oriented to college computer curriculum. In *2016 11th International Conference on Computer Science & Education (ICCSE) Proceedings* (pp. 439–442). IEEE.

*Liu, B., & He, J. (2014). Teaching mode reform and exploration on the University computer basic based on computational thinking training in Network Environment. In *2014 9th International Conference on Computer Science & Education (ICCSE) Proceedings* (pp. 59–62). IEEE.

*Liu, X., & Xu, H. (2016). Reform on college fundamentals of computer course based on mobile computing. In *2016 11th International Conference on Computer Science & Education (ICCSE) Proceedings* (pp. 890–893). IEEE.

Lye, S. Y., & Koh, J. H. L. (2014). Review on teaching and learning of computational thinking through programming: What is next for K-12? *Computers in Human Behavior, 41*, 51–61.

Mannila, L., Dagiene, V., Demo, B., Grgurina, N., Mirolo, C., Rolandsson, L., & Settle, A. (2014). Computational thinking in K-9 education. In A. Clear & R. Lister (Eds.), *Proceedings of the working group reports of the 2014 on innovation & technology in computer science education conference* (pp. 1–29). New York: ACM.

*Miller, L. D., Soh, L. K., Chiriacescu, V., Ingraham, E., Shell, D. F., Ramsay, S., & Hazley, M. P. (2013). Improving learning of computational thinking using creative thinking exercises in CS-1 computer science courses. In *2013 I.E. Frontiers in Education Conference Proceedings* (pp. 1426–1432). IEEE.

*Miller, G., Hermans, F., & Braun, R. (2016). Gradual structuring: Evolving the spreadsheet paradigm for expressiveness and learnability. In *2016 15th International Conference on Information Technology Based Higher Education and Training (ITHET) Proceedings* (pp. 1–8). IEEE.

*Moreno, J., & Robles, G. (2014). Automatic detection of bad programming habits in scratch: A preliminary study. In *2014 I.E. Frontiers in Education Conference (FIE) Proceedings* (pp. 1–4). IEEE.

*Moreno-León, J., & Robles, G. (2016). Code to learn with Scratch? A systematic literature review. In *Global Engineering Education Conference (EDUCON), 2016 IEEE* (pp. 150–156). IEEE.

*Moreno-León, J., Robles, G., & Román-González, M. (2016). Comparing computational thinking development assessment scores with software complexity metrics. In *2016 I.E. Global Engineering Education Conference (EDUCON) Proceedings* (pp. 1040–1045). IEEE.

*Nesiba, N., Pontelli, E., & Staley, T. (2015). DISSECT: Exploring the relationship between computational thinking and English literature in K-12 curricula. In *2015 I.E. Frontiers in Education Conference (FIE) Proceedings* (pp. 1–8). IEEE.

Papert, S. (1996). An exploration in the space of mathematics educations. *International Journal of Computers for Mathematical Learning, 1*(1), 95–123.

*Park, C. J., Hyun, J. S., & Heuilan, J. (2015). Effects of gender and abstract thinking factors on adolescents' computer program learning. In *2015 I.E. Frontiers in Education Conference (FIE) Proceedings* (pp. 1–7). IEEE.

*Peel, A., Fulton, J., & Pontelli, E. (2015). DISSECT: An experiment in infusing computational thinking in a sixth grade classroom. In *2015 I.E. Frontiers in Education Conference (FIE) Proceedings* (pp. 1–8). IEEE.

*Prieto-Rodriguez, E., & Berretta, R. (2014). Digital technology teachers' perceptions of computer science: It is not all about programming. In *2014 I.E. Frontiers in Education Conference (FIE) Proceedings* (pp. 1–5). IEEE.

Riley, D. D., & Hunt, K. A. (2014). *Computational thinking for the modern problem solver*. Boca Raton: CRC Press.

*Rodrigues, R. S., Andrade, W. L., & Campos, L. M. S. (2016). Can computational thinking help me? A quantitative study of its effects on education. In *2016 I.E. Frontiers in Education Conference (FIE) Proceedings* (pp. 1–8). IEEE.

Román-González, M., Pérez-González, J. C., Jiménez-Fernández, C. (2016). Which cognitive abilities underlie computational thinking? Criterion validity of the Computational Thinking Test. *Computers in Human Behavior,* 1–14.

*Sabitzer, B., & Pasterk, S. (2015). Modeling: A computer science concept for general education. In *2015 I.E. Frontiers in Education Conference (FIE) Proceedings* (pp. 1–5). IEEE.

Sáez-López, J. M., Román-González, M., & Vázquez-Cano, E. (2016). Visual programming languages integrated across the curriculum in elementary school: A two year case study using "scratch" in five schools. *Computers & Education, 97*, 129–141.
*Scott, A., & Barlowe, S. (2016). How software works: Computational thinking and ethics before CS1. In *2016 I.E. Frontiers in Education Conference (FIE) Proceedings* (pp. 1–9). IEEE.
*Se, S., Ashwini, B., Chandran, A., & Soman, K. P. (2015). Computational thinking leads to computational learning: Flipped class room experiments in linear algebra. In *2015 International Conference on Innovations in Information, Embedded and Communication Systems (ICIIECS) Proceedings* (pp. 1–6). IEEE.
Selby, C. (2013). *Computational thinking: The developing definition*. Paper presented at The 18th Annual Conference on Innovation and Technology in Computer Science Education, Canterbury, UK.
*Shell, D.F., Hazley, M.P., Soh, L.K., Ingraham, E., & Ramsay, S. (2013). Associations of students' creativity, motivation, and self-regulation with learning and achievement in college computer science courses. In *2013 I.E. Frontiers in Education Conference Proceedings* (pp. 1637–1643). IEEE.
*Shell, D. F., Hazley, M. P., Soh, L. K., Miller, L. D., Chiriacescu, V., & Ingraham, E. (2014). Improving learning of computational thinking using computational creativity exercises in a college CSI computer science course for engineers. In *2014 I.E. Frontiers in Education Conference (FIE) Proceedings* (pp. 1–7). IEEE.
Snodgrass, M. R., Israel, M., & Reese, G. C. (2016). Instructional supports for students with disabilities in K-5 computing: Findings from a cross-case analysis. *Computers & Education, 100*, 1–17.
*Standl, B. (2016). A case study on cooperative problem solving processes in small 9th grade student groups. In *2016 I.E. Global Engineering Education Conference (EDUCON) Proceedings* (pp. 961–967). IEEE.
Swaid, S. I. (2015). Bringing computational thinking to STEM education. *Procedia Manufacturing, 3*, 3657–3662.
*Taylor, K. (2013). Can utilizing social media and visual programming increase retention of minorities in programming classes?. In *2013 I.E. Frontiers in Education Conference Proceedings* (pp. 1046–1048). IEEE.
The Royal Society. (2012). *Shut down or restart? The way forward for computing in UK schools*. London: The Royal Society Retrieved from http://royalsociety.org/uploadedFiles/Royal_Society_Content/education/policy/computing-in-schools/2012-01-12-Computing-in-Schools.pdf.
*Toedte, R. J., & Aydeniz, M. (2015). Computational thinking and impacts on K-12 science education. In *2015 I.E. Frontiers in Education Conference (FIE) Proceedings* (pp. 1–7). IEEE.
*Towhidnejad, M., Kestler, C., Jafer, S., Nicholas, V. (2014). Introducing computational thinking through stealth teaching. In *2014 I.E. Frontiers in Education Conference (FIE) Proceedings* (pp. 1–7). IEEE.
*Vieira, C., & Magana, A. J. (2013). Using backwards design process for the design and implementation of computer science (CS) principles: A case study of a Colombian elementary and secondary teacher development program. In *2013 I.E. Frontiers in Education Conference Proceedings* (pp. 879–885). IEEE.
Voogt, J., Fisser, P., Good, J., Mishra, P., & Yadav, A. (2015). Computational thinking in compulsory education: Towards an agenda for research and practice. *Education and Information Technologies, 20*(4), 715–728.
Wing, J. M. (2006). Computational thinking. *Communications of the ACM – Self managed systems, 49*(3), 33–35.
Wing, J. M. (2008). Computational thinking and thinking about computing. *Philosophical Transactions of the Royal Society of London A: Mathematical Physical and Engineering Sciences, 366*(1881), 3717–3725.
Wing, J.M. (2010). *Computational thinking: What and why?*. Retrieved from http://www.cs.cmu.edu/~CompThink/resources/TheLinkWing.pdf.

*Zhao, Y., Zhang, C., Chen, X., & Luo, X. (2013a). Innovative practice teaching mode, pay attention to students' practice ability training Reforming the practice teaching modes of "The Fundamentals of Computer". In *2013 8th International Conference on Computer Science & Education (ICCSE) Proceedings* (pp. 1117–1122). IEEE.

*Zhao, Y. J., Zhang, C. Y., Li, M. J., & Wang, J. (2013b). Innovative practices teaching mode research of the fundamentals of computer. In *2013 8th International Conference on Computer Science & Education (ICCSE) Proceedings* (pp. 1154–1159). IEEE.

Zhong, B., Wang, Q., & Chen, J. (2016). The impact of social factors on pair programming in a primary school. *Computers in Human Behavior, 64*(C), 423–431.

Chapter 3
Microworlds, Objects First, Computational Thinking and Programming

Greg Michaelson

3.1 Overview

Teaching of programming has long been dominated by language-oriented approaches, complemented by industrial design techniques, with little attendant pedagogy. An influential alternative has been Papert's constructivism, through playful exploration of constrained microworlds. The archetypal microworld is based on turtle graphics, as exemplified in Papert's Logo language. Here, students compose and repeat sequences of operations to steer and move a turtle that leaves a trail behind it. Contemporary graphical environments, like Alice and Scratch, augment the turtle world with colourful interacting animated avatars.

However, the microworld approach scales poorly to systematic programming driven by problem-solving. Many students find the transition from novice coding to problem-solving-oriented programming problematic (Moors and Sheenan 2017). Furthermore, microworld languages tend to be relatively impoverished, lacking types and data structures.

Objects first is a contemporary approach to teaching programming through object orientation, which seeks to bridge microworlds and systematic programming. Here, students explore, modify and extend pre-formed objects analogous to microworlds, in constrained subsets of full strength languages, typically Java. However, there is growing evidence that, as with the original microworlds, some students find the transition to problem-solving-based programming difficult.

Computational thinking (CT), as popularised by Wing, offers an approach to problem-solving in which programming is the final stage. CT has been widely heralded as a new pedagogy of programming. However, interpretations of CT vary widely from a loose assemblage of techniques to a systematic discipline.

G. Michaelson (✉)
School of Mathematical & Computer Sciences, Heriot-Watt University, Edinburgh, Scotland
e-mail: G.Michaelson@hw.ac.uk

M. S. Khine (ed.), *Computational Thinking in the STEM Disciplines*,
https://doi.org/10.1007/978-3-319-93566-9_3

In this chapter, I will argue that both microworlds and objects first build superficial programming skills at the expense of deeper competences in problem-solving. I will further argue that systematic CT, driven by seeking patterns in concrete instances, offers a way to refocus on problem-solving for programming.

What follows may seem a bit disjointed, but it gets there in the end. My approach is a mix of pedagogy, history and opinion: I trust it's clear which is which.

3.2 Microworlds and Logo

Since Wing's highly influential intervention (Wing 2006), there has been worldwide interest in computational thinking (CT) as a pedagogy of problem-solving. Tedre and Denning (2016) offer a succinct account of CT before and after Wing. In particular, they draw attention to the key role of Papert in envisioning CT.

Papert (1993) was a proponent of Piaget's constructivist model of cognitive development. Here, a child's transition, from the preadolescent concrete to the adult abstract operational stages of thought, is informed by learning by discovery, termed *bricolage*, which is the exploration of different assemblies of available skills. Thus, Papert expounded a notion of problem-solving based on *combinatorial thinking*, i.e. systematic exploration, in some *microworld*, i.e. a constrained domain, *through thinking about thinking*, i.e. debugging an incorrect solution to find a better one.

This approach derives from early Artificial Intelligence Research. In their report on activities in the MIT AI laboratory, Minsky and Papert (1971) say that:

> ...we see solving a problem often as getting to know one's way around a "micro-world" in which the problem exists. (Minsky and Papert 1971) (cited in (Weir 1987, p.12))

and that:

> *We think that learning to learn is very much like debugging complex computer programs.* To be good at it requires one to know a lot about describing processes and manipulating such descriptions. (italics in original) (Minsky and Papert 1971)

Hence, a notation for describing microworld processes, and an environment for manipulating process descriptions, could facilitate both combinatorial thinking and thinking about thinking.

Papert was highly enthused by the educational possibilities offered by mass access to computers with graphical capabilities. He envisaged how what we now call personal computing facilities might be used to embody and animate microworlds for teaching children.

Primarily concerned with mathematics education, Papert's Logo programming language was intended for manipulating a geometric microworld of turtle graphics. His idea was that, by acting out the behaviours they wished the turtle to perform, a neophyte could learn how to assemble rules characterising those behaviours in the microworld on the computer. Rule assemblies could then be falsified by comparing the behaviour of the on-screen turtle with the intended behaviour, for reformulation.

Papert promoted the benefits of:

> …*deliberately* thinking like a computer, according, for example, to the stereotype of a computer program that proceeds in a step-by-step, literal, mechanical fashion. There are situations where this style of thinking is appropriate and useful. Some children's difficulties in learning formal subjects such as grammar or mathematics derive from their inability to see the point of such a style. (italics in original) (Papert 1993, p. 27)

While Papert is somewhat vague about his pedagogy, his collaborator Weir (1987) thoroughly explored how best to use computers to support "a kind of messing about". For Weir, a microworld is a:

> *small, coherent domain of objects and activities implemented as a computer program* and corresponding to an interesting part of the real world. (italics in original) (Weir 1987, p. 12)

She went on to elaborate the key features of computer-based environments for exploring a microworld, of which I think the following are the most pertinent:

> 1. …as the learner interacts with environment (sic), she "manipulates" concrete embodiments of the concepts to be learned; she "experiences" the concepts directly, not through language about them.
>
> 2. … The idea is to juxtapose experience in the real world with experience in the computational world, so that each complements the other.
>
> 3. A cluster of activities is made available so that any one concept is met in several different contexts and in different combinations.
>
> …
>
> 5. It is useful to choose aspects of the environment about which the learner is likely to have intuitions, to have naïve theories arising from her "street sense"…
>
> …
>
> 7. There should be several levels of formal description, so that a student can move backward and forward from experiential to formal modes of operation.
>
> ’’’
>
> 9. It is to be expected that there will be some things you CAN'T do easily in a particular microworld. (Weir 1987, p. 105)

We can see in this characterisation the essential elements of the objects-first approach, which we will explore below.

Logo's design was strongly influenced by its MIT stablemate LISP (Berkley and Bobrow 1964). Logo (McArthur 1973) is procedural rather than declarative and has a richer syntax than LISP, making it more suitable for use with beginners. Nonetheless, like LISP, Logo is untyped and uses a list form for all data structures. Like LISP, while Logo is Turing complete, its expressiveness depends heavily on a substantial vocabulary of symbolic operators.

The typical use of Logo for teaching is to start with sequences of concrete turtle operations, introduce fixed repetition and then explore abstraction through naming sequences as procedures and generalising operation arguments as variables. Weir notes that:

> Using variables in this way provides a concrete way of approaching the abstract notion of an algebraic variable as it occurs in school mathematics. (Weir 1987, p. 23)

I think that abstraction through the introduction of variables is the key link from coding to programming. However, neither Papert nor Weir offers any guidance for how to do so. We will return to this below.

It is important to remember that Logo was intended to support a constructivist pedagogy that integrated combinatorial thinking and thinking about thinking, not as a programming language per se. Weir seemed to regret the latter use:

> The books that have appeared since tend to regard "doing Logo" as learning to program. But what about acquiring aesthetic and scientific concepts? (Weir 1987, p. 11)

Reflecting on his widely read book *Mindstorms*, Papert (1993) acknowledged that:

> *Mindstorms* unquestionably has a bug for giving prominence to structured programming as a model for thinking about thinking ... although *Mindstorms* emphatically proposes the idea of "bricolage" as a model for general scientific theorizing, this idea comes late in the book and is not developed as an alternative style of programming.... (Papert 1993, p. xv)

There is an enormous literature about Logo and its deployment, much supportive and much critical. There is also an enormous literature about constructivist pedagogy, again much supportive and much critical. I will not explore either here.

Nonetheless, it is salient to reiterate that, in its constructivist conception, Logo was developed to facilitate learners making the transition through bricolage from preadolescent concrete thinking to adult abstract thinking. Little consideration was given to its use with adolescents and adults, like senior secondary and undergraduate students, who might be expected to have attained abstract thought, and for whom the bricolage learning style may not be appropriate.

3.3 From Logo to Objects

From the outset, Logo was widely adopted for school use, at least by schools that could afford the required computing and support infrastructure. For example, an early report for Alberta Education (Kieren 1984) recommends its deployment across schools, at all levels, for a variety of topics. However, the report is guarded about the use of Logo in computer literacy and computer science, beyond early years, and recommends avoiding premature use of more advanced features like:

> procedures, sub-procedures, variables, recursion, or top-down programming. (Kieren 1984, p. 23)

That is, the use of abstraction mechanisms is to be delayed.

Within mainstream computing education in higher education (HE), Logo had considerably less purchase. Programming courses had become universal from the mid-1960s onwards, with content and pedagogy driven largely by changing industrial practices (Michaelson 2015).

Nonetheless, the motivations behind Logo were taken seriously by both computing educationalists and language developers. In particular, Logo influenced the evolution of the general-purpose object-oriented Smalltalk, at Xerox PARC in the early 1970s (Kay 1993).

Object orientation (OO) long predates Logo: the Simula languages (Dahl 2004) were developed in the 1960s for discrete event simulation, and Simula 67 was the first widely used OO language. Kay, Smalltalk's designer, acknowledged Simula as a key influence on Smalltalk (Kay 1993).

Kay, having encountered Piaget's and Papert's pedagogies through Minsky, visited Papert's team in 1968 and was impressed by Logo's use with children in local schools. His group started teaching Smalltalk to school children from 1973, deploying what were effectively microworlds to underpin learning (Kay 1993).

Smalltalk is a far more expressive language than Logo but, like Logo, was implemented from the start in a visual environment. Indeed, one of the first classes constructed in Smalltalk was for Logo turtle graphics (Kay 1993).

However, Kay observed that non-programmer adults found the transition from very simple to somewhat more complex problems very hard. He analysed a program he thought straightforward to construct and found 17 "non-obvious ideas":

> And some of them were like the concept of the arch in building design: very hard to discover, if you don't already know them. (Kay 1993, p. 82)

To address this, Kay's team decided that design should be taught explicitly. His collaborator Goldberg introduced design templates as intermediary forms, to aid decomposing a problem into classes and messages; this approach proved successful.

We could view the use of design templates as a form of *scaffolding*. This constructivist concept was elaborated by Wood et al. (1976), building on Vygotsky's notion of *zones of proximal development* (Vygotsky 1978), in turn developed as a critique of Piaget:

> *...the zone of proximal development...is the distance between the actual developmental level as determined by independent problem solving and the level of potential development as determined by problem solving under adult guidance or in collaboration with more capable peers* (italics in original). (Vygotsky 1978, p. 86)

Thus, Wood et al. argued that traversing a zone of development requires scaffolding to facilitate it:

> This scaffolding consists essentially of the adult "controlling" those elements of the task that are initially beyond the learner's capacity, thus permitting him to concentrate upon and complete only those elements that are within his range of competence ... It may result, eventually, in development of task competence by the learner at a pace that would far outstrip his unassisted efforts. (Wood et al. 1976, p. 90)

We will return to this notion below.

3.4 OO and Objects First

Through the 1970s and 1980s, mainstream higher education (HE) computing continued to be led by industrial imperatives. Impressionistically, while COBOL, and Algol descendants like Pascal, still predominated, we can observe a steady transition to the OO language C++, driven by the rapid uptake of the free, platform-independent UNIX operating system from Bell Laboratories.

UNIX was written in C and ran on minicomputers and workstations, offering relatively low-cost multi-access, appealing to HE, and a robust software engineering environment, appealing for industrial computing. In turn, C++ was derived from C and, like Smalltalk, influenced by Simula 67 (Stroustrup 1987).

In the same period, emerging industrial approaches to OO design, for example, Rumbaugh et al.'s (1990) object modelling technique (OMT), were also adopted in HE. Typically, however, a C++ subset was used to teach traditional procedural programming prior to, and independently of, OO. Smalltalk was not widely adopted.

A fundamental change in undergraduate computing education came in the mid-1990s, with the widespread adoption of SUN Microsystem's OO language Java (Gosling et al. 1996), eventually displacing COBOL for commercial computing and rapidly gaining use for wider software engineering alongside C++.

Java is disconcertingly like C in appearance, but had OO as the key language design principle from the outset. Unlike Smalltalk and C++, Java supports single rather than multiple inheritance, making it simpler for OO teaching.

Java was complemented by the integration of similar but competing industrial OO methodologies, including OMT, into the Unified Modelling Language (UML) (Booch et al. 2005) which rapidly became, and remains, the standard OO design and education practice.

Herein lie the roots of objects first (OF). Many practitioners noted that students who had already been exposed to procedural programming found the subsequent adoption of OO difficult. It was thought that this might be circumvented by starting with OO. See, for example, (Wallace and Martin (1997). Thus, Java and OO with UML quickly displaced procedural programming as the initial undergraduate teaching methodology.

Nonetheless, constructivist thinking, and Papert's pedagogy, were still current in computing education. For example, Brusilovsky et al. (1997) critiqued general-purpose languages as being too large, over-oriented to numeric and symbolic computation and lacking visualisation. Citing Logo as a major influence, they advocated the use of mini-languages oriented to constrained domains, much like microworlds, for teaching programming principles, in visual environments. They also advocated the use of subsets of full languages, which they term sublanguages.

Echoing Papert and Weir, Brusilovsky et al. commented that:

> Note that the application of a mini-language is never the goal itself, but a method of mastering a set of notions and skills. If this set contains not only programming concepts, but also some concepts from another domain, a mini-language might be useful to learn this domain (as Logo is used to learn geometry). (Brusilovsky et al. 1997)

Now, just before the emergence of Java, Kölling et al. (1995) critiqued extant OO languages, like C++, Smalltalk and Eiffel, as too complex, and their development environments as too unwieldy, for initial teaching. Subsequently, Kölling and Rosenberg built the Blue OO teaching language (Kölling and Rosenberg 1996a) and development environment (Kölling and Rosenberg 1996b). Serendipitously, Blue shares two key characteristics of Brusilovsky et al.'s (1997) recommendations: a small language intended to aid learning, implemented in a simple visual environment to support experimentation.

As Kölling acknowledges (Kölling 2016), academics face an uphill struggle to broaden the use of novel in-house teaching languages, to compete with industrial languages. For example, in 2017, Cass (2017) reported that the top ten languages for IEEE Spectrum readers were Python, C, Java, C++, C#, R, JavaScript, PHP, Go and Swift. For comparison, Murphy et al. (2017) reported that in UK universities in 2016, the top ten teaching languages were Java, Python, C++, C, JavaScript, Haskell, C#, Processing, Matlab and PHP.

Thus, the promising outcomes from experiences with Blue were subsequently revisited by Barnes and Kölling (2003), in the BlueJ environment for teaching initial OO in Java.

BlueJ was explicitly developed to support OF (Kölling 2016, p. 13). In their guidelines for teaching with BlueJ, Kölling and Rosenberg (2001) observe that:

> With BlueJ we can really interact with objects as the very first thing we do. Since objects can be created interactively, the first activity for students should be to open an existing project, create a few objects, make method calls on these objects and inspect the objects' state. Here, we really interact with objects before introducing any other concept. Objects come truly first. (Kölling and Rosenberg 2001, p. 2)

This serendipitously met Brusilovsky et al.'s (1997) third key recommendation: a microworld like approach to programming education.

In a typical curriculum based on OF, as exemplified say by Barnes and Kölling (2003), students are presented with a programming environment for visualising objects, pre-primed with some simple class. Students start by constructing instances of the class and invoking methods to learn what changes they cause. Next, they start to solve problems that involve modifying instances in specified ways, by sequencing method invocations. Students then explore the code behind the methods, and, with appropriate guidance, modify methods to change their behaviours, and construct new methods with new behaviours.

This approach is far more general than the original microworld conception. In principle, the scope and properties of the introductory class of objects are limited solely by the instructor's ingenuity and skill, rather than being constrained by a domain-oriented notation like the original Logo.

OF in general and BlueJ specifically have proved highly popular, and both are still widely used in HE. Nonetheless, the efficacy of OF is controversial.

3.5 Critiques of Objects First

Hu (2004) provides a thorough survey of limitations to OF. In particular, he observes that OF gives an inadequate grasp of basic algorithms, because arrays are covered late.

Of interest from a constructivist perspective, and contrary to my suggestion above, Hu argues that most young adults lack appropriate abstract thinking capabilities for OO:

> When addressing the reasons students were unable to see the forest in their programming activities, Reek (1995) pointed out that "the problem isn't that the students are stupid, but rather that at age eighteen their thinking maturity is still at the concrete level". Teaching objects-first ignores this fact and thus creating an environment that forces students to think at a higher level of abstraction, which is often the point where the confusion starts. (Hu 2004, p. 212)

To me, this suggests that OF does not adequately scaffold abstraction. However, Hu recommends a return to procedural programming for beginners, followed by OO.

Lister et al.'s (2006) comprehensive, but dense, analysis of the SIGCSE mailing list discussion of OF, which considered 99 postings by 39 people, reached no definitive conclusions:

> There is a fairly strong consensus that programming is hard both to teach and to learn, but the case that objects-early is harder (or easier) than objects-late has not yet been made conclusively. (Lister et al. 2006, p. 150)

However, they make the interesting observation that:

> A key distinction is evident from the debate, however. Two computing sub-disciplines are contending over the role of programming as:
> 1. A manipulative tool for the conduct of algorithmic thought experiments in a purely scientific CS model, as opposed to
> 2. The centrality of design in the construction of large scale software systems by professional software engineers. (Lister et al. 2006, p. 159)

Finally, in their well-designed study, Ehlert and Schulte (2009) compared programming learning outcomes for 17-year-olds following either an OF or a procedural approach in a first programming course. Both cohorts covered the same topics overall, but in different orders. For OF, students began with instance experimentation, before moving on to variables and control structures. For the procedural approach, students began with variables and control structures, meeting OO much later. The evaluation found that, at the end of the study, the OF cohort had somewhat higher attainment of overall concepts than the procedural cohort, but, in a follow-up 8 weeks later, there was no significant difference in attainment.

Curiously, both cohorts found arrays and association difficult: the procedural cohort met arrays three topics before the OF cohort, and both met association as the final topic. To me, this suggests that neither approach provided appropriate scaffolding.

3.6 Computational Thinking, Papert and Objects First

Let us now return to computational thinking (CT). In popular discourse, we can discern a spectrum from what we might call weak to strong CT.

At one end, weak CT is toolkit of general-purpose, domain-independent problem-solving techniques that derive from computing. We have seen that for Papert (1993), problem-solving involves "thinking like a computer". Alternatively, CT may be presented as thinking like a computer scientist. For example, for Google (2017):

> The basic skills of computer scientists and the way they think are computational thinking. The area in which you apply CT can be any subject area or topic, even the subject area or topic you teach.

Denning et al. (2017, pp. 32–33) offer a robust, if scattergun, criticism of the alleged primacy and all-embracing nature of CT.

Contrariwise, Denning (2017) critiques what he sees as too narrow a conception of CT as problem-solving:

> Underlying all the claims is an assumption that the goal of computational thinking is to solve problems. Is everything we approach with computational thinking a problem? No. We respond to opportunities, threats, conflicts, concerns, desires etc. by designing computational methods and tools – but we do not call these responses problem-solutions. It seems overly narrow to claim that computational thinking, which supports the ultimate goal of computational design, is simply a problem solving method. (Denning 2017, p. 39)

Between weak and strong CT, Yadav et al. (2017) list nine core concepts from the Computer Science Teachers Association/International Society for Technology in Education (CSTA/ISTE):

> data collection, data analysis, data representation, problem decomposition, abstraction, algorithms and procedures, automation, parallelization, and simulation. (Yadav et al. 2017, p. 57)

They also list six CT practices from a College Board/National Science Foundation stand-alone course:

> connecting computing, creating computational artefacts, abstracting, analyzing problems and artefacts, communicating, and collaborating. (Yadav et al. 2017, p. 58)

While this offers more precision, nonetheless, on this basis just about any educational activity could be claimed to embody CT.

At the other end of the spectrum, strong CT is a systematic discipline of problem-solving. Thus, Kao (2011), following Wing (2006), presents the four key aspects of CT as:

> ...
>
> - Decomposition: the ability to break down a problem into subproblems.
> - Pattern recognition: the ability to notice similarities,differences, properties, or trends in data.
> - Pattern generalization: the ability to extract unnecessary details and generalize those that are necessary in order to define a concept or idea in general terms.
> - Algorithm design: the ability to build a repeatable, step-by-stepprocess to solve a particular problem. (Kao 2011, p. 6)

BBC Bitesize (2017) repeats these almost verbatim, identifying Kao's (2011) pattern generalisation as abstraction:

> There are four key techniques (cornerstones) to computational thinking:
>
> - Decomposition – breaking down a complex problem or system into smaller, more manageable parts
> - Pattern recognition – looking for similarities among and within problems
> - Abstraction – focusing on the important information only, ignoring irrelevant detail
> - Algorithms – developing a step-by-step solution to the problem, or the rules to follow to solve the problem. (BBC Bitesize 2017)

Note that these aspects lie at the core of the CSTA/ISTE characterisation. Curiously, Google (2017) includes a variant of Kao's definition, while also repeating the CSTA/ISTE list.

Both Papert's approach and OF are based on some pre-given decomposition of a domain, and scaffold algorithm formation through combinatorial thinking, that is by exploring the efficacy of different groupings of rules (Papert) or methods (OF) against some problem requirement.

Here, I think that OF offers little advance on Papert. In particular, there seems to be little principled consideration of what constitute key problem-solving and programming concepts or how their acquisition should be staged. Thus, there are no explicit criteria for choosing an initial object class or deploying it to scaffold concept acquisition beyond method sequencing.

Furthermore, neither approach:

- Scaffolds reflection on the efficacy of potential solutions found by discovery, that is, thinking about thinking
- Has anything to say about pattern identification or abstraction
- Offers any guidance on decomposition of a novel problem domain from scratch

3.7 Scaffolding Programming with Patterns and Computation Structures

While I have critiqued Kao's characterisation as not taking adequate account of information in problem-solving (Michaelson 2015), I am very much of the strong CT persuasion.

I also see problem-solving as the heart of programming. However, this view is by no means universal.

For example, Guizdal (2017) suggests that starting with problem-solving in teaching computer science may be counter-productive. Citing Sweller (1988), he asserts that:

> Problem-solving creates enormous cognitive load that interferes with learning to understand. (Guizdal 2017, pp. 11)

Instead, Guizdal proposes starting with program comprehension:

> To teach for understanding, we would give students worked examples and ask them questions about the examples, ask students to predict outcomes or next steps in a visualisation.... (Guizdal 2017, pp. 11)

While I acknowledge that program comprehension is a vital component of learning to program, I think that problem-solving is key to becoming an effective programmer, right from the start. Furthermore, I think that understanding and comparing code fragments is fundamental to abstraction in CT-based problem-solving.

As I have argued elsewhere (Michaelson 1992, 2015), I think that problem-solving should be driven by abstraction from concrete instances of a specific problem, to identify the constructs appropriate to a general solution. In particular, we should use abstraction to identify the types and variables that an algorithm may manipulate to solve an arbitrary instance.

This approach is in keeping with Vygotsky's (1962) characterisation of the final stages in the development of concept formation in young persons:

> Only the mastery of abstraction, combined with advanced complex thinking, enables the child to progress to the formation of genuine concepts. A concept emerges only when the abstracted traits are synthesized anew and the resulting abstract synthesis becomes the main instrument of thought. The decisive role in this process, as our experiments have shown, is played by the word, deliberately used to direct all the part processes of advanced concept formation. (Vygotsky 1962, p. 78)

That is, as Berger (2005) suggests for mathematics, I see the role of the variable in problem-solving as comparable to the role of the word in Vygotsky's notion of advanced concept formation.

Here, I think that the CT notion of pattern identification offers a very practical bridge from concrete instances to abstractions with variables. However, the term "pattern" may be misleading. We normally understand a pattern to have some repetitive fixed structure, for example, in tiling a floor as a black and white chequerboard. But, for problem-solving, we actually want to identify *the underlying regularities in differences*. So a pattern is *some property that lots of different instances share*; ideally one that enables us to *generate new instances* that also share the property.

I also think that the notion of identifying a pattern should be more than just a metaphor for some informed intuition that comes with experience. Rather, pattern identification should be taken literally, as an approach that involves directly comparing structural and operational features in solving concrete instances of problems, to identify how they are similar and how they differ.

This generalises Papert's combinatorial thinking, where the learner behaves within a constrained domain to write down sequences of actions for the computer to perform. In an arbitrary domain, we proceed by solving lots of concrete instances of a problem, analysing what we did with the concrete data to get to the concrete results and then looking for patterns across instances in both data and actions. We can do this by decomposing our data and actions at finer and finer levels of detail until we can identify and generalise the patterns amongst and across elementary operations on elementary data. This is reminiscent of Turing's analysis of someone doing arithmetic with pencil and squared paper, going to right down to symbol-by-symbol operations (Turing 1936).

Furthermore, I think we should view patterns as templates, encompassing both information and computation:

- Patterns in data lead to variables and expressions to calculate with them.
- Patterns in computations lead to assignments, structured programming constructs and subprograms.

Thus, I envisage patterns as skeletal forms in some well-defined notation, for example, a pseudocode or reference language (Scottish Qualifications Authority 2015) or, indeed, some programming language. That is, ultimately, patterns are rendered as chunks of syntax with holes in them.

Note that, here, there are fundamental differences to OO design patterns (Gamma et al. 1995) which are used to structure complex programs constructed from classes with well-defined interfaces. This usually involves retrofitting extant components to a pattern. In my conception, control patterns have closer analogies with higher-order function from functional programming (Michaelson 1992). These are used to abstract common patterns of recursion for instantiation with context-specific functions to control repetition and computation.

3.7.1 Finding Variables and Expressions

An elementary expression typically has variables for values that change and constants for values that don't change, combined by unchanging operations. We can find expressions by abstracting over concrete calculations.

For example, suppose we want to travel 400 km. If our vehicle goes at 100 km per hour, the journey takes:

$$400/100 == 4 \text{ h}$$

And if our vehicle goes at 80 km per hour, the journey takes:

$$400/80 == 5 \text{ h}$$

Comparing the expressions, they have "400/" in common so we can abstract where they differ:

$$400/? \text{ h}$$

Now, we can introduce a variable called "speed", which has to be an integer like 100 and 80, and then for unknown values find:

400/speed

We all know how to do this. How about making it explicit?

3.7.2 Finding Computation Patterns

Let's next consider three archetypal computation structures and their corresponding programming language forms: choice, repetition and iteration.

First of all, an *if* statement typically has a choice condition, with branches depending on whether it is true or false. We can view this as a pattern for discriminated computation. In turn, this may depend on being able to identify a pattern to divide data into two groups, where each group is processed in the same manner. That pattern then frames the condition.

Next, a *while* statement typically has a termination condition and a body. We can view this as a pattern that determines when to stop performing a repeated computation. In turn, this may depend on being able to identify a pattern in either data or actions, characterising how each step determines the next.

Finally, a *for loop* typically has a control variable with initialisation and incrementation expressions, a termination condition, and a body. We can view this as a pattern for counted or indexed computation. In turn, this may depend on being able to identify a pattern characterising a sequence of data, where each element is processed or generated in the same manner, or processing each element depends on some property of the previous element.

In all three cases, our understanding of the computation pattern guides our abstractions. That is, when we explore data looking for computation patterns, we should ask of it questions that enable us to choose amongst the possibilities.

All three computation patterns involve:

- A condition that determines what to do next
- What is done next
- What it is done to

So maybe we might interrogate our data by asking:

- What have we got?
- What are we doing to it?
- Why or when are we doing it?

3.7.3 Example

Suppose we're organising a sports club outing to the chariot racing, and we want to work out how many full price and how many concession price tickets we need to purchase. Suppose we know everyone's age:

12 44 23 63 13 69 10 12 61 . . .

and that anyone under 16 or over 60 can get a concession.

To begin with, let's write down what we do with each concrete age in turn:

Data	Action	Condition
12	Count 1 concession	Under 16
44	Count 1 full price	Between 16 and 60
23	Count 2 full price	Between 16 and 60
63	Count 2 concession	Over 60
13	Count 3 concession	Under 16
69	Count 4 concession	Over 60
. . .	. . .	. . .

We've already implicitly abstracted out accumulation variables for the concession and full price counts, just by describing what we're doing.

On closer inspection, we can see that the action for each concrete data item involves incrementing a count depending on which condition is met:

Data	Action	Condition
12	Increment concession	Under 16
44	Increment full price	Between 16 and 60
23	Increment full price	Between 16 and 60
63	Increment concession	Over 60
13	Increment concession	Under 16
69	Increment concession	Over 60
. . .	. . .	. . .

We can regroup by the three conditions:

Data	Action	Condition
12	Increment concession	Under 16
13	Increment concession	Under 16
44	Increment full price	Between 16 and 60
23	Increment full price	Between 16 and 60
63	Increment concession	Over 60
69	Increment concession	Over 60
. . .	. . .	. . .

Now we can spot the choice structure pattern, abstract over the age in the three conditions and introduce an if statement for each:

```
IF age under 16 THEN
increment concession
ELSE
IF age over 60 THEN
increment concession
ELSE
increment full
END IF
END IF
```

Alternatively, we can look for a pattern based on grouping abstracted actions:

Data	Action	Condition
12	Increment concession	Under 16
13	Increment concession	Under 16
63	Increment concession	Over 60
69	Increment concession	Over 60
44	Increment full price	Between 16 and 60
23	Increment full price	Between 16 and 60
...	...	...

Again, we can abstract over the age but introduce two if statements are having combined conditions for common actions:

```
IF age under 16 OR over 60 THEN
increment concession
ELSE
increment full price
END IF
```

We can now think about how processing the sequence of data, rather than each data item, is structured. Let’s make the process of dealing with each item in turn more explicit:

Data	Action	Condition
	Look for more data	More data
12	...	...
	Look for more data	More data
44	...	...
	Look for more data	More data
23	...	...
...	...	...
	Look for more data	No more data
	Stop	

This looks like a repetition structure of the form:

```
look for more data
WHILE more data DO
IF ... THEN ... ELSE ...
look for more data
END WHILE
```

Again, we all know how to do this sort of blow-by-blow analysis, and with experience, we come to do so in big "intuitive" steps. But beginners don't know where to start. Again, how about making this analysis explicit?

3.7.4 Finding Information Structures

Note that the notion of a type encompasses structured information, for example, linear sequences and nested structures, not just base types like numbers. However, identifying how information is structured need not lead us immediately to storing data in such a structure. That depends on whether processing the information requires it to be stored for subsequent access.

For example, it is tempting to automatically identify a linear sequence of numbers with an array. However, finding the average of a sequence of numbers can be performed by taking each number in turn. In contrast, finding all the numbers less than the average requires them all to be held until the average is known. Again, this is something we can tell by systematic and detailed working with concrete instances.

3.8 Conclusions

Teaching programming is hard. It seems deeply unsatisfactory that some people seem to get it and some don't, and that we don't understand why.

After critiquing the bricolage/microworld and objects-first approaches to teaching programming, I've argued that we might better deploy a pedagogy based on strong CT. That is, I think that we can teach beginners how to systematically analyse concrete instances of problems, to tease out the essential patterns in data and computations, leading to abstractions that can form the basis of algorithms and, ultimately, programs.

Acknowledgements I would like to thank Nancy Falchikov for valuable discussions about constructivism, development and cognition.

References

Barnes, D., & Kölling, M. (2003). *Objects First with Java: A practical introduction using BlueJ*. Upper Saddle River: Prentice-Hall.

Berger, M. (2005). Vygotsky's theory of concept formation and mathematics education. In H. L. Chick & J. L. Vincent (Eds.), *Proceedings of the 29th conference of the International Group for the Psychology of Mathematics Education* (Vol. 2, pp. 153–160). Melbourne: PME.

BBC Bitesize. (2017). *Introduction to computational thinking*. http://www.bbc.co.uk/education/guides/zp92mp3/revision (inspected 16/5/17).

Berkley, E. C., & Bobrow, D. G. (Eds.). (1964). *The programming language LISP: Its operation and applications*. Cambridge, MA: MIT Press.

Booch, G., Rumbaugh, J., & Jacobson, I. (2005). *The Unified Modeling Language user guide* (2nd ed.). Upper Saddle River: Addison-Wesley.

Brusilovsky, P., Calabrese, E., Hvorecky, J., Kouchnirenko, A., & Miller, P. (1997). Mini-Languages: A way to learn programming principles. *Education and Information Technologies, 2*(1), 65–83.

Cass. S. (2017). The 2017 top programming languages, *IEEE Spectrum*, 18th July, 2017.

Dahl, O. J. (2004). The birth of object orientation: the Simula languages. In *From object-orientation to formal methods*. Berlin: Springer.

Denning, P. J. (2017). Remaining trouble spots with computational thinking. *CACM, 60*(6), 33–39.

Denning, P. J., Tedre, M., & Yongpradit, P. (2017). Misconceptions about computer science. *CACM, 60*(3), 31–33.

Ehlert, A., & Schulte, C. (2009). Empirical comparison of objects-first and objects-later. In *Proceedings of the Fifth International Workshop on Computing Education Research* (ICER'09), ACM (pp. 15–26).

Gamma, E., Helm, R., Johnson, R., & Vlissides, J. (1995). *Design patterns: Elements of reuseable object-oriented software*. Boston: Addison- Wesley.

Google. (2017). What is computational thinking? In computational thinking for educators. https://computationalthinkingcourse.withgoogle.com/unit?lesson=8&unit=1 (inspected 16/5/17).

Gosling, J., Joy, B., & Steele, G. L., Jr. (1996). *The Java language specification*. Boston: Addison Wesley.

Guizdal, M. (2017). Balancing teaching CS efficiently with motivating students. *CACM, 60*(6), 10–11.

Hu, C. (2004). Rethinking of teaching objects first. *Education and Information Technologies, 9*(3), 209–218.

Kao, E. (2011). Exploring computational thinking at Google. *CSTA Voice, 7*(2), 6.

Kay, A. C. (1993). The early history of Smalltalk. *ACM SIGPLAN Notices, 28*(3), 69–95.

Kieren, T. E. (1984). *LOGO in education: What, how, where, why and consequences*. Planning Services, Alberta Education.

Kölling, M. (2016). Lessons from the design of three educational programming environments: Blue, BlueJ and Greenfoot. *International Journal of People-Oriented Programming, 4*(1), 25–32.

Kölling, M., & Rosenberg, J. (1996a). Blue – A language for teaching object oriented programming. In *Proceedings of 27th SIGCSE technical symposium on computer science education, Philadelphia, Pennsylvania, USA* (pp. 190–194). *SIGCSE Bulletin*, *28*(1).

Kölling, M., & Rosenberg, J. (1996b). An object oriented development environment for the first programming course. In *Proceedings of 27th SIGCSE technical symposium on computer science education, Philadelphia, Pennsylvania, USA* (pp. 83–87). *SIGCSE Bulletin*, *28*(1).

Kölling, M., & Rosenberg, J. (2001). Guidelines for teaching object orientation with Java. In *Proceedings of the 6th Conference on Information Technology in Computer Science Education (ITiCSE 2001)*, Canterbury (pp. 33–36).

Kölling, M., Koch, B., & Rosenberg, J. (1995). Requirements for a first year object-oriented teaching language. *ACM SIGCSE Bulletin, 27*(1), 173–177.

Lister, R., Berglund, A., Clear, C., et al. (2006). Research perspectives on the objects-early debate. In *Working group reports on Innovation and Technology in Computer Science Education (ITiCSE-WGR '06), ACM* (pp. 146–165).

McArthur, C. D. (1973). LOGO user's guide and reference manual, Bionics Research Reports: No. 14, Bionics Research Laboratory, School of Artificial Intelligence, University of Edinburgh.

Michaelson, G. (1992). *Elementary Standard ML.* UCL Press.

Michaelson, G. (2015). Teaching programming with computational and informational thinking. *Journal of Pedagogic Development, 5*(1), 51–65.

Minsky, M., & Papert, S. (1971). Artificial Intelligence Research Report, Memo AIM-252, AI Laboratory, MIT.

Moors, L., & Sheenan, R. (2017). Aiding the transition from novice to traditional programming environments. In *Proceedings of IDC 2017: ACM interaction design and children conference* (pp. 509–514). Stanford University.

Murphy, E., Crick, T., Davenport, J.H. (2017). An analysis of introductory programming courses at UK Universities, *The Art, Science, and Engineering of Programming*, Vol. 1, No. 2, 2017, Article 18; 23 pages.

Papert, S. (1993). *Mindstorms.* (2nd edn). Basic Books.

Reek, M. (1995). A top-down approach to teaching programming. *SIGSCE Bulletin, 27*(1), 6–9.

Rumbaugh, J., Blaha, M., Premerlani, W., Eddy, F., & Lorensen, W. (1990). *Object-oriented modeling and design.* New Jersey: Prentice Hall.

Scottish Qualifications Authority. (2015). *Reference language for advanced higher computing science question papers.* http://www.sqa.org.uk/files_ccc/ComputingScienceReflanguageSpecificationsSQPAH.pdf. (consulted 17/5/17).

Stroustrup, B. (1987). *The C++ programming language*. Addison-Wesley.

Sweller, J. (1988). Cognitive load during problem solving: Effects on learning. *Cognitive Science, 12*, 257–285.

Tedre, M., & Denning, P. (2016). The long quest for computational thinking. In *Proceedings of the 16th Koli calling conference on computing education research, November 24–27, 2016* (pp. 120–129). Koli.

Turing, A. (1936). On computable numbers, with an application to the Entscheidungs problem. *Proceedings of the London Mathematical Society, 42*(1), 230–265.

Vygotsky, L. S. (1962). *Thought and language*. Cambridge, MA: MIT Press.

Vygotsky, L. S. (1978). *Mind in society*. Cambridge: MIT Press.

Wallace, C., & Martin, P. (1997). Not **whether** Java but **how** Java. In *Proceedings of Java in the computing curriculum conference (JICC 1)*, South Bank University.

Weir, S. (1987). *Cultivating minds: A Logo casebook.* New York: Harper and Row.

Wing, J.M. (2006). Computational thinking. *CACM viewpoint*, March, pp. 33–35.

Wood, D., Bruner, J. S., & Ross, G. (1976). The role of tutoring in problem solving. *Journal of Child Psychology and Psychiatry, 17*, 89–100.

Yadav, A., Stephenson, C., & Hong, H. (2017). Computational thinking for teacher education. *CACM, 60*(4), 55–62.

Chapter 4
Toward a Phenomenology of Computational Thinking in STEM Education

Pratim Sengupta, Amanda Dickes, and Amy Farris

4.1 Introduction

In this chapter, we argue for an epistemological shift from viewing coding and computational thinking as mastery over computational logic and symbolic forms to viewing them as a more complex form of *experience.* Rather than viewing computing as regurgitation and production of a set of axiomatic computational abstractions, we argue that computing and computational thinking should be viewed as discursive, perspectival, material, and embodied experiences, among others. These experiences include, but are not subsumed by, the use and production of computational abstractions. We illustrate what this paradigmatic shift toward a more phenomenological account of computing can mean for teaching and learning STEM in K-12 classrooms by presenting a critical review of the literature, as well as by presenting a review of several studies we have conducted in K-12 educational settings grounded in this perspective.

Papert (1987) famously referred to *technocentrism* as the fallacy of referring all questions about technology to the technology itself. A critical look at the history of educational computing tells us that the research in this field has also been predominantly technocentric in nature. Calls for taking into account the learners' experiences as building blocks for deeper learning and the development of disciplinary

Partial support from NSF CAREER Award #115230 and the Imperial Oil Foundation is gratefully acknowledged. All opinions are the author's and not endorsed by funding agencies.
Version 5.0 (6 March 18)

P. Sengupta (✉)
University of Calgary, Calgary, AB, Canada
e-mail: pratim.sengupta@ucalgary.ca

A. Dickes
Harvard University, Cambridge, MA, USA

A. Farris
The Pennsylvania State University, University Park, PA, USA

M. S. Khine (ed.), *Computational Thinking in the STEM Disciplines*,
https://doi.org/10.1007/978-3-319-93566-9_4

expertise in STEM certainly have been made (e.g., Papert 1980; DiSessa 2000). However, the predominant effect of this call has also been technocentric in the sense that it has resulted in the creation of a new genre of programming languages (e.g., LOGO, Scratch, NetLogo, StarLogo TNG, AgentSheets, ViMAP, CTSiM, etc.) and microcontrollers (e.g., Arduino) designed to be easily usable for the "novice programmer." The technocentric focus is also evident in the learning objectives and assessment of computational thinking, which predominantly focus on the production and use of computational abstractions (e.g., see the studies reviewed by Grover and Pea (2013a)). Only a few, recent examples have focused on phenomenological aspects of computational thinking, such as the centrality of discourse (Grover and Pea 2013b; Farris and Sengupta 2014), the role of embodied reasoning (Francis et al. 2016), aesthetic experiences (Farris and Sengupta 2016) and the importance of managing, rather than ignoring uncertainty (Farris et al. 2016) in the development of computational thinking in STEM curricular contexts. And while recent arguments have been made for an increased awareness for paying attention to sociological dimensions of computing such as computing in public spaces (Sengupta and Shanahan 2017), virtual communities (e.g., online Scratch communities) and out-of-school, DIY makerspaces (Kafai and Burke 2013), our focus here is on the K-12 public school classroom.

Our chapter is an argument for deepening and broadening the focus on the phenomenology of computing and computational thinking in K-12 STEM curricular contexts and classrooms. Our concerns are both epistemological and pedagogical and are grounded historically as well as in the pragmatics of K-12 classrooms with the focus on sustaining computing as a long-term practice. The first part of the chapter presents a critical and synthetic review of the literature and argues for a phenomenological approach toward developing an epistemology of computational thinking that foregrounds the uncertainty and complexity in the experience of computing and science, in professional practice and in STEM classrooms. The second part of the chapter presents a set of pedagogical approaches for sustaining computing and computational thinking through computational modeling in the STEM classroom. This is presented in the form of a critical review of studies that are conducted by our research group in K-12 classrooms in the USA, including studies that were conducted in the form of partnerships with teachers.

4.2 The Need for a Phenomenology of Computational Thinking

Since the phrase "computational thinking" has been popularized by Wing (2006), there have been a plethora of studies on computational thinking in education. Yet, beyond the early work on computational literacy by Papert (1980) and diSessa (2001), the *epistemology* of computational thinking has received very little attention in the literature. In this section, we examine core beliefs and assumptions about the nature of knowledge and knowing that are and should be involved in thinking computationally, by adopting a historical perspective as well as by reviewing recent research in and relevant to educational computing, from a phenomenological perspective. We highlight the importance of grounding computational thinking in representational and epistemic practices that are central to *knowing* and *doing* in science and, more broadly, in STEM education. The phenomenologist Merleau-Ponty (1962) defined *sense experience* as "that vital communication with the world which makes it present as a familiar setting of our life" (Merleau-Ponty 1962, pp 61). We believe that thinking carefully in terms of these practices can help us understand the materiality, uncertainty and subjectivity inherent in the students' and teachers' *sense experiences* of computational thinking in STEM classrooms, for reasons we explain in more detail next.

4.2.1 Inseparability of Abstractions and Practices in Computing and Science

Citing a definition coined together with Jan Cuny of the National Science Foundation and Larry Snyder of the University of Washington, Wing (2011) defined "computational thinking" to indicate the "thought process involved in formulating problems and their solutions so that the solutions are represented in a form that can be effectively carried out by an information-processing agent [CunySnyderWing10]" (Wing 2011, p 20). According to Wing, the "essence of computational thinking is *abstraction*" (Wing 2008, pp 3717). She argued that computational thinking involves dealing with abstractions in the following ways: (a) defining abstractions, (b) working with multiple layers of abstraction, and (c) understanding the relationships among the different layers (Wing 2008). Abstractions, according to Wing, give computer scientists the power to scale and deal with complexity. She noted:

> Abstraction is used in defining patterns, generalizing from instances, and parameterization. It is used to let one object stand for many. It is used to capture essential properties common to a set of objects while hiding irrelevant distinctions among them. (Wing 2011, p 20)

Wing's conceptualization of abstraction, as the excerpt above shows, therefore, emphasizes the notion of generalization. Abstractions, in her view, are generalized computational representations that can be used (i.e., applied) in multiple situations or contexts. In this sense, as Sengupta et al. (2013) pointed out, her definition of abstraction is similar to Locke's. In Locke's view, abstraction is the process in which "ideas taken from particular beings become general representatives of all of the same kind" (Locke 1690/1979).

However, a phenomenological interpretation of Wing's notion of abstractions is incomplete without a deeper understanding of the contextualization that necessitates and grounds computational abstractions in professional practice. For example, the computer scientist and software engineering researcher Douglas C. Schmidt (2006) points out that software researchers and developers typically engage in creating abstractions that help them program in terms of their contextualized design goals – e.g., the specific problem that they are solving, which is often in a different field (domain) of professional practice. The abstractions that "need" to be created are essential because the end user must be shielded from avoidable complexities, such as the CPU, memory, and network devices, and, instead, interact directly with the domain-specific problem (Schmidt 2006). Similarly, it is important to note that even Wing (2011) acknowledges the complexity of computing systems as resulting from the material and physical constraints underlying the information-processing agent and its operating environment. She argues that while considering what computational thinking is, we must also "worry about boundary conditions, failures, malicious agents and the unpredictability of the real world" (Wing 2011, pp 20).

We therefore believe that the term "thinking" in computational thinking is a semantic reduction of its intended meaning. Phenomenologically, computational thinking involves both representational and epistemic work that are also grounded disciplinarily and materially. It is in this light that Sengupta et al. (2013) argued that when the notion of computational abstractions is grounded *in use*, it could be understood as a practice that draws upon concepts that are fundamental to computing and computer science, and it also includes practices such as problem representation, abstraction, decomposition, simulation, verification, and prediction that are also central to modeling, reasoning, and problem-solving in a large number of scientific, engineering, and mathematical disciplines (National Research Council 2007; NGSS 2015).

Sociologists and philosophers of science have also identified the inseparability of abstractions and practice in the work of scientists. It is rarely the case that the transformation of an initial idea to a successful scientific experiment or a model is a simple and linear process that relies on solely the invention and use of abstractions. The philosopher Andrew Pickering pointed out that scientists are always enmeshed in a "mangle of practice" (Pickering 1995). That is, scientists struggle continuously in order to get theories and instruments on one hand and the natural world on the other to perform in the ways that their investigations require. The creation of scientific knowledge can therefore be understood as a dynamical process of interactive stabilization of material and human agency – a process that Pickering termed as the *dance of agency* (Pickering 1995; see also Lehrer 2009). Uncertainty, and managing uncertainty are unavoidable aspects in this work, even though the most popular image of scientific work tends to be one of the certitude of accurate predictions (Duschl 2008).

A central focus of the scientific work is the invention, reproduction, and modification of scientific inscriptions – such as graphs, equations, computer code, etc. – which tend to amplify certain aspects of the phenomena under investigation while reducing emphasis on other, less relevant aspects (Latour 1990). This is similar and synergistic to the work of defining and using contextually relevant computational abstractions, as we pointed out earlier. Additionally, computational models can also bring to light new, unexpected ways of thinking about the phenomena by bringing different disciplinary perspectives in contact with one another (MacLeod and Nersessian 2015). The process of creation of these inscriptions – which are collectively termed "modeling" – involves both representational and epistemic work in a deeply intertwined manner (Giere 1988; Pickering 1995; Lehrer 2009). This perspective is known as the "science as practice" perspective and is now regarded as a cornerstone of science education research (NGSS 2015). In the following subsections, we consider the subjective and perspectival nature of the work involved in modeling, and in particular, computational modeling.

4.2.2 Subjectivity in Representational Work

Studies of scientists and their production of scientific inscriptions reveal a rather amorphous nature of scientific knowledge and work (Pickering 1995; Ochs et al. 1996; Latour 1999; Daston and Galison 2007). For Pickering (1995), as we mentioned in the previous section (Sect. 4.2.1), subjectivity arises from the dance of agency between theory formulation and the materiality of the physical world. Latour (1999) argues that while a common image of science implies objectivity and certitude, viewing science as *research* can help us see it as a much more complex experience – one that is uncertain and subjective, and both human and non-human. Ochs et al. (1996) highlighted the central role that interpretive work, including negotiation between scientists, plays in dealing with uncertainty during a research project. They also demonstrated that the interpretive nature and uncertainty of this work – an epistemic phenomenon – are deeply tied to the representational infrastructure (Ochs et al. 1996). This is echoed by Daston and Galison (2007), who pointed out that as representational technologies evolve and new representational technologies emerge, they necessitate new forms of uncertainty and interpretive work.

Daston and Galison (2007) argued that with the introduction of photographic technology and the printing press, the epistemic stance of scientific work shifted from a falsely "objectivitist" stance to "trained judgment." This was evident in their comparison between the nineteenth century introduction of photographic technology where the machinic nature of photography created an impression that scientist could "get out of the way" and let the photograph produce what became perceived as bare, uninterpreted, objective "facts." In contrast, beginning in the early to mid-twentieth century, with the advent of the printing press that in turn widened the audience for scientific works such as atlases, the production of scientific images became necessarily more interpretive on the part of the scientist, with a clear goal of *enhancing the communicativity* of the images, which Daston and Galison (2007) termed "trained judgment."

Building on this work, Farris et al. (2016) have argued that the advent of computing as a *key* mode and medium of scientific inquiry further amplifies this epistemic stance of "trained judgment." A case to point, they argued, is that recent, long-term ethnographic studies of biomedical engineering labs illustrate how the malleability and inherent interdisciplinary of the practice of computational modeling results in new conceptual innovations in scientific practice (Nersessian 2012;

Chandrasekharan and Nersessian 2015). Nersessian and colleagues showed that computational modeling can be particularly helpful for creating new scientific knowledge in the field of complex systems, by (a) bridging the gap between theorization, dynamic visualization, and experimental work, (b) bringing together multiple disciplinary perspectives, (c) using stochastic modeling techniques in cases where clear mechanistic accounts are difficult to obtain, and (d) making it possible to communicate directly with colleagues about complex, predictive visualizations of the target phenomena.

4.2.3 Computational Modeling as Perspectival Work

In his seminal book, *Mindstorms*, Papert argued that working with the LOGO turtle is a "model for what it is to get to know an idea the way you get to know a person" (Papert 1980, pp 136). Papert argued that it involves *getting to know* the turtle, through exploring what it can or cannot do. He cautioned that this should not mean that all ideas be reduced to computational terms; rather, the early experience with turtles is a good model of learning. That is, "... it is a good way to 'get to know' subject by 'getting to know' its powerful ideas" (Papert 1980, p 138). As an illustrative case, he noted that when children learn Newtonian mechanics using LOGO, they do so through modeling changing velocities, i.e., by specifying how fast the turtle should move. The propositional forms of these phenomena are represented in the form of physical laws in the form of linear mathematical equations, and the fallacy of education is that these laws which are the products of complex work (i.e., Pickering's mangle of practice) in which qualitative thinking that is less completely specified and seldom stated in propositional form play an important role. Therefore, it is the qualitative experience of *thinking like the turtle* and *thinking with the turtle* that makes the experience of learning a powerful and a deep one and one that is quite antithetical to *learning as usual* in K-12 science (and beyond). These forms of reasoning enable the learner to engage in embodied and intuitive reasoning (Papert 1980; Wilensky and Reisman 2006; Dickes et al. 2016b; Sengupta and Wilensky 2009).

The early success of LOGO has led to the development of several LOGO-like programming languages and modeling environments such as NetLogo (Wilensky 1999), Scratch (Resnick et al. 2009), AgentSheets (Repenning and Sumner 1995), CTSiM (Sengupta et al. 2013; Basu et al. 2016), and ViMAP (Sengupta et al. 2015b). Computational models developed in such languages are more generally known as agent-based models (ABMs). When users develop ABMs, they construct programs by providing simple rules to a computational object or agent (e.g., the sprite in Scratch, the turtle in LOGO, etc.), which then enacts the rules through movement in computational space. These agent-level actions are repeated over time and/or across multiple agents. In the former case, it enables learners to generate models of continuous movement (Newtonian mechanics) from temporal aggregations of discrete actions (Sengupta and Farris 2012; Sengupta et al. 2012). In the latter case, it enables learners to model dynamical systems (e.g., ecological interdependence) in which multiple agents are simultaneously interacting with each other (Dickes and Sengupta 2013; Dickes et al. 2016b).

Because the agent-level interactions, attributes, and behaviors are often *body-syntonic* (i.e., can be explained and understood through simple embodied actions of the child), young children can model complex scientific phenomena using such forms of computing (Papert 1980; Danish 2014; Dickes et al. 2016b; Levy and Wilensky 2008). As Dickes et al. (2016b) demonstrated, by engaging in agent-based modeling, even young learners can investigate and develop explanations of system-level, emergent behaviors from the perspective of agents within the system. They key argument supported by these studies is that thinking like the agent provides learners an intuitive pathway in exploring emergent outcomes of the system (Wilensky and Reisman 2006; Levy and Wilensky 2008). Evelyn Fox Keller's biography of the biologist Barbara McClintock supports this claim, citing evidence that thinking like the agent (e.g., a chromosome) enabled McClintock to make significant advances in her research on human genetic structures (Keller 1984). Similarly, Ochs et al. (1996) also identified that scientists' sensemaking in the domain of physical sciences also involves such mental projections of the self into the phenomenon of inquiry.

4.3 Phenomenological Approaches for Sustaining Computing in STEM Classrooms

What does the theoretical review in the preceding section mean for the praxis of computing in STEM education? We argue that the *experience* of coding in STEM, from the perspective of the learners and teachers, especially over a long period of time, is inherently heterogeneous. That is, dealing with computational abstractions in the context of STEM disciplinary contexts and classrooms involves engaging with multiple forms and genres of representations beyond coding, and often translating between these representations requires interpretive judgments. This stands in contrast to the views that have been more traditionally supported by educational researchers, where the goal is to "apply" algorithmic thinking and computational abstractions to determine the correct answer. This complexity is left out in technocentric images of coding, even when they apparently focus on computational productions by participants.

In the remainder of this section, we propose some phenomenological approaches that can help us address these issues in the K-12 STEM classroom. We will review a set of studies conducted in partnership with K-12 teachers and students. Participants in these studies used coding in order to design and develop models in science and math on a long-term basis, throughout the academic year. We present a close examination of the nature of the experience through which teachers and students appropriated coding and computational thinking as the language of doing scientific work in their classrooms. We begin with an argument for adopting a particular *genre* of programming and modeling (*agent-based* programming and modeling) for modeling across disciplines, which is essential for long-term curricular integration. We then suggest a set of pedagogical guidelines for integrating programming in the K-12 STEM curricula, grounded in the perspectives of teachers and learners in K-12 classrooms.

4.3.1 Agent-Based Computational Modeling as a Transdisciplinary Practice

Scientific practices like modeling develop only over the long term, both historically within the sciences and ontogenetically within the lifetime of individuals. This is because modeling is a rather nuanced and complex form of epistemology, even though most educational texts and curricula do not directly address these complexities (Lehrer 2009). The *yearlong science classroom* is a better context for engaging children in such extended forms of practice, rather than the predominant tradition in educational research to conduct intervention studies where children engage in modeling (including computational modeling) spanning a few hours to a few days. But, in order to support such long-term curricular integration, we must take into consideration how to integrate computational modeling and programming *across* disciplinary contexts.

Different forms of phenomena lend themselves to different forms of modeling (Lehrer and Schauble 2007), and we have found that at the elementary, middle, and high school levels, the categories of *linear continuity* and *emergent aggregation* can be helpful guides for us in selecting scientific phenomena across disciplines that can lend themselves well to computational modeling and programming. An example of modeling *linear continuity* would be modeling motion as a continuous change in position, where the behavior of a single "agent" (e.g., a ball rolling on a ramp) can be modeled as a temporal series of changes of position and/or other variables such as speed and acceleration that obey linear mathematical relationships (Sherin et al. 1993; Sengupta and Farris 2012). An example of modeling *emergent aggregation* would be modeling ecological interdependence, where multiple agents simultaneously interact with each other and the environment, which in turn result in aggregate-level outcomes, e.g., the dynamical relationship between the predator and prey populations in an ecosystem (Wilensky and Reisman 2006; Dickes and Sengupta 2013; Wagh et al. 2017). Such aggregate-level behaviors or outcomes are known as *emergent*, because although linear relationships between individual agents (objects) produce these behaviors, these behaviors are not apparent in the description of either the individual objects or the relationships (Lehrer and Schauble 2007; Wilensky and Resnick 1999). Other examples of emergent phenomena that have been successfully adopted by teachers and students through the use of agent-based computational modeling and programming include electrical conduction (Sengupta and Wilensky 2011), crystallization (Blikstein and Wilensky 2009), molecular chemistry (Stieff and Wilensky 2003), evolution (Dickes and Sengupta 2013; Wagh et al. 2017), ethnocentrism (Hostetler et al. 2018), etc. This suggests that adopting agent-based modeling and programming as the form of computing can make it possible for educators to use the same genre of modeling and programming across multiple disciplines.

It has also been argued that students' conceptual difficulties in understanding both linear continuity and both emergent aggregation have similar origins (Reiner et al. 2000; Chi 2005). For example, Reiner et al. (2000) argued that physics novices tend to use substance-based knowledge when reasoning about concepts like force, heat, light, and electric current (e.g., force as a property of an object). For example, the misconception that continuing motion implies a continued force in the direction of the movement is generated from a more primitive idea (called phenomenological primitives or p-prims) called "continuous force," which can be abstracted from common everyday experiences of needing constant effort to keep an object in motion (DiSessa 1993). Note that these novice intuitive ideas about physics have an underlying structure of a direct schema – one that involves an agent either acting on another agent or an agent being acted upon by an impetus (Talmy 1983). On the other hand, an expert-like understanding of kinematics involves being able to conceptualize a situation in terms of more complex interactions – e.g., situations involving lack of motion, or constant speed could be conceptualized as forms of dynamic equilibrium between interacting systems (Clement 1993; Greeno and Van De Sande 2007). Similarly, in the domain of ecology, researchers have argued that commonly noted misconceptions are indicative of direct schema or event schema, which imply a direct cause-effect relationship (such as "A" causes "B") or an event that has a finite duration of time (as opposed to being continuous), whereas the expert conception of ecological phenomena involves a more complex cognitive structure involving the dynamic and decentralized nature of emergent phenomena in terms of a myriad of simultaneous interactions (Chi 2005). However, studies have also shown that pedagogical approaches based on agent-based models and modeling can act as productive learning environments, using which novice learners can develop deep understandings of dynamic, aggregate-level phenomena by bootstrapping, rather than discarding their agent-level intuitions (Dickes and Sengupta 2013; Dickes et al. 2016b; Wilensky and Reisman 2006; Levy and Wilensky 2008).

This body of research also provides useful guidelines for the sequence of learning activities in each domain, and our general pedagogical approach explicitly adopts the perspective that expert-like scientific knowledge can result through building upon and refining existing naive intuitive knowledge (Dickes et al. 2016b; Danish 2014; Sengupta et al. 2015b). For example, the initial learning activities leverage a naive conceptualization of the domains and progressively scaffold them toward refinement. In kinematics, learners begin by inventing representations of motion in terms of measures of speed (how fast an object is moving) and inertia (innate tendency of an object to continue its current state of rest or motion, which often takes an anthropomorphic form in novice reasoning), and gradually move to a force-based, more canonical description of motion in subsequent activities (Sengupta and Farris 2012; Farris et al. 2016). Similarly, in ecology, students begin with programming the behavior of single agents in the ecosystem and gradually develop more complex programs for modeling the behavior and interaction of multiple species within the ecosystem (Wilensky and Reisman 2006; Danish 2014; Sengupta et al. 2013; Dickes et al. 2016b).

4.3.2 Framing Programming as Designing Mathematical Measures of Change

Our studies have demonstrated that framing programming as "mathematizing" in the science classroom can serve as a productive pedagogical approach for integrating programming in the K-12 science classroom (Sengupta et al. 2013, Sengupta et al. 2015a, b, 2018; Dickes et al. 2016a; Farris et al. 2016). In this approach, programming is used in the context of creating computational models of scientific phenomena through designing discrete mathematical representations of units of change, for representing change over time. That is, the computational code created by students serve to define a "unit" of measurement, which would then get repeated as the program was "run" to produce the desired motion.

From the perspective of praxis in the K-12 science classroom in North American public schools, this form of activity is of critical importance for classroom integration of computational modeling and programming. Teachers in US and Canadian public schools who we have worked with have reported that interpreting and constructing mathematical measures (e.g., units of measurement and graphs) are areas where most of their students experience difficulties (e.g., see Sengupta et al. 2018). This is also of importance for US and Canadian public schools because manipulating units is emphasized in standardized assessments (in the USA) and the program of studies (in Canada), and therefore, teachers acknowledge this as an important learning goal in their regular science classroom.

We see this as a great opportunity for integration of computational modeling and programming in K-12 science classrooms. Our studies show that agent-based programming and modeling can help students overcome conceptual challenges in understanding linear continuity (e.g., kinematics; see Sengupta and Farris 2012) and emergent aggregation (e.g., ecology; see Dickes et al. 2016b), through the iterative design of measures of change over time. This is because the activity of programming the behavior of agents requires the learners to define the event in discrete measures (Sengupta et al. 2015b). The state of the simulation, at any instant, represents a single event in the form of spatialized representations of agent actions and interactions. To "run" the simulation, these events are repeated a number of times specified by the user. By engaging in iterative cycles of building, sharing, refining, and verifying computational models, students refine their understanding of what actions and interactions of agents represent an "event," which are then displayed on graphs. This enables students to define and explore different kinds of units and see their simulation measured in those units (Farris et al. 2016) and even merge computational modeling with artistic design (Sengupta et al. 2012).

4.3.3 Supporting Perspectival Work Through Embodied Modeling

Research in science education suggests that the integration of ABMs in elementary classrooms also benefits greatly from the use of other synergistic forms of modeling such as embodied and physical modeling. Programming an agent involves learning to think like the agent, because it can help students understand the relationship between their code and the simulated output. In our studies, all teachers saw embodied modeling as a valuable activity for teaching students how to think like an agent. Embodied modeling introduces the students to the relevant computational rules represented by the agent-based programming commands through embodied interactions with the material world, and in doing so, helps them debug their programs and deepens their understanding of the graphs in the simulations (Dickes et al. 2016a, b).

Why are these different forms of modeling necessary? Science educators and cognitive scientists have argued that embodied thinking is central to the development of agent-based thinking and representational practices (Papert 1980; Goldstone and Wilensky 2008; Wilensky and Reisman 2006). For example, in a recent study conducted in a third-grade classroom, students began with an embodied modeling activity of foraging behavior, followed by the generation of mathematical inscriptions based on their embodied actions, and finally, conducted further inquiry of interdependence in an ecosystem using two separate ABMs (Dickes et al. 2016b). We found that the students recalled and built upon their embodied modeling experiences as butterflies foraging for nectar (see Fig. 4.1), during their subsequent interactions with the agent-based simulation of a butterfly-bird-flower ecosystem (see Fig. 4.2). We also found that creating mathematical inscriptions (bar graphs) to represent the data collected during the embodied modeling activity provided a representational continuity between the embodied modeling activities and the ABMs, as well as with previous representational forms that students used and developed in their science and math classes prior to the study. And finally, we also found that embodied modeling activities, especially in the case of modeling interactions between different types of agents, must be designed so that students are able to take on the perspectives of different types of agents, rather than prompting students to take on the perspective of only one type of agent.

As students engaged iteratively in cycles of embodied modeling and graphing by taking on the perspective of the agents in the system, and then modeled the same phenomena using multi-agent-based NetLogo simulations, we found that they were

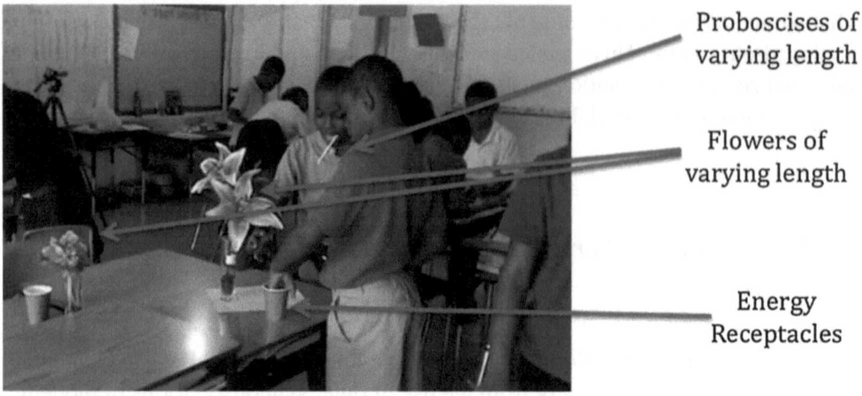

Fig. 4.1 Students participating in phase I's embodied modeling activity

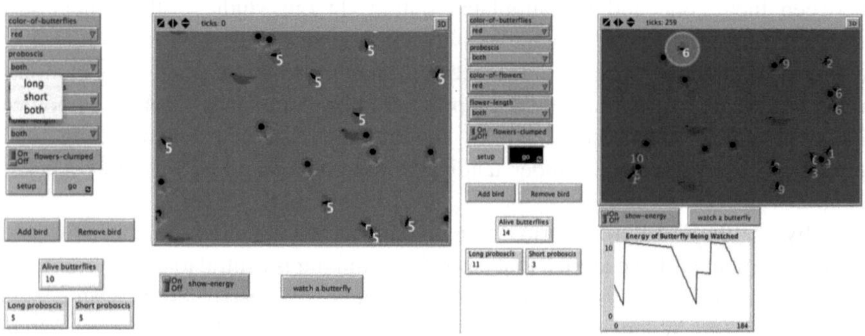

Fig. 4.2 Screenshots of the predator ABM (left) and watched energy ABM (right). Both models were designed to actively recruit students' previous embodied modeling experiences shown in Fig. 4.1

able to develop progressively more complex forms of mechanistic explanations of emergence. Mechanistic explanations focus on the processes that underlie cause-effect relationships and thereby take into account how the activities of the constituent components affect one another (Russ et al. 2008). In particular, we found that learners were able to engage in a particular form of mechanistic reasoning that Russ et al. (2008) termed *chaining*. During chaining, learners use knowledge about the causal structure of the phenomena to make claims about what must have happened previously to bring about the current state of things (backward chaining) or what will happen next given that certain entities or activities are present now (forward chaining).

This is an important finding from the perspective of computational thinking in the context of science education, because this suggests that event-based programming and modeling can support children in developing deep conceptual understandings of complex scientific phenomena. Furthermore, this also suggests that focusing on supporting the growth of students' mechanistic reasoning through modeling may be helpful for integrating computational thinking in science classrooms. As Sengupta et al. (2013) identified, mechanistic reasoning in the domain of science education is well aligned with *algorithm design* and *complexity analysis* in the domain of computational thinking.

4.3.4 Refining Computational Modeling Through Disciplinarily Grounded Classroom Norms

Our studies also illustrate that emphasizing mathematizing and measurement as key forms of learning activities can help teachers meaningfully integrate programming as a "language" of science (Dickes et al. 2016a; Sengupta et al. 2018). Long-term studies of classroom integration of computional modeling and programming in the science curricula has further shown that teachers can seamlessly accomplish this by supporting the development of disciplinarily grounded classroom norms for developing and refining mathematical measures (Dickes et al. 2016a). Science educators have shown that the iterative design of mathematical measures can result in deep conceptual growth of students in elementary science, especially when these activities are integrated throughout the curriculum over several months (Lehrer 2009). Lehrer and colleagues have also shown that asking the question *what counts as a "good" model* also needs to be established in classroom instruction as a norm, in order to deepen students' engagement with scientific modeling in elementary grades. Furthermore, similar to Cobb and his colleagues' work in the mathematics classroom explained in the next paragraph (McClain and Cobb 2001; Yackel and Cobb 1996; Cobb et al. 1992), these norms also follow shifts toward deeper disciplinary warrants over time (Lehrer and Schauble 2006; Ford and Forman 2006; Lehrer et al. 2008). In such classrooms, mathematical modeling becomes a meaning-making lens through which the natural world can be systematized and described (Lehrer et al. 2001).

The specific genre of classroom norms that we have found to be at work in our studies has been termed *sociomathematical norms* (McClain and Cobb 2001; Yackel and Cobb 1996; Cobb et al. 1992). In a recent paper, we outlined and demonstrated how the emphasis on developing and refining sociomathematical norms pertaining to the design of mathematical measures of motion can help teachers seamlessly integrate programming with science education in a third-grade classroom and how they are taken up in students' work (Dickes et al. 2016b). Sociomathematical norms differ from general social norms that constitute the classroom participation structure in that they concern the normative aspects of classroom actions and interactions that are specifically mathematical. These norms regulate classroom discourse and influence the learning opportunities that arise for both the students and the teacher. As in the work of Cobb and his colleagues (Yackel et al. 1991; Cobb et al. 1992), we also found that it was the classroom teacher who initiated and guided the development of these norms in order to foster and sustain classroom microcultures characterized by explanation, justification, and argumentation.

In our study (Dickes et al. 2016b), an important and rather fundamental sociomathematical norm that began as the central guiding question posed by the teacher at the beginning of the class was "what counts as a *good* model." Similar to Yackel and Cobb (1996), we found that this norm typically originated as a socially defined norm and shifted over time to a more sociomathematically defined norm. That is, students' initial warrants were decided on the basis on how many of their peers "liked" a particular model during class discussion and sharing of models rather than thinking more deeply about how their ViMAP code represented the relevant phenomenon they were modeling. However, over time, these warrants became progressively more grounded in the mathematically warrants of how representative their code were of the relevant phenomena being modeled. The class jointly took up normative ways of thinking about and representing motion (walking) through designing and refining *approximate* and *predictive* measures of change over time, using embodied modeling activities, drawings of their embodied modeling activities that represented "step-sizes", and their ViMAP code and graphs (Dickes et al. 2016b).

Overall, we found that students' use of the ViMAP programming commands became increasingly sophisticated as they held their models accountable to the sociomathematical norms (Dickes et al. 2016a). Over a 6-week period, we scored each student's final ViMAP model at the end of each class period in terms of whether they used appropriate computational abstractions identified by Sengupta et al. (2013) as being relevant to computational thinking such as variables, loops, and initialization. Students' code was scored on the appropriate and non-redundant use of variables and loops in their models and whether their graphs represented appropriate element(s) of the phenomenon being simulated using their ViMAP code. The growth in students' computational fluency is evident in Fig. 4.3. For example, a score of zero meant none of the variables used were appropriate, whereas a score of 3 meant no

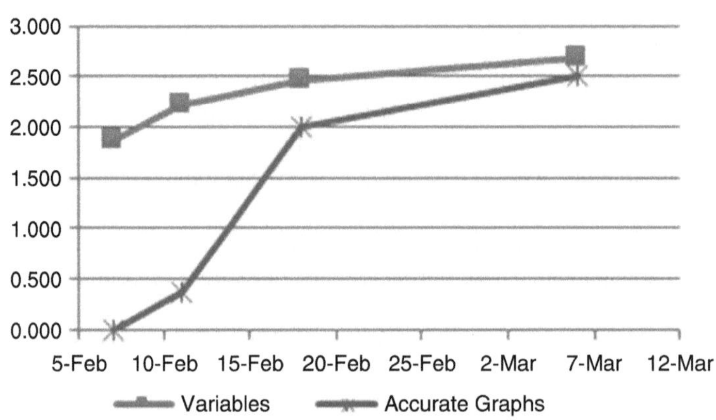

Fig. 4.3 Improvement in computational thinking supported by sociomathematical norms

use of redundant or incorrect variables. The accuracy of the graphs in students' later models were indicative of the appropriate use of the "repeat" command (i.e., loops) and order of placement of the "place measure" command. This in turn relied on their conceptual understanding of when to initialize the measurement (i.e., initialization) and how often the desired measurement had to be repeated in order to generate the graph (loops).

4.3.5 Framing Coding as Designing for Authentic Use

In a study conducted in a fourth-grade classroom in a low-income (90% free lunch), public charter school in Nashville, we investigated how collaboratively designing computational machines for authentic users could support the integration of coding in STEM education (Sengupta et al. 2015b). The first phase of the study focused on introducing students to agent-based programming through creating geometric shapes (e.g. squares, circles, spirals) using the ViMAP programming language (Sengupta et al. 2015b). ViMAP uses the NetLogo modeling platform (Wilensky 1999) as its simulation engine and enables learners to design, program and graph NetLogo simulations using both programming blocks and text-based programming (see Fig. 4.6). This phase lasted for eight class periods. For the next 18 class periods, students worked in dyads on a STEM design challenge (capstone activity), i.e., constructing mathematical machines and user guides for generating geometric shapes using a distributed computing infrastructure.

During the capstone learning activity, learners worked in dyads and constructed a mathematical machine for generating geometric shapes. Each machine consists of two components: virtual and physical. The virtual component was a ViMAP program that learners constructed using visual programming primitives selected from the ViMAP programming library. The physical component consisted of two physical control interfaces, each designed to control the reading on one of the distance sensors. Each sensor controlled a distinct turtle variable (e.g., color, speed, rotation). This was an activity that required intersubjective collaboration (Sengupta et al. 2015b), because while each member of the dyad independently designed one of these physical control structures using Lego bricks, the dyad was responsible for jointly designing the ViMAP program. Figure 4.4 shows an example of student work.

We specified that other fourth-grade teachers in Nashville would use these machines, so that students had a specific image of user(s) in mind. To ensure authenticity of the users, we also invited three graduate students in education with prior math teaching experience in elementary grades, but unaffiliated with our study, to serve as "users." The user testing took place twice: first in mid-March (user testing 1) and in late April (user testing 2). During both the user testing events, each user interacted with a dyad's machine for about 20 min and provided them written and verbal feedback. After user testing 1, students improved their machines and user

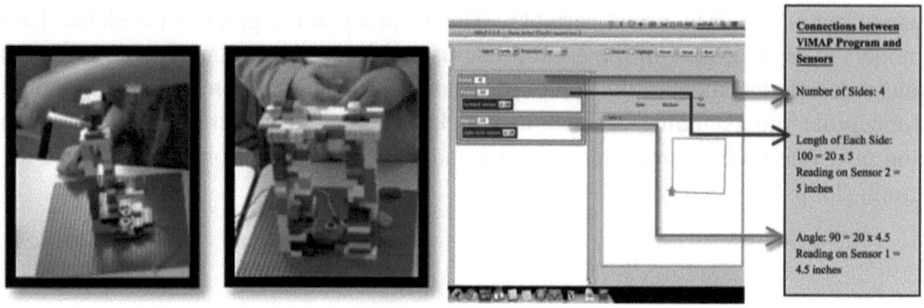

Fig. 4.4 (**a**) (Left) Jerry's pulley mechanism for controlling turn of the turtle via 1. (**b**) (Middle) Chuck's machine for controlling the speed of the turtle via sensor 2. (**c**) (Right) is a screenshot of their ViMAP program for generating a square, and our annotation makes explicit the multiplicative reasoning involved in generating *angles* and *sides* of the square

guides in order to address the issues highlighted in the feedback. User testing 2 was also the capstone activity.

We compared the work of each dyad at two stages: user testing 1 (UT1) and user testing 2 (UT2). In terms of children's mechanistic explanations (Russ et al. 2008), we found that compared to UT1, attending to what the user needs to know resulted in improving greatly the quality of students' mathematical explanations during UT2. Their explanations, as evident both in their user guides and verbal explanations during the user testing process, made explicit the mathematical relationships between algorithmic elements (e.g., number of loops in their ViMAP program) and variables in their ViMAP programs, and the actions of the turtle in every step (e.g., right turn), which in turn was directly effected by the users' actions (e.g., sensor reading generated by the user). The greater emphasis on identifying and representing the relationships between computational abstractions (algorithms and variables), mathematical relationships, and the mechanics of the physical setup resulted from the need to create designs that were more *communicative* (Sengupta et al. 2015b). A sample comparison is shown in Fig. 4.5a.

The phenomenological lesson here is that when coding is embedded in an authentic design activity intended for and tested by an authentic audience, paying attention to the needs and the perspective of the user can deepen the coders' conceptual understanding of the relationship of computational abstractions with disciplinarily grounded knowledge and representations.

4.3.6 Support Transition from Visual to Text-Based Programming

Another important issue for sustaining programming in K-12 STEM classrooms, especially in the higher grades (middle school or high school), is that although visual

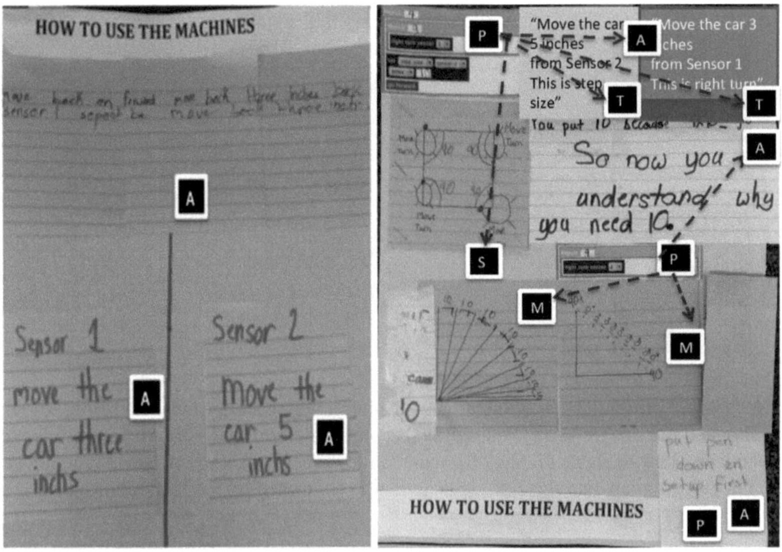

Fig. 4.5a Jacinda and Tom's user guides in user testing 1 (left) and user testing 2 (right). We annotated their user guides using the schematic shown in Fig. 4.5b

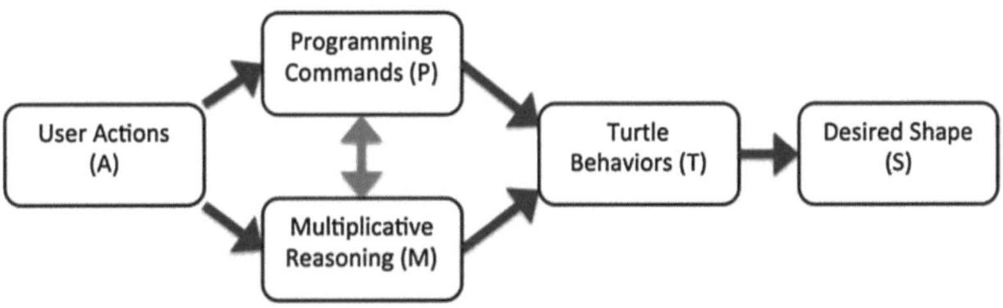

Fig. 4.5b A schematic for mechanistic explanations used by all groups in user testing 2

programming lowers the barrier for entry into programming, learners who intend to pursue careers in computing may find the drag-and-drop nature of visual programming inauthentic or find it difficult to transition to text-based programming (DiSalvo 2014). In a recent study conducted in an eighth-grade classroom, we investigated this issue (Sengupta et al. 2015b). We used ViMAP, because ViMAP is a dual-mode programming language that enables users to engage in both blocks and text-based programming. Visual programming commands in ViMAP are defined as short NetLogo procedures (see Fig. 4.6), which students can easily access and modify using text-based NetLogo code. In our study, after engaging in visual programming with ViMAP for approximately 2 months to build simulations of interdependence in ant ecosystems, the teacher and the students wanted to make deeper changes in the underlying text-based NetLogo code. But, given the limited instructional time, the teacher found it challenging to help students create new simulations in NetLogo

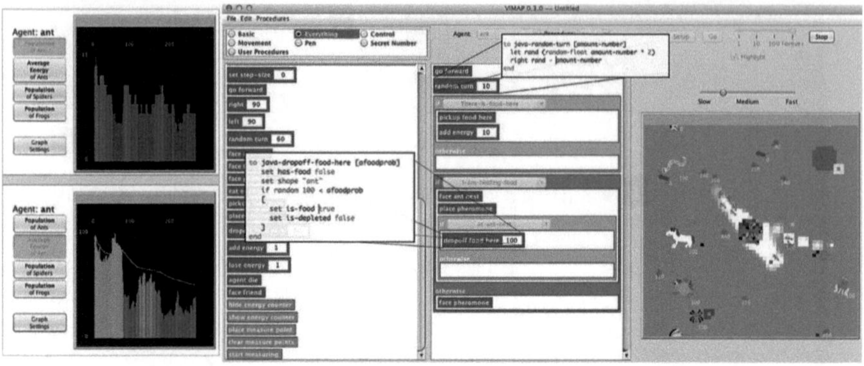

Fig. 4.6 ViMAP-Ant-Foodweb simulation and programming commands developed by eighth graders. Popped-out images show NetLogo procedures underlying the ViMAP commands created by the students. Left to right, graphs of population and energy, library of ViMAP commands, a sample ViMAP program, and the NetLogo simulation controlled by the ViMAP program

using text-based programming. This required a lot of "overhead," because the language syntax was often disconnected from the relevant scientific concepts.

To address this issue, the teacher then decided to return to the ViMAP-Ants unit (see www.vimapk12.net for the curricular activities) and asked the students to work in small groups to create new ViMAP commands by modifying and extending the underlying NetLogo code. For the eighth graders, this work was motivated by a capstone project of designing and creating a version of ViMAP-Ants in order to teach fourth graders about food webs, which is a required curricular standard in fourth grade. The teacher introduced the students to relevant "chunks" (procedures) in the NetLogo code pertaining to specific ViMAP commands they were already familiar with. She led class discussions in which the students collaboratively interpreted and explained the significance of the computational abstractions in NetLogo code in terms of relevant scientific concepts represented in the ViMAP commands. Learning the syntax and new forms of abstractions (such as classes) in text-based programming therefore became deeply intertwined with the relevant concepts in ecology (e.g., hierarchy of organisms in food webs). While students' previous work using visual programming introduced them to computational abstractions such as loops, variables, and conditionals, text-based programming further deepened their computational thinking because it involved creating computational objects or classes, declaring new local variables, creating and modifying conditionals, editing and repurposing lists, and using random numbers. Students' growth in computational thinking was further evident in a post-assessment activity, in which they successfully created new commands for a NetLogo simulation of a different ecosystem without teacher-provided assistance (Sengupta et al. 2015b).

4.4 Discussion: Computational Thinking as *Experience* in K-12 STEM

In *Quest for Certainty*, John Dewey argued against empiricist ontology that substitutes data for objects (and inquiry). Data, he argued, signifies a phenomenon for further inquiry; but instead, empiricism often represents data as being self-evident (Dewey 1929 (1984)). In a similar vein, the persistent fallacy of the predominant epistemology in educational computing, especially as it pertains to computational thinking in education, is the normative notion that knowledge is some *antecedent reality* (Dewey 1929 (1984)), reified in terms of learners' use of computational abstractions used commonly by professional coders. That is, for researchers, the experience of learners is substituted by canonical assessments of the "computational abstractions" that the learners used in their computer programs. Certainly, there are efforts, especially by constructionist scholars, to demonstrate how computing can take on diverse and personally meaningful forms (Resnick et al. 2000; Farris and Sengupta 2016), but the hallmark of the experience of coding, as reported in nearly all research articles on computational thinking (including some of our own previous work), remains the deft use of computational abstractions by learners who haven't had much prior experience with coding. This is the danger of technocentrism (Papert 1987) realized – where the questions about technology are being answered by referring the questions to the technology itself.

In this chapter, we have argued for paradigmatic shift in the epistemology and pedagogy of computing and computational thinking, especially for K-12 STEM education. Our position is that we must shift away from empiricist ontology that Dewey argued against (Dewey 1929 (1984)), and technocentrism that Papert argued against (Papert 1987), toward more phenomenological perspectives, in terms of trying to both understand and support the development of computational thinking as *experience* in the context of K-12 STEM education. Epistemologically, we have argued that computational thinking must be reconceptualized more appropriately as an intersubjective experience, as opposed to more cognitivist and technocentric images of learning and reasoning that can be assessed through the production of symbolic code. Contextualizing computational abstractions in K-12 science classrooms is a complex experience that can be better understood as a "phenomenal field" (Merleau-Ponty 1962), rather than by simply focusing on a cognitivist image of "thinking". This experience is rife with uncertainty and involves significant instructional work. For example, even in short-term studies, Sengupta et al. (2013) and Basu et al. (2016) acknowledge and highlight the importance of extensive scaffolding provided by facilitators in order to help students in overcoming challenges in designing and using the necessary computational abstractions for modeling kinematics and ecology. However, despite such efforts, a commonly used approach of assessing computational thinking relies primarily (and in many cases, only) on evaluating structural elements of learners' computer programs (e.g., Grover et al. 2018; Weintrop et al. 2018; Dasgupta et al. 2016). Pedagogically, we have argued that addressing this issue necessitates careful considerations of the complexities of K-12 classrooms, without ignoring teachers' and students' experiences in which computing and coding are situated. In particular, we have proposed the following pedagogical guidelines for sustaining computational thinking in the K-12 classroom:

1. Reframing programming and coding as "modeling," in particular, as the design of mathematical units of measurement of change over time, for the K-12 science classroom;
2. Highlighting transdisciplinary representational and epistemic practices such as design and modeling to support continuity in learning experiences across disciplines;
3. Designing complementary activities that use embodied modeling and non-computational materials as representational and cognitive amplifications of computational code;
4. Focusing on disciplinarily grounded, normative instructional approaches (e.g., sociomathematical norms) during classroom instruction for refining computational modeling;
5. Reframing coding and modeling as designing for an *authentic* audience; and
6. Using both visual (block-based) and text-based programming languages for longer-term curricular integration.

This list is far from exhaustive. However, given the context in which most of our studies have been carried out – high-poverty, predominantly nonwhite classrooms in public schools with limited resources – we believe that these guidelines can help us focus our attention on issues that can make a difference in terms of helping teachers and students adopt computing and computational thinking as a "language" of STEM, especially on a long-term basis.

References

Basu, S., Biswas, G., Sengupta, P., Dickes, A., Kinnebrew, J. S., & Clark, D. (2016). Identifying middle school students' challenges in computational thinking-based science learning. *Research and Practice in Technology Enhanced Learning, 11*(1), 13.

Blikstein, P., & Wilensky, U. (2009). An atom is known by the company it keeps: A constructionist learning environment for materials science using agent-based modeling. *International Journal of Computers for Mathematical Learning, 14*(2), 81–119.

Chandrasekharan, S., & Nersessian, N. J. (2015). Building cognition: The construction of computational representations for scientific discovery. *Cognitive Science, 39*, 1727–1763.

Chi, M. T. (2005). Commonsense conceptions of emergent processes: Why some misconceptions are robust. *The Journal of the Learning Sciences, 14*(2), 161–199.

Clement, J. (1993). Using bridging analogies and anchoring intuitions to deal with students' preconceptions in physics. *Journal of Research in Science Teaching, 30*(10), 1241–1257.

Cobb, P., Wood, T., Yackel, E., & McNeal, B. (1992). Characteristics of classroom mathematics traditions: An interactional analysis. *American Educational Research Journal, 29*(3), 573–604.

Danish, J. A. (2014). Applying an activity theory lens to designing instruction for learning about the structure, behavior, and function of a honeybee system. *The Journal of the Learning Sciences, 23*(2), 100–148.

Dasgupta, S., Hale, W., Monroy-Hernández, A., & Hill, B. M. (2016, February). *Remixing as a pathway to computational thinking*. In Proceedings of the 19th ACM Conference on Computer-Supported Cooperative Work & Social Computing (pp. 1438–1449). ACM.

Daston, L., & Galison, P. (2007). *Objectivity*. New York: Zone Books.

Dewey, J. (1929 (1984)). *The later works of John Dewey 1929: The quest for certainty*. SIU Press.

Dickes, A. C., & Sengupta, P. (2013). Learning natural selection in 4th grade with multi-agent based computational models. *Research in Science Education, 43*(3), 921–953.

Dickes, A. C., Farris, A. V., & Sengupta, P. (2016a). *Integrating agent-based programming with elementary science: The role of socio-mathematical norms*. In: Proceedings of the 12th International Conference of Computers in Education (ICCE 2016), pp 128 – 138.

Dickes, A., Sengupta, P., Farris, A. V., & Basu, S. (2016b). Development of mechanistic reasoning and multi-level explanations in 3rd grade biology using multi-agent based models. *Science Education, 100*(4), 734–776.

DiSalvo, B. (2014). Graphical qualities of educational technology: Using drag-and-drop and text-based programs for introductory computer science. *IEEE Computer Graphics and Applications, 34*(6), 12–15.

DiSessa, A. A. (1993). Toward an epistemology of physics. *Cognition and Instruction, 10*(2–3), 105–225.

DiSessa, A. A. (2001). *Changing minds: Computers, learning, and literacy*. Cambridge, MA: MIT Press.

Duschl, R. (2008). Science education in three-part harmony: Balancing conceptual, epistemic, and social learning goals. *Review of Research in Education, 32*(1), 268–291.

Farris, A. V., & Sengupta, P. (2014). Perspectival computational thinking for learning physics: A case study of collaborative agent-based modeling. In: *Proceedings of the 12th international conference of the learning sciences* (pp. 1102–1107). ICLS. 2014.

Farris, A. V., & Sengupta, P. (2016). Democratizing children's computation: Learning computational science as aesthetic experience. *Educational Theory, 66*(1–2), 279–296.

Farris, A. V., Dickes, A. C., Sengupta, P. (2016). Development of disciplined interpretation using computational modeling in the elementary science classroom. In *Proceedings of the 13th international conference of the learning sciences* (pp. 282–290). ICLS. 2016.

Ford, M. J., & Forman, E. A. (2006). Redefining disciplinary learning in classroom contexts. *Review of Research in Education, 30*(1), 1–32.

Francis, K., Khan, S., & Davis, B. (2016). Enactivism, spatial reasoning and coding. *Digital Experiences in Mathematics Education, 2*, 1–20.

Giere, R. N. (1988). *Explaining science: A cognitive approach*. Chicago: University of Chicago Press [RNG].

Goldstone, R. L., & Wilensky, U. (2008). Promoting transfer by grounding complex systems principles. *The Journal of the Learning Sciences, 17*(4), 465–516.

Greeno, J. G., & Van De Sande, C. (2007). Perspectival understanding of conceptions and conceptual growth in interaction. *Educational Psychologist, 42*(1), 9–23.

Grover, S., & Pea, R. (2013a). Computational thinking in K–12: A review of the state of the field. *Educational Researcher, 42*(1), 38–43.

Grover, S., & Pea, R. D. (2013b). Using a discourse-intensive pedagogy and android's app inventor for introducing computational concepts to middle school students. In *Proceedings of the 44th ACM technical symposium on computer science education* (SIGCSE '13). New York: ACM.

Grover, S., Basu, S., & Schank, P. (2018, February). *What we can learn about student learning from open-ended programming projects in middle school computer science*. In Proceedings of the 49th ACM Technical Symposium on Computer Science Education (pp. 999–1004). ACM.

Hostetler, A., Sengupta, P., & Hollett, T. (2018). Unsilencing critical conversations in social-studies teacher education using agent-based modeling. *Cognition and Instruction, 36*(2), 139–170.

Kafai, Y. B., & Burke, Q. (2013). The social turn in K-12 programming: Moving from computational thinking to computational participation. In *Proceeding of the 44th ACM technical symposium on computer science education* (pp. 603–608). ACM.
Keller, E. F. (1984). *A feeling for the organism, 10th aniversary edition: The life and work of Barbara McClintock*. London: Macmillan.
Latour, B. (1990). Drawing things together. In M. Lynch & S. Woolgar (Eds.), *Representation in scientific practice* (pp. 19–68). Cambridge: MIT Press.
Latour, B. (1999). *Pandora's hope: essays on the reality of science studies*. Cambridge: Harvard University Press.
Lehrer, R. (2009). Designing to develop disciplinary dispositions: Modeling natural systems. *American Psychologist, 64*(8), 759.
Lehrer, R., & Schauble, L. (2006). Cultivating model-based reasoning in science education. In R. K. Sawyer (Ed.), *The Cambridge handbook of the learning sciences* (pp. 371–387). New York: Cambridge University Press.
Lehrer, R., & Schauble, L. (2007). Contrasting emerging conceptions of distribution in contexts of error and natural variation. In M. Lovett & P. Shah (Eds.), *Thinking with data* (pp. 149–176). Mahwah: Lawrence Erlbaum Associates.
Lehrer, R., Schauble, L., Strom, D., Pligge, M. (2001). Similarity of form and substance: Modeling material kind. In *Cognition and instruction: Twenty-five years of progress* (pp. 39–74).
Lehrer, R., Schauble, L., & Lucas, D. (2008). Supporting development of the epistemology of inquiry. *Cognitive Development, 23*(4), 512–529.
Levy, S. T., & Wilensky, U. (2008). Inventing a "mid level" to make ends meet: Reasoning between the levels of complexity. *Cognition and Instruction, 26*(1), 1–47.
Locke, J. (1690/1979). *An essay concerning human understanding*. New York: Oxford University Press.
MacLeod, M., & Nersessian, N. J. (2015). Modeling systems-level dynamics: Understanding without mechanistic explanation in integrative systems biology. *Studies in History and Philosophy of Science Part C: Studies in History and Philosophy of Biological and Biomedical Sciences, 49*, 1–11.
McClain, K., & Cobb, P. (2001). An analysis of development of sociomathematical norms in one first-grade classroom. *Journal for Research in Mathematics Education, 32*(3), 236–266.
Merleau-Ponty, M. (1962). *Phenomenology of perception*. New York: Routledge.
National Research Council. (2007). *Taking science to school: Learning and teaching science in grades K-8*. Washington, DC: National Academies Press.
Nersessian, N. J. (2012). Modeling practices in conceptual innovation: An ethnographic study of a neural engineering research laboratory. In U. Feest & F. Steinle (Eds.), *Scientific concepts and investigative practice* (pp. 245–269). Berlin: DeGruyter.
Next Generation Science Standards. (2015). www.nextgenscience.org/next-generation-science-standards
Ochs, E., Gonzales, P., & Jacoby, S. (1996). "When I come down I'm in the domain state": Grammar and graphic representation in the interpretive activity of physicists. *Studies in Interactional Sociolinguistics, 13*, 328–369.
Papert, S. (1980). *Mindstorms: Children, computers, and powerful ideas*. New York: Basic Books.
Papert, S. (1987). Information technology and education: Computer criticism vs. technocentric thinking. *Educational Researcher, 16*(1), 22–30.
Pickering, A. (1995). *The mangle of practice: Time, agency, and science*. Chicago: University of Chicago Press.
Reiner, M., Slotta, J. D., Chi, M. T., & Resnick, L. B. (2000). Naive physics reasoning: A commitment to substance-based conceptions. *Cognition and Instruction, 18*(1), 1–34.
Repenning, A., & Sumner, T. (1995). Agentsheets: A medium for creating domain-oriented visual languages. *Computer, 28*(3), 17–25.

Resnick, M., Berg, R., & Eisenberg, M. (2000). Beyond black boxes: Bringing transparency and aesthetics back to scientific investigation. *The Journal of the Learning Sciences, 9*(1), 7–30.

Resnick, M., Maloney, J., Monroy-Hernández, A., Rusk, N., Eastmond, E., Brennan, K., Millner, A., Rosenbaum, E., Silver, J., Silverman, B., & Kafai, Y. (2009). Scratch: programming for all. *Communications of the ACM, 52*(11), 60–67.

Russ, R., Scherr, R., Hammer, D., & Mikeska, J. (2008). Recognizing mechanistic reasoning in student scientific inquiry: A framework for discourse analysis developed from philosophy of science. *Science Education, 93*, 875–891.

Schmidt, D. C. (2006). Model-driven engineering. *Computer, 39*(2), 25–31.

Sengupta, P., & Farris, A. V. (2012). Learning kinematics in elementary grades using agent-based computational modeling: a visual programming-based approach. In *Proceedings of the 11th international conference on interaction design and children* (pp. 78–87). ACM.

Sengupta, P., & Shanahan, M. C. (2017). Boundary play and pivots in public computation: New directions in STEM education. *International Journal of Engineering Education, 33*(3), 1124–1134.

Sengupta, P., & Wilensky, U. (2009). Learning electricity with NIELS: Thinking with electrons and thinking in levels. *International Journal of Computers for Mathematical Learning, 14*(1), 21–50.

Sengupta, P., & Wilensky, U. (2011). Lowering the learning threshold: Multi-agent-based models and learning electricity. In *Models and modeling* (pp. 141–171). Cham: Springer.

Sengupta, P., Farris, A. V., & Wright, M. (2012). From agents to continuous change via aesthetics: Learning mechanics with visual agent-based computational modeling. *Technology, Knowledge and Learning, 17*(102), 23–40.

Sengupta, P., Kinnebrew, J., Basu, S., Biswas, G., & Clark, D. (2013). Integrating computational thinking with K12 science education using agent-based modeling: A theoretical framework. *Education and Information Technologies, 18*, 351–380.

Sengupta, P., Krishnan, G., Wright, M., & Ghassoul, C. (2015a). Mathematical machines & integrated STEM: An intersubjective constructionist approach. *Communications in Computer and Information Science, 510*, 272–288.

Sengupta, P., Dickes, A. C., Farris, A. V., Karan, A., Martin, K., & Wright, M. (2015b). Programming in K12 science classrooms. *Communications of the ACM, 58*(1), 33–35.

Sengupta, P., Brown, B., Rushton, K., & Shanahan, M. C. (2018). Reframing coding as "Mathematization" in the K12 classroom: Views from teacher professional learning. *Alberta Science Education Journal, 45*(2), 28–36.

Sherin, B., diSessa, A. A., & Hammer, D. (1993). Dynaturtle revisited: Learning physics through collaborative design of a computer model. *Interactive Learning Environments, 3*(2), 91–118.

Stieff, M., & Wilensky, U. (2003). Connected chemistry—Incorporating interactive simulations into the chemistry classroom. *Journal of Science Education and Technology, 12*(3), 285–302.

Talmy, L. (1983). How language structures space. In H. Pick & I. Acredelo (Eds.), *Spatial orientation: Theory, research, and application*. New York: Plenum Press.

Wagh, A., Cook-Whitt, K., & Wilensky, U. (2017). Bridging inquiry-based science and constructionism: Exploring the alignment between students tinkering with code of computational models and goals of inquiry. *Journal of Research in Science Teaching, 54*(5), 615–641.

Weintrop, D., Afzal, A., Salac, J., Francis, P., Li, B., Shepherd, D. & Franklin, D. (2018). Evaluating CoBlox: A Comparative Study of Robotics Programming Environments for Adult Novices. In *Proceedings of the 2018 CHI Conference on Human Factors in Computing Systems (CHI'18)*. pp. 366:1–12. Montreal QC, Canada: ACM Press.

Wilensky, U. (1999). NetLogo. http://ccl.northwestern.edu/netlogo/. Center for Connected Learning and Computer-Based Modeling, Northwestern University, Evanston.

Wilensky, U., & Resnick, M. (1999). Thinking in levels: A dynamic systems approach to making sense of the world. *Journal of Science Education and technology, 8*(1), 3–19.

Wilensky, U., & Reisman, K. (2006). Thinking like a wolf, a sheep, or a firefly: Learning biology through constructing and testing computational theories – an embodied modeling approach. *Cognition and Instruction, 24*(2), 171–209.

Wing, J. M. (2006). Computational thinking. *Communications of the ACM, 49*(3), 33–35.

Wing, J. M. (2008). Computational thinking and thinking about computing. *Philosophical Transactions of the Royal Society A: Mathematical, Physical and Engineering Sciences, 366*(1881), 3717–3725.

Wing, J. (2011). Computational thinking—What and why? *The Link: News from the School of Computer Science., 6*, 20–23.

Yackel, E., & Cobb, P. (1996). Sociomathematical norms, argumentation, and autonomy in mathematics. *Journal for Research in Mathematics Education, 22*(4), 390–408.

Yackel, E., Cobb, P., & Wood, T. (1991). Small-group interactions as a source of learning opportunities in second-grade mathematics. *Journal for Research in Mathematics Education, 22*(5), 390–408.

Chapter 5
Strictly Objects First: A Multipurpose Course on Computational Thinking

Johannes Krugel and Peter Hubwieser

5.1 Introduction

In 2006, Janet Wings presented her groundbreaking concept of *computational thinking* (CT), which seems to be broadly accepted these days. She defined CT as "… the process of abstraction, choosing the right abstractions, operating in terms of multiple layers of abstraction simultaneously, defining the relationships between layers." Two years before, the German state of Bavaria had introduced a new compulsory subject (from 2 to 6 years, depending on the school type) of computer science (CS) in its grammar schools (Gymnasium) with a very similar direction (Hubwieser 2012). The teaching approach for this subject was developed in the years 1995 – 2000. Up to now, more than 700,000 students have attended this course at least in its shortest version of 2 years. The course is based on the *strictly objects-first* approach, introducing the concepts *object*, *attribute*, and *method* just before *class* and long before any programming activity. Based on the object-oriented paradigm, the course covers data structures of common software types (graphic and text processors, hypertext, spreadsheets) as well as data bases and object-oriented programming.

Later, this concept was applied to an introductory CS lecture for engineering students at the Technical University of Munich, which was conducted over 8 years with a maximum of 400 participants per year.

In Germany, the implementation of computer science education at school is very diverse, unregulated, and inconsistent in many states. To compensate the resulting differences in prerequisite knowledge among the freshmen of our university, we developed a massive open online course (MOOC) called "LOOP: Learning Object-Oriented Programming" that introduces computational thinking and object-oriented

J. Krugel (✉) · P. Hubwieser
School of Education, Technical University of Munich, Munich, Germany
e-mail: johannes.krugel@tum.de

M. S. Khine (ed.), *Computational Thinking in the STEM Disciplines*,
https://doi.org/10.1007/978-3-319-93566-9_5

concepts before any programming activity, very similar to our original teaching concept. The course includes various interactive exercises to enable the learners to experiment with the presented concepts. Furthermore, we implemented programming exercises with constructive feedback for the learners using a web-based integrated development environment and additionally an automatic grading system. The target group of this online course are prospective students of science or engineering that are due to attend CS lessons in their first terms. The course was conducted as a prototype with a limited number of participants. In a concluding survey, the participants submitted textual feedback on the course; some of them proposed specific improvements for the employed interactive exercises. In addition, we conducted a cluster analysis to find out how the participants behaved over the course.

In this paper, we discuss the relationships between the CT concept of Wing, its operationalization elaborated by others, and our object-oriented view on the basic concepts of computation. We describe the school subject and the university lecture, before presenting details of our MOOC, its didactical approach, and first results of a pilot study.

5.2 Background and Related Work

In the following, we describe the ideas of *computational thinking* and *object orientation*, as well as the *objects-first* principle. We furthermore describe the theoretical background of the teaching methods, motivational theory, and some related MOOCs for introductory CS.

5.2.1 Computational Thinking

Janet Wings' concept of *computational thinking* (CT) seems to be broadly accepted today, also outside the CS community (see, e. g., Mishra et al. (2013)). Recently, a comprehensive review of the field of computational thinking in K12 was published in Seehorn et al. (2011).

Yet, for defining or identifying competencies in the field of CT, we need an operationalization of its definition. The K12 standards of the CSTA from 2011 contain (among four others) a strand called "computational thinking." The 48 standards of this strand can be seen as an operationalization of CT. In Table 5.1, we list several of those standards that might be relevant for our teaching concept.

Table 5.1 Relevant CSTA standards of the strand *computational thinking*

No.	The students shall be able to	Level	Grade
4	Recognize that software is created to control computer operations	1	K-3
6	Understand and use the basic steps in algorithmic problem-solving (e. g., problem statement and exploration, examination of sample instances, design, implementation, and testing)	1	3–6
7	Develop a simple understanding of an algorithm (e. g., search, sequence of events, or sorting) using computer-free exercises	1	3–6
46	Use the basic steps in algorithmic problem-solving to design solutions (e. g., problem statement and exploration, examination of sample instances, design, implementing a solution, testing, evaluation)	2	6–9
48	Define an algorithm as a sequence of instructions that can be processed by a computer	2	6–9
50	Act out searching and sorting algorithms	2	6–9
51	Describe and analyze a sequence of instructions being followed (e. g., describe a character's behavior in a video game as driven by rules and algorithms)	2	6–9
52	Represent data in a variety of ways including text, sounds, pictures, and numbers	2	6–9
57	Use abstraction to decompose a problem into subproblems	2	6–9
88	Use predefined functions and parameters, classes, and methods to divide a complex problem into simpler parts	3.A	9–10
90	Explain how sequence, selection, iteration, and recursion are building blocks of algorithms	3.A	9–10
94	Describe how various types of data are stored in a computer system	3.A	9–10
141	Compare and contrast simple data structures and their uses (e. g., arrays and lists)	3.B	10–12
145	Decompose a problem by defining new functions and classes	3.B.	10–12

5.2.2 *Object Orientation*

The object-oriented paradigm was defined, e. g., in 1990 by Rosson and Alpert as a combination of the four aspects: *communicating objects*, *abstraction*, *shared behavior*, and *designing with objects* (Rosson 1990). In order to define the concept of object orientation, James Rumbaugh stated in 1991 that object orientation would have to encompass four aspects: *identity of objects*, *classification*, *polymorphism*, and *inheritance* (Rumbaugh and Blaha 1991). In 1992, Henderson-Sellers identified the following three main conceptual components: *encapsulation/information hiding*, *abstraction/class/object*, and *inheritance/polymorphism* (Henderson-Sellers 1992).

Armstrong conducted a very interesting review of 239 publications and presented a list of the 20 aspects of object orientation that were addressed most frequently in the definitions of object orientation. The five most frequently used were *inheritance*, *object*, *class*, *encapsulation*, and *method* (Armstrong 2006).

5.2.3 Objects-First Principle

Discussing the *objects-first* paradigm, Lewis stated: "*A distinction must quickly be made between initially writing classes that define objects, and using objects defined by preexisting classes. It is sometimes suggested that if students do not write multiple classes and methods initially, they are not being indoctrinated into an object-oriented approach. Most educators agree, however, that using objects is the appropriate first step*" (Lewis 2000).

In 2008, Bennedsen and Schulte conducted a very interesting survey about the understanding and implications of *objects first* among introductory programming teachers (Bennedsen and Schulte 2010). They contacted about 700 authors, teachers and SIGCSE members and received 298 at least partly filled out questionnaires. The content analysis led to the suggestion of the three categories *using objects*, *creating classes*, and *concepts* ("involves the teaching of the general principles and ideas of the object-oriented paradigm, focusing not just on programming but on creating object-oriented models in general"). The authors deduced three common sequences of *objects-first* courses (Bennedsen and Schulte 2010):

1. *Using objects, followed by:*
   (a) *Creating classes, followed by concepts*
   (b) *Concepts, followed by creating classes*
2. *Creating classes, followed by concepts.*

In her doctoral thesis, Ira Diethelm defined the strategy *strictly models and objects first* that should start with the usage and manipulation of objects before the class concept is introduced (Diethelm 2007), corresponding to the sequences *1.a* and *1.b* accordingly (Bennedsen and Schulte 2010).

5.2.4 Teaching Methods

According to Biggs (1999), the highest level of teaching focuses on "what the student does":

> *This implies a view of teaching that is not just about facts, concepts and principles to be covered and understood, but about:*
>
> 1. *What it means to understand those concepts and principles in the way we want them to be understood.*
> 2. *What kind of TLAs (teaching/learning activities) are required to reach those kinds of understandings.*

In fact it was our main concern during the development of this course to propose the actual learning activities that students should perform as related to the different curriculum topics. We did not regard pseudo-activities like "listen," "understand," or "read" as satisfying in this respect. Instead, we tried to suggest observable active learning operations, similar to Anderson and Krathwohl (2001) in relation to their cognitive process concept, e. g., "program," "implement," "present," "explain," or "calculate." In summary, we tried to follow as closely as possible the principles of constructivism (see, e. g., Ben-Ari 1998).

5.2.5 *Motivation*

Based on his empirical investigations, J. M. Keller developed the ARCS Model of Instructional Design as a "method for improving the motivational appeal of instructional materials" (Keller 1987a, b). The model contains four conceptual categories that "subsume many of the specific concepts and variables that characterize human motivation" (Keller 1987b):

- *Attention* has not only to be directed to the appropriate stimuli but also sustained during the learning process.
- *Relevance* provides the answer to the question "Why do I have to study this?"
- *Confidence* can influence a student's persistence and accomplishment. Confident people tend to believe that they can effectively accomplish their goals by means of their actions, while unconfident people want to impress others and worry about failing.
- *Satisfaction* makes people feel good about their accomplishments.

5.2.6 *Related MOOCs*

There are many online courses for learning the basics of computer science. In the following, we sketch some introductory MOOCs (massive open online courses) and SPOCs (small private online courses) that explicitly cover computational thinking or object-oriented programming (OOP) and that were recently published in the scientific literature.

Liyanagunawardena et al. describe the experiences with a MOOC for the introduction to programming where the learners have the possibility to build an Android game. They report on a good community experience and that one difficulty for the learners was to install the development software (Liyanagunawardena et al. 2014).

Piccioni et al. describe a SPOC used to complement an existing course for the introduction to programming. As gamification element, badges are awarded to learners (Piccioni et al. 2014).

Falkner et al. developed a MOOC in which the participants learn programming by producing animations and digital artwork (Falkner et al. 2016).

Alario et al. developed a MOOC with interactive exercises using, among others, the software Greenfoot (Alario-Hoyos et al. 2016; Kölling 2010).

Kurhila et al. describe a MOOC for the introduction to CS offered in Finish schools (Kurhila and Vihavainen 2015; Vihavainen et al. 2012).

5.3 Course Origin and History

Before presenting our MOOC in more detail, we describe the design of the school subject and the introductory CS lecture for STEM (science, technology, engineering, and mathematics) students at the university.

5.3.1 Compulsory Subject CS in Bavarian Grammar Schools

In the year 2000, the government of the German state of Bavaria decided to introduce a new compulsory subject of *computer science (CS)* in all its 400 *Gymnasiums*, which represent the most demanding of its three types of secondary schools (Hubwieser 2012). This subject was introduced simultaneously with the reduction of the Bavarian Gymnasium from nine to eight grades that was put into operation in grade 5 and 6 in 2004.

The new CS subject comprises mandatory courses in grades 6/7 for all students of Gymnasium, followed by courses that are compulsory for the students of the science and technology track (the largest of the four education tracks of this school type, attended by about 50% of an age group) in grade 9 and 10. In grades 11 and 12, there are elective courses that qualify for an optional graduation exam in informatics.

Regarding the learning content, the CS course is based on a *strictly objects-first* approach. The courses start in grade 6 with the introduction of the concepts *object*, *attribute*, *method*, and *class* in the context of vector graphics. In a second step, the concept of *aggregation* is introduced in the context of word processing. Following this, the students work with recursive aggregation, applied to file systems. This leads to the construction of folder trees as a representation of hierarchical structures. In the next step, the trees are generalized to graphs by the application of references in the context of hypertext systems. At the end of grade 7, the students work out their first programs, using a virtual robot, e. g., the *Robot Karol*. "Programming" is understood at this point as creating new methods for the class *Robot*.

In grade 9, the students learn to apply the concept of *function*. They construct functional models (data flow diagrams), where the information processing units are restricted to functions. The models are implemented on spreadsheet systems. The rest of the school year in grade 9 is dedicated to (object-oriented) data modeling. The students work with the basic concepts of data base systems – record, table, query – using a relational data base system. They design entity relationship models of more complex systems that consist of several tables, connected by relationships, and implement their models using relational data base systems. The students of grade 10 consolidate their object-oriented knowledge by "real" object-oriented programming, designing object, class, and state models and implementing them with a suitable object-oriented programming language, which currently will be Java in most of the schools. Additionally, they learn to apply the concepts of *sub-* and *superclass*, *inheritance*, and *polymorphism.*

In the elective courses in grade 11, the object-oriented modeling and programming concepts are extended by the recursive data structures: *list*, *tree*, and *graph.* Afterward, the students are introduced to basic concepts of software engineering, which they apply within the context of a large programming project. In the second year of the elective course, the students are introduced in several important subject areas of computer science in grade 12: formal languages, parallel programming, assembler programming, and limitations of computability. For more details about this course, see Hubwieser (2012).

We had derived the following global learning objectives for our project (Hubwieser 2007). The students should:

- Acquire the capability of independent opinion and responsible acting in the information society.
- Be able to act responsibly and efficiently in a world of work and profession that is ubiquitously penetrated by information technology (IT).
- Master efficiently the tools and understand the limitations, chances, and risks of information technology.
- Learn the responsible, efficient usage of information technology based on knowledge of the theoretical foundations and basic principles of the systems.
- Master complex systems, particularly being able to describe their structure and behavior and communicate about them in a competent way.
- Be prepared for the application of information technology in other school subjects.
- Be able to choose their career based on a sufficient knowledge of the possibilities and principal limitations of future IT developments.

Concerning the level of educational objectives, we have explained also in Hubwieser (2007) that our CS lessons are very demanding, e. g., the students have to:

- Analyze problems in order to represent them properly by an object-oriented model.
- Evaluate alternative models in order to choose one of them.
- Create models and programs out of their models.

This shows that the educational objectives reach the most difficult cognitive process dimension *create* according to Anderson and Krathwohl (2001).

Regarding the lowest category of learning objectives, I have demonstrated in Hubwieser (2007, 2008a) that the instructional objectives of even quite simple object-oriented programs easily demand 40 or even more instructional objectives in order to be understood by the students.

The intended knowledge is a very crucial aspect still, even if it gains its importance mainly as a component of learning objectives or, following Hartig (2001), as

Table 5.2 Object-oriented framework of knowledge areas

Knowledge field	Object-oriented concept	Examples, tools, and software
Data structures of standard software	Object and class models of documents, attributes, methods	Graphics, text, hypertext, presentations
Algorithm	State and activity models, methods	Karol, Scratch, etc.
Function	Functional models	Spreadsheets
Data bases	Records as objects, tables as classes	Relational data base systems
Computer networks, Internet	Parallel collaboration of objects, defined by protocols	E-mail, client-server-systems, network protocols
Formal languages	Syntax of programming languages and protocols	Java, BNF
Machine level programming	Object-oriented models of processors and computer systems	Assembly programming, class "PROCESSOR"

one of several facets of *competence*. After all, the relevant subject matter knowledge determines the substantial and logical structure of the teaching process.

As it is impossible to teach *all* the knowledge of CS in secondary schools, we had to select the knowledge elements that seem to be particularly valuable for our target group, with respect to the intentions we had defined specifically for our project. In particular, we had to respect the primary objective of extended general education. For this purpose, we applied the concept of Fundamental Ideas of CS of A. Schwill (1994).

Regarding the overall logical structure of the course, we decided to follow the *objects-first* paradigm, described by Lewis (2000).

During this process of the curriculum development, it turned out that the object-oriented approach could even serve as a conceptual framework that contained all other knowledge fields that we had decided to be encompassed in the curriculum (see Table 5.2).

As far as we have evidence about the success of this conception, it seems quite successful now, 10 years after the start of the first course.

5.3.2 Introductory CS Lectures for STEM Students at the University

We also designed and gave introductory lectures in CS for STEM students at the university. The main goal of these courses was to introduce freshmen of engineering (major in geodesy) and business administration into the fundamentals of object-oriented programming (OOP). The course comprised 2 weekly hours of lecturing and 2 more hours of practice in groups of 20 students. It ran over one semester each and was taught in German language.

Following a *strictly objects-first* strategy (Gries 2008), we distributed the learning objectives over the parts of the course that precede the "serious" programming part and thereby avoiding to confront the students with too many unknown concepts at the point they have to write their first program. Basically, we suggest the students to look at an object as a state machine (Hubwieser 2006). In order to realize this in a student-oriented way, the students need to be able to understand a simulation program of a typical state machine, e. g., a traffic light system. The curriculum of the course was structured in the following seven chapters:

1. *Modeling.* Informatics (main subject areas, typical working methods), functional modeling (data flow diagrams), modeling techniques in computer science
2. *Object-Oriented Modeling.* Objects in documents (object, class, attribute, method, class card, object card); artificial languages (grammars, BNF); states of objects (state, transition, state diagram, real and program objects); object diagram, association, class diagram, multiplicity of associations, compound objects (creation of objects as values of attributes)
3. *Algorithms.* The concept of algorithm (programming languages), class definition (definition and declaration, signature of methods, access modifier, attribute declaration, definition of methods); structure of algorithms (graphical representation of algorithms, structural components of algorithms, nesting of components, input and output of algorithms); properties of algorithms (terminating, deterministic, determined)
4. *Object-Oriented Programming.* Definition of classes (structure of object-oriented programs, definition and declaration, signature of methods, access modifier, attribute declaration, definition of methods); assignment statement, ring exchange, assignment in constructor methods, encapsulation, equality; translation of computer programs, compiler vs. interpreter, execution of programs, course of events of a program; communication by methods (input, output, side effects, local and global variables/attributes); creating objects at runtime, constructor method, references, removal of objects; implementation of algorithms (structure elements in programming languages – sequence, conditional statement, repetition); arrays, index.
5. *State Modeling.* Finite automatons, triggering and triggered action, state chart; implementation of automatons (switch statement); conditional transitions (complete state modeling, implementation of conditional transitions).
6. *Interaction and Recursion.* Implementation of associations (unidirectional, bidirectional, 1:1, 1:*n*, *m*:*n* multiplicities, association class); sequence charts (calling of methods, sequence charts); recursive algorithms (linear and cascading recursion)
7. *Generalization.* Sub- and super-classes, specialization, inheritance; implementation of specialization, overriding of methods, generalization, class hierarchies; polymorphism (calling methods of foreign classes, abstract classes)

5.4 MOOC

Making our approach for learning object-oriented accessible to an even wider audience, we decided to develop an online course called "LOOP: learning object-oriented programming" (Krugel and Hubwieser 2017). In the following, we give the motivation for creating the course, describe the course elements, and present the details of the course design.

5.4.1 *Motivation for Creating the Course*

CS education in school is varying strongly in many countries. In consequence, the prerequisite knowledge of freshmen at universities is very inhomogeneous (Hubwieser et al. 2015). To investigate this at our university, we performed a survey in a CS1 lecture in October 2015, applying a specific instrument by Linck et al. (2013), as proposed, e. g., by Hering et al. (2014). The results of the 874 participants revealed huge differences in the students' prior knowledge.

As students cannot be expected to be present at the university before lecturing starts, MOOCs (massive open online courses) seem to represent potential solutions to compensate or reduce these differences. Yet, as learning to program is a substantial cognitive challenge (Hubwieser 2008b), such MOOCs run in danger to overstrain the students, frustrating them already before their studies.

To meet this challenge, we designed our course LOOP that starts with a gentle introduction to computational thinking (Wing 2006) and object-oriented concepts before the programming part to avoid excessive cognitive load, following the concept *strictly objects first* (Gries 2008).

The target group of the course are prospective students of science or engineering that are due to attend CS lessons in their first terms. We created our course on the learning platform *edX Edge*[1].

5.4.2 *Videos*

All topics of the course are presented in short videos with an average length of 5 min. The videos were produced based on the suggestions of Guo et al. (2014) and similar to the suggestions by Alonso-Ramos et al. (2016) published shortly after our recording.

Each of the 24 videos begins with a short advance organizer to help the learners focus on the relevant aspects. This is augmented with the talking head of the

[1]https://edge.edx.org

respective instructor (using chroma key compositing) facilitating the learners to establish a personal and emotional connection (Alonso-Ramos et al. 2016).

For the actual content of the videos, we decided to use a combination of slides and tablet drawing. The background of the video consists of presentation slides and the instructor uses a tablet to draw and develop additional aspects or to highlight important part of the slides ("Khan-style"). This turned out to yield quite engaging videos with a reasonable effort for preparing and recording. All slides are provided for download and we additionally added audio transcripts for the videos. By such video, audio, and textual representations, several senses are addressed simultaneously, making the content accessible to learners with different learning preferences or impairments.

5.4.3 *Quizzes*

After each video, the course contains quizzes as formative assessment. The main purpose is to provide the learners with direct and instant feedback on the learning progress. The quizzes use the standard assessment types offered by the MOOC platform, e. g., single- / multiple-choice questions, drop-down lists, drag-and-drop problems, or text input problems. Depending on the answer, the learner gets a positive feedback or, otherwise, for example, hints which previous parts of the course to repeat in more detail.

5.4.4 *Interactive Exercises*

The videos introduce new concepts to the learners, and the quizzes test the progress, which is, however, in general not sufficient to acquire practical competencies (Alario-Hoyos et al. 2016). Following a rather constructivist approach, we intend to let the learners experiment and interact with the concepts directly. Considering that, we include interactive exercises or programming task for all learning steps throughout the course. Special care was devoted to the selection and development of those interactive exercises to enable the learners to experiment and interact directly with the presented concepts. It can be a major obstacle for potential participants having to install special software (Liyanagunawardena et al. 2014; Piccioni et al. 2014), which is especially problematic in an online setting without a teacher who could help in person. We therefore decided to use only purely web-based tools. There are already many web-based tools for fostering computational thinking and learning OOP concepts available on the web. We selected the most suitable tools to support the intended learning goals. Where necessary, we adapted or extended them to meet our needs. All tools are integrated seamlessly into the learning platform resulting in a smooth user experience.

5.4.5 Programming Exercises

While in several introductory CS MOOCs the learners have to install an integrated development environment (IDE) for writing their first computer programs, we decided to rely on web-based tool also for this purpose (like Piccioni et al. 2014). We chose to use *Codeboard*[2] (Estler and Nordio) (among several alternatives, Derval et al. 2015; Skoric et al. 2016; Staubitz et al. 2016) because of the usability and seamless integration into the *edX* platform using the Learning Tools Interoperability (LTI) standard.

The programming assignments are graded automatically, and the main purpose is to provide helpful feedback to the learner. We therefore implemented tests for each assignment that make heavy use of the Java reflection functionality. While standard unit tests would fail with a compile error if, e. g., an attribute is missing or spelled differently. Reflection makes it possible to determine for a learner's submission if, e. g., all attributes and methods are defined with the correct names, types, and parameters. Writing the tests requires more effort than for standard unit tests but can give more detailed feedback for the learners in case of mistakes.

Additionally we integrated the automatic grading and feedback system *JACK* (Striewe and Goedicke 2013) using the external grader interface of the *edX* platform. Apart from static and dynamic tests, *JACK* also offers the generation and comparison of traces and visualization of object structures; however, we do not use this extended functionality yet.

5.4.6 Course Design

Computational thinking (CT) as introduced by Wing (2006) is a universal personal ability that can be used in many disciplines. Since the target group of our course comes from various different fields of study, we incorporated CT as integral part of the course. CT is on the one hand intended to facilitate learning programming and on the other hand a sustainable competency that can be used also outside of our course.

As pointed out in Hubwieser (2008b), there is a fundamental didactical dilemma in teaching OOP: on the one hand, modern teaching approaches postulate to teach in a "real-life" context (Cooper and Cunningham 2010), i. e., to pose authentic problems to the students. Therefore, it seems advisable to start with interesting, sufficiently complex tasks that convince the students that the concepts they have to learn are helpful in their professional life. However, if we start with such problems, we might ask too much from the students, because they will have to learn an enormous amount of new, partly very difficult concepts at once (Hubwieser 2008b).

Following a *strictly objects-first* approach (Gries 2008) and similar to the design of the school subject and the introductory lecture, we solved this problem by

[2]https://codeboard.io

distributing the learning objectives over the parts of the course that precede the "serious" programming part. This avoids to confront the learners with too many unknown concepts when they have to write their first program. Among others, we suggest to the students to look at an object as a state machine (Hubwieser 2008b). In order to realize this in a learner-oriented way, the students need to be able to understand a simulation program of a typical state machine, e. g., a traffic light system.

Concerning the choice of the examples, we set the emphasis on the relevance for the everyday life, which leads, for instance, to banking or domestic appliances.

LOOP consists of the following five chapters, which are described in more detail below:

1. Object-oriented modeling
   - 1.1 Objects
   - 1.2 Classes
   - 1.3 Methods and parameters
   - 1.4 Associations
   - 1.5 States of objects
2. Algorithms
   - 2.1 Concept of algorithm
   - 2.2 Structure of algorithms
3. Classes in programming languages
   - 3.1 Class definition
   - 3.2 Methods
   - 3.3 Creation of objects
4. Object-oriented programming
   - 4.1 Implementing algorithms
   - 4.2 Arrays
5. Associations and references
   - 5.1 Aggregation and references
   - 5.2 Managing references
   - 5.3 Communication of objects
   - 5.4 Sequence charts

5.4.6.1 Chapter 1: Object-Oriented Modeling

Following the concept *strictly objects first*, Chapter 1 is based on object-oriented modeling of standards software documents like graphics, texts, or hypertexts as already proposed in Hubwieser (2000). This idea was inspired by the prior work of U. Freiberger about the object-oriented modeling of word processors (Freiberger

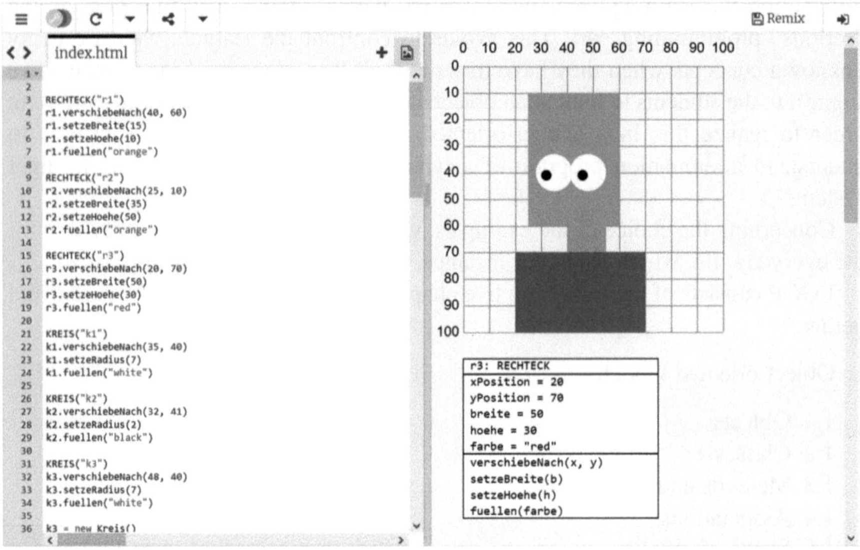

Fig. 5.1 Interactive exercise: construction of a graphic with simple commands (using *trinket.io*)

1988–1990). As described in Hubwieser (2003), the course starts with the application of the concepts *object* and *attribute* in the context of vector graphics, worked out by the students using web-based vector graphic drawing tool *SVG-edit*.[3] The learners have the task to draw a simple graphic using rectangles, circles, and lines (which implicitly also already prepares for the idea of *classes*). The learners are then asked to publish their drawing in the discussion forum of the course and to introduce themselves to the community.

To let the learners experience that objects have a state, that the state can change, and that this is usually achieved by method calls, we developed a new interactive exercise. The learners can draw a picture on a canvas in the web browser by using simple commands in a restricted pseudo programming language, which only allows the creation of graphical objects (e. g., circles and rectangles) and method calls on those objects (e. g. *move()*, setWidth(), *setFillColor()*, . . .). By this way, the students rediscover the same objects, classes, attributes, and methods that they have learned at the beginning of the course in the context of vector graphics. To implement this exercise, we combined the tool *trinket*[4] (providing an online code editor connected to a canvas) and the JavaScript library *SVG.JS*[5] (providing an interface for drawing objects) (see Fig. 5.1). We adapted and extended this such that the learners can inspect the drawn objects by showing the Unified Modeling Language (UML) object diagram when hovering over an object. As an assignment in the course, we prompted

[3]https://github.com/SVG-Edit

[4]https://trinket.io

[5]http://svgjs.com

the learners to draw an animal or a cartoon figure and to share it in the discussion forum.

As the execution of an algorithm in an object-oriented program is realized by a sequence of object states (represented by the values of the attributes), the concept *state* is essential to understand this. We start by defining the state of an object by the combination of the states of all its attributes, each of them represented by the actual value of the attribute. This means that an object changes its state if one of its attributes changes its value. The transition between two states is triggered by a method that changes the values of an attribute. In order to describe the states of an object, we introduce state charts.

We then turn our attention toward the states of real-world objects and model them in further state charts, for example, the state chart of (German) traffic lights. The traffic light system uses three indicators (red, yellow, and green light, each with the states "on" and "off") to display four different system states, which the road users interpret usually as "stop" (red), "prepare for stopping if not very close" (yellow), "go" (green), and "prepare for forthcoming go" (red-yellow). We discover that the real technical system "traffic light" uses only four of its eight technically possible states (defined by all possible combinations of the two states of each of the three lamps). For example, the state "on-off-on" (in colors: red-green) is technically possible, but not used. In an interactive assignment, the students are asked to complete a start chart for a hair dryer (see Fig. 5.2).

As a consequence of the abovementioned considerations, from now on every object is perceived as a state machine.

5.4.6.2 Chapter 2: Algorithms

To understand the concept of algorithms, Chapter 2 begins with a definition and some examples and counterexamples of algorithms. Enabling the learners to interact with and to visualize simple algorithms, we integrated the geometric web framework *CindyJS*[6] written in JavaScript (von Gagern et al. 2016). As example, we use the Euclidian algorithm for calculating the greatest common divisor of two numbers. The learner can modify the input by moving a point in the plane and observe at the same time the steps of the algorithm.

Syntax can be a major obstacle when learning to program (Kölling 2010). We therefore tried to reduce the cognitive load and make the first steps easier by providing a gentle introduction. To facilitate the understanding for the structure of algorithms, we included a gamification element using block-based programming (see Fig. 5.3). We integrated a series of maze riddles from *Blockly-Games*,[7] which can be solved by combining move operations with structural elements like loops and conditional statements.

---

[6]http://cindyjs.org

[7]https://blockly-games.appspot.com

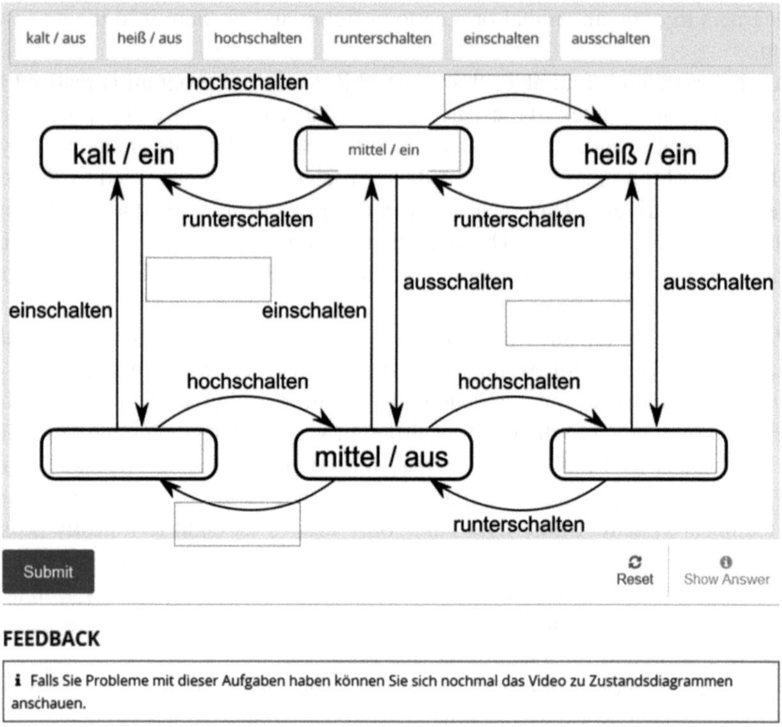

Fig. 5.2 Interactive exercise: completing a state chart for a hair dryer (using drag-and-drop)

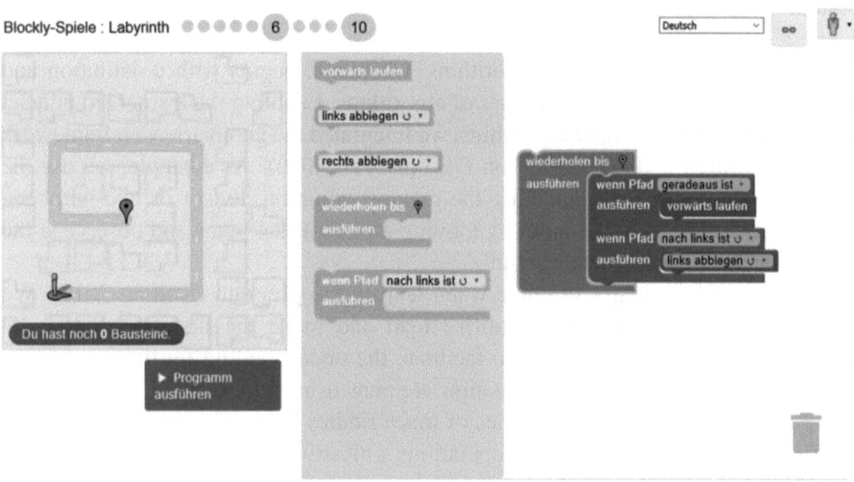

Fig. 5.3 Interactive exercise: programming with blocks (using *Blocky Games*)

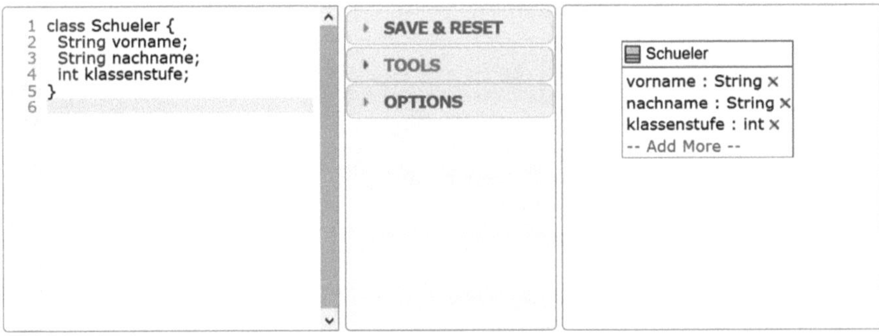

Fig. 5.4 Interactive exercise: connection between class card and program code (*UmpleOnline*)

5.4.6.3 Chapter 3: Classes in Programming Languages

After this introductory step, we transfer the concept of a *class* (including the concepts of attributes and methods) into a programming language (here: *Java*) by comparing a class card with the definition of the same class in the programming language. Before the learners start to implement their first class in from scratch, we let them experience the connection between the class card and the corresponding Java implementation using the web-based tool *UmpleOnline*[8] (Lethbridge 2014). This tool enables the learner to modify a class diagram and simultaneously observe the changes in the Java implementation and the other way around (see Fig. 5.4).

On this occasion, we also introduce some new concepts like data types and constructors. We, however, try to reduce the cognitive load and hide "advanced" aspects (like access modifiers) to let the learners focus on the essential parts of the class definitions.

One of the big challenges of informatics education is the proper introduction of the assignment command (e. g., "assign the value 5 to the variable *number*"), particularly its distinction from the equality statement (e. g., "the value of *number* is equal to 5"). In the syntax world of C and its successors (e. g., Java), this is made even more difficult by the unlucky choice of the "=" sign for the assignment operator, which is well known by the students from the subject of mathematics, symbolizing equality there. The semantics of the assignment command will be explored by closely looking at the values of the attributes during sequences of assignments. Proper understanding might be tested by the task of exchanging the values of two attributes. This leads directly to the state concept of attributes (respectively, of variables) which is essential to understand an assignment like "*counter* = *counter* + *1*" where the attribute (variable) *counter* appears in two different states, represented by its values before and after the execution of the assignment.

[8]http://try.umple.org

Fig. 5.5 Interactive exercise: visualizing the program execution (using *Java-Tutor*)

5.4.6.4 Chapter 4: Object-Oriented Programming

Now that the structural elements of algorithms (elementary, sequential, conditional, and repetitive processing structures), the concept of *states* of objects, and also the syntax of class definitions are known, it is not a big step for the learners to start implementing their first algorithms.

For visualizing the execution of a program, we chose to use the tool *Java-Tutor*[9] (based on the very similar *Python-Tutor*) (Guo 2013). The learners can run a program step-by-step with the possibility to navigate forward and backward while observing the control flow (see Fig. 5.5). It also includes a graphical representation of the memory contents, preparing the learner to understand related concepts such as references.

Chapter 5 also introduces the concept of *arrays*, as a rather simple but very instructive element, especially in combination with repetitions.

5.4.6.5 Chapter 5: Associations and References

One of the most challenging tasks in teaching informatics is the introduction of pointers or references, which is essential for the understanding of certain effects that arise, for example, if two references point toward the same object. Fortunately, our students already know the aggregation as a special type of association (marked with "contains"), for example: an object of the class *Folder* may contain an object of the

[9]http://www.pythontutor.com/java.html

class *File*. This leads us exactly to the right concept: if an object *obj1* of *Class1* contains an object *obj2* of *Class2*, the latter will be implemented as an attribute of the type *Class2* in *Class1*, whose value is actually a reference toward the external object *obj2*. It is crucial for the proper perception that the students keep in mind that the aggregation association is always connecting objects, although it is drawn in the class diagram.

Chapter 6 additionally introduces sequence diagrams as a way to visualize the behavior of a system with several objects and the communication between the objects.

5.5 Pilot Conduction

We prepared the online course on the platform *edX Edge* and conducted it during the summer holidays 2016 with a limited number of participants as prototype for a MOOC. The course was announced internally at our university as preparation course for CS basics. Participation was voluntary and did not count toward a grade, but we issued informal certificates for successful participation (= obtaining at least 50% of the possible points in at least 12 of 16 course units).

In an introductory online questionnaire, we asked the participants about their gender, major, and previous programming experience. Additionally we asked about the intentions to complete the course, providing four options (see Table 5.3).

The course took 5 weeks (1 week for each chapter) and the targeted workload of the learners was 10 h per week. The communication among the learners and with the instructors took place entirely in the discussion forum. The main task of the instructor during the conduction of the course was to monitor the forum and to react accordingly, e. g., answer questions or fix problems with the grading system.

In a concluding questionnaire distributed after the course, we asked for positive and negative textual feedback regarding the course.

Table 5.3 Participant's intentions to complete the course

Option	Answers
I just want to have a look at the course	14
I want to study some topics that are relevant for me	18
I want to study most topics of the course	12
I want to complete the whole course	29
(No answer)	4
Total	77

5.5.1 First Results

The course attracted 187 registrations. For the introductory questionnaire, we received 77 responses (female, 21; male, 52; no answer, 4) with a very diverse study background (33 different majors, including biology, business studies, engineering, and mathematics). Regarding programming, 10 participants had no experience, 35 had basic knowledge, and 27 participants had already written a "bigger" program of at least 100 lines of code (no answer: 5 participants).

The discussion forum contained in total 178 posts at the end of the course. However, there was not a lot of discussion and communication among the participants themselves, and most posts were answers to the exercises as required by the assignments (see Sect. 3.4), presumably because we did not actively focus on initiating lively discussions in this prototypical conduction of the course.

From the 77 responses of the introductory questionnaire, 41 stated that they want to complete most topics or the whole course (see Table 5.3). In general, MOOCs have a rather high dropout (Delgado Kloos et al. 2014; Garcia et al. 2014; Piccioni et al. 2014). At the end of the course, we were happy that 40 participants gained the course certificate (however, not necessarily the same learners as the 41 from the questionnaire).

In the concluding survey distributed after the course, we received 11 answers. The participants proposed specific improvements for the employed interactive exercises, among others to use a more user-friendly web-based drawing tool (or to additionally allow the use of offline software) and to include more difficult exercises. Yet, the overall feedback was encouragingly positive. The learners stated to like the videos, the explanations, the interactive exercises, and the overall alignment of the course elements.

The scores of the assignments reflect a picture similar to the responses of the participants. The average scores are quite low, which is caused mainly by the high frequency of score 0, in other words by the participants that did not really try to solve the tasks (see Table 5.4). The average scores decline monotonously over all course sections, except the tasks in Sects. 1.5 and 5.1, where the average scores increase relative to the preceding tasks.

5.5.2 Cluster Analysis

To investigate the learning progress of the participants even further, we conducted a hierarchical cluster analysis of their scores. We calculated the average of the achieved relative scores over each of the 16 course sections for each of the 187 participants. For example, if a person scored 3 points in a section with a maximum score of 12 points over all assignments, we entered the value 0.25. This resulted in a 187×16 matrix of real values (from 0.00 to 1.00) with one line per participant and one column per section. For the clustering, we regarded the columns of this matrix as

Table 5.4 Results of assignments per course section

Section	Average over all participants	Frequency of average score 0
1.1	0.35	55.9 %
1.2	0.34	63.8 %
1.3	0.32	66.5 %
1.4	0.26	69.1 %
1.5	0.37	62.8 %
2.1	0.25	72.9 %
2.2	0.24	71.8 %
3.1	0.24	74.5 %
3.2	0.21	76.6 %
3.3	0.19	79.8 %
4.1	0.14	81.4 %
4.2	0.12	84.6 %
5.1	0.14	83.5 %

dimensions. Thus, the set of 16 average scores of each participant could be regarded as the definition of a certain position in a 16-dimensional space. By this way, the pairwise distance between the positions of all participants could be calculated in a specific distance matrix with 187 columns and 187 rows, applying 2 different distance metrics (*Maximum* and *Euclidian*). Finally, a hierarchical clustering was performed on this distance matrix, starting with one cluster per person and combining successively more and more persons to larger clusters according to their relative distance, applying several different clustering strategies (*ward.D*, *Complete*, *Average*, *and McQuitty*; for details, see Everitt et al. 2001). The calculation was performed in the statistical programming language *R* by applying the function *hclust*.

As hierarchical clustering is a local heuristic strategy, the results have to be inspected according to their plausibility. For this purpose, we looked for plausible dendrograms that represented a proper hierarchy. We found that the Euclidean distance metrics produced the best results in combination with the *ward.D* algorithm. Figure 5.6 shows the result. To find a suitable number of clusters, we inspected these dendrograms from the top down to a level where we found as many clusters as possible, but avoiding too small clusters with less than five members. We found that the best height to cut would be at five branches (see Table 5.5).

Finally, we calculated the average performance over all course sections for each of these clusters. The results are displayed in Fig. 5.7.

Obviously, the clustering reflects the dropout behavior of the participants. It reveals that the 90 students of *cluster 2* did not have any success in the whole course, presumably just playing around with the MOOC. The 29 students of *cluster 1* achieved some relevant score points in Sect. 1.1 (Objects) and Sect. 1.5 (States of objects). The 22 participants of *cluster 3* started quite well, but dropped out during the second part of Chapter 1 and Chapter 2. Interestingly, they also show a relative good performance in Sect. 1.5, similar to *cluster 1*. This seems to be a last but in the end unsuccessful trial to catch up again. The 20 students of *cluster 4* decline from

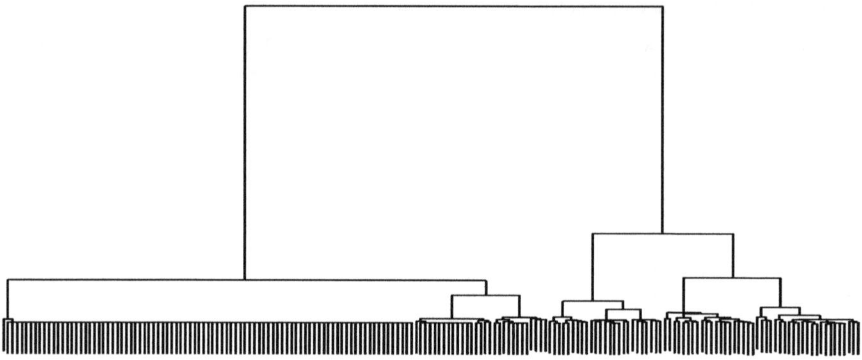

Fig. 5.6 Exemplary dendrogram, clustering by Euclidean distance with the *ward.D* algorithm

Table 5.5 Results of clustering over the performance in the course sections

Cluster	Students
1	29
2	90
3	25
4	20
5	23

Sects. 4.1 to 5.2, failing in the last two exercises. Finally, the 23 students of *cluster 5* made it successfully to the end, although having some troubles with Sect. 4.3 (Arrays), Sect. 5.2 (Managing references), and Sect. 5.4 (Sequence charts), but catching up again in Sect. 5.1 (Aggregation and references) and Sect. 5.3 (Communication of objects).

5.6 Discussion

A very similar teaching concept is applied in three introductory CS courses: a school subject, a lecture at the university, and a MOOC. The circumstances of the courses are very different, concerning, e.g., the audience (school pupils vs. STEM students), the setting (classroom vs. online) and the course duration (several years vs. 6 weeks). Our underlying ideas for the design of all three courses are, however, very similar: the learning objectives are distributed across the course to reduce the cognitive load of the learners. This is achieved by teaching concepts from computational thinking before the actual programming tasks.

Based on the results of the pilot conduction of the MOOC, we are going to improve and optimize our course. In particular, we will look carefully to the results of the cluster analysis and optimize especially the sections that represent local minima in Fig. 5.7 (e. g., Sects. 1.4, 2.2, 4.3, 5.2, and 5.4) because the assignments in these sections might represent potential hurdles for the learners.

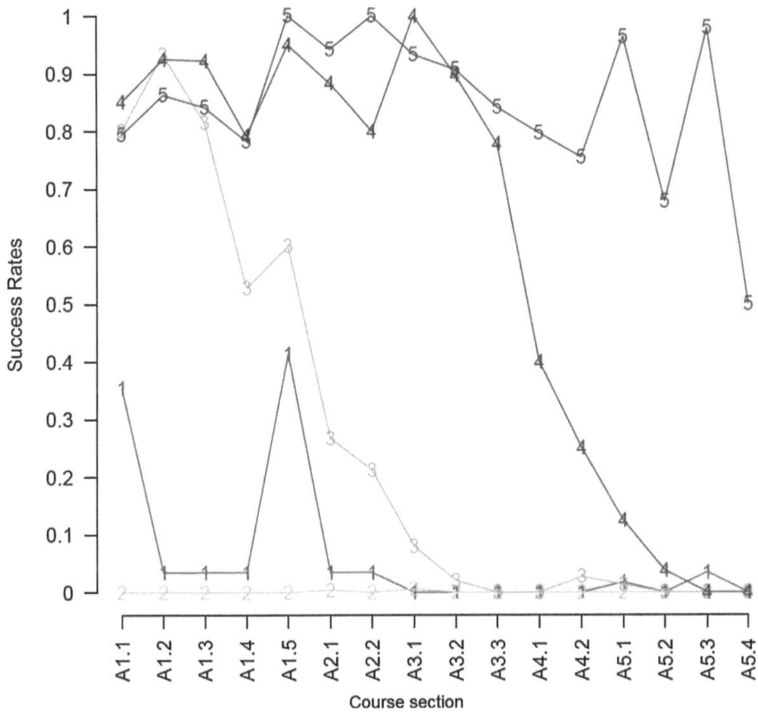

Fig. 5.7 Average performance of the student clusters in the course sections

While we focused so far mostly on the design and creation of the material and exercises, we are going to shift the focus more toward the communicational aspects and incorporate more collaborative elements and peer-grading features into the course. We are going to offer the course to the public as a MOOC and aim to evaluate the learning processes by mining all data produced by the participants. We furthermore plan to measure the effect of the MOOC on the knowledge of the freshmen when entering university.

In the long term, we aim to use LOOP as a general tool to analyze learning processes in object-oriented programming. The online setting allows to perform experiments and analysis on a scale much larger than in a regular classroom course and, furthermore, poses new research questions (Settle et al. 2014).

Acknowledgments We thank Alexandra Funke and Marc Berges for developing parts of the course and Elias Hoffmann, Elisabeth Eichholz, and Simon Zettler for testing and evaluating a preliminary version of the course.

References

Alario-Hoyos, C., Delgado Kloos, C., Estévez-Ayres, I., Fernández-Panadero, C., Blasco, J., Pastrana, S.,... Villena-Román, J. (2016). Interactive activities: The key to learning programming with MOOCs. In *European stakeholder summit on experiences and best practices in and around MOOCs (EMOOCS'16).* Norderstedt: Books on Demand.

Alonso-Ramos, M., Martin, S., Albert Maria, J., Morinigo, B., Rodriguez, M., Castro, M., Assante, D. (2016). Computer science MOOCs: A methodology for the recording of videos. In *IEEE global engineering education conference (EDUCON'16)* (pp. 1115–1121). https://doi.org/10.1109/EDUCON.2016.7474694.

Anderson, L. W., & Krathwohl, D. R. (2001). *A taxonomy for learning, teaching, and assessing: A revision of Bloom's taxonomy of educational objectives* (4th ed.). New York: Longman.

Armstrong, D. J. (2006). The quarks of object-oriented development. *Communications of the ACM, 49*, 123–128. https://doi.org/10.1145/1113034.1113040.

Ben-Ari, M. (1998). Constructivism in computer science education. *ACM SIGCSE Bulletin, 30*(1), 257–261. https://doi.org/10.1145/274790.274308.

Bennedsen, J., & Schulte, C. (2010). BlueJ visual debugger for learning the execution of object-oriented programs. *ACM Transactions on Computing Education, 10*(2), 1–22. https://doi.org/10.1145/1789934.1789938.

Biggs, J. B. (1999). What the student does: Teaching for enhanced learning. *Higher Education Research and Development, 18*(1), 57–75.

Cooper, S., & Cunningham, S. (2010). Teaching computer science in context. *ACM Inroads, 1*, 5–8. https://doi.org/10.1145/1721933.1721934.

Delgado Kloos, C., Munoz-Merino, P. J., Munoz-Organero, M., Alario-Hoyos, C., Perez-Sanagusti, M., Parada G., H.A., … Luis Sanz, J. (2014). Experiences of running MOOCs and SPOCs at UC3M. In *IEEE global engineering education conference (EDUCON'14)* (pp. 884–891). https://doi.org/10.1109/EDUCON.2014.6826201.

Derval, G., Gego, A., Reinbold, P., Frantzen, B., Van Roy, P. (2015). Automatic grading of programming exercises in a MOOC using the INGInious platform. In *European stakeholder summit on experiences and best practices in and around MOOCs (EMOOCS'15)* (pp. 86–91).

Diethelm, I. (2007). Strictly models and objects first – Ein Unterrichtskonzept für OOM. In S. E. Schubert (Ed.), *Vol. 112. LNI, Didaktik der Informatik in Theorie und Praxis (INFOS 2007)* (pp. 45–56). GI.

Estler, C., & Nordio, M. C. (n.d.). Retrieved from http://codeboard.io/.

Everitt, B. S., Landau, S., & Leese, M. (2001). *Cluster analysis*. London: Arnold.

Falkner, K., Falkner, N., Szabo, C., Vivian, R. (2016). Applying validated pedagogy to MOOCs. In *ACM conference on innovation and technology in computer science education (ITiCSE'16)* (pp. 326–331). New York: ACM. https://doi.org/10.1145/2899415.2899429.

Freiberger, U. (1988-1990). *Standardsoftware, Didaktische Hilfen. Teil 1–3: Arbeitskreis Standardsoftware*. Augsburg.

von Gagern, M., Kortenkamp, U., Richter-Gebert, J., & Strobel, M. (2016). Cindy JS. In G. M. Greuel, T. Koch, P. Paule, & A. Sommese (Eds.), *Lecture notes in computer science, 5th international congress on mathematical software (ICMS'16)* (pp. 319–326). Cham: Springer New York. https://doi.org/10.1007/978-3-319-42432-3_39.

Garcia, F., Diaz, G., Tawfik, M., Martin, S., Sancristobal, E., Castro, M. (2014). A practice-based MOOC for learning electronics. In *IEEE global engineering education conference (EDUCON'14)* (pp. 969–974). https://doi.org/10.1109/EDUCON.2014.6826217.

Gries, D. (2008). A principled approach to teaching OO first. *ACM SIGCSE Bulletin, 40*(1), 31. https://doi.org/10.1145/1352322.1352149.

Guo, P. J. (2013). Online Python Tutor. In T. Camp, P. Tymann, J. D. Dougherty, K. Nagel (Eds.), *44th ACM technical symposium on computer science education (SIGCSE'13)* (p. 579). https://doi.org/10.1145/2445196.2445368.

Guo, P. J., Kim, J., & Rubin, R. (2014). How video production affects student engagement. In M. Sahami, A. Fox, M. A. Hearst, & M. T. H. Chi (Eds.), *1st ACM conference on Learning@Scale (L@S'14)* (pp. 41–50). New York: ACM. https://doi.org/10.1145/2556325.2566239.

Hartig, J. (2001). Concept of competence: A conceptual clarification. In D. S. Rychen & L. Salganik (Eds.), *Defining and selecting key competencies* (pp. 45–65). Seattle: Hogrefe and Huber.

Henderson-Sellers, B. (1992). *A book of object-oriented knowledge: Object-oriented analysis, design and implementation: A new approach to software engineering, Prentice-Hall object-oriented series*. Englewood Cliffs: Prentice-Hall.

Hering, W., Huppertz, H., Krämer, B. J., Schreier, S., Magenheim, J., Neugebauer, J. (2014). On benefits of interactive online learning in higher distance education. In *6th international conference on mobile, hybrid, and on-line learning (eLmL'14)* (pp. 57–62).

Hubwieser, P. (2000). *Didaktik der Informatik: Grundlagen, Konzepte, Beispiele, Springer-Lehrbuch* (1st ed.). Berlin: Springer.

Hubwieser, P. (2003). Object models of IT-systems: Supporting cognitive structures in novice courses of informatics. In T. J. van Weert & R. K. Munro (Eds.), *IFIP/International Federation for Information Processing, informatics and the digital society: Social, ethical and cognitive issues; IFIP TC3/WG 3.1 & 3.2 open conference on social, ethical, and cognitive issues of informatics and ICT* (pp. 129–140). Boston: Kluwer Academic Publishers.

Hubwieser, P. (2006). Functions, objects and states: Teaching informatics in secondary schools. In R. T. Mittermeir (Ed.), *Lecture notes in computer science, informatics education – the bridge between using and understanding computers*. Springer.

Hubwieser, P. (2007). A smooth way towards object oriented programming in secondary schools. In D. Benzie, & M. Iding (Eds.), *Informatics, mathematics and ICT: A golden triangle: Proceedings of the working joint IFIP conference: WG3.1 Secondary Education, WG3.5 Primary Education; College of Computer and Information Science*. Boston.

Hubwieser, P. (2008a). Analysis of learning objectives in object oriented programming. In R. T. Mittermeir, & M. M. Syslo (Eds.), *Lecture notes in computer science, informatics education – supporting computational thinking, 3rd international conference on informatics in secondary schools – evolution and perspectives (ISSEP'08)* (pp. 142–150). Springer.

Hubwieser, P. (2008b). Analysis of learning objectives in object oriented programming. In R. T. Mittermeir, & M. M. Syslo (Eds.), *Lecture notes in computer science, informatics education – supporting* computational thinking, *3rd international conference on informatics in secondary schools – evolution and perspectives (ISSEP'08)* (pp. 142–150). Springer. https://doi.org/10.1007/978-3-540-69924-8_13.

Hubwieser, P. (2012). Computer science education in secondary schools – The introduction of a new compulsory subject. *ACM Transactions on Computing Education, 12*(4), 1–41. https://doi.org/10.1145/2382564.2382568.

Hubwieser, P., Giannakos, M. N., Berges, M., Brinda, T., Diethelm, I., Magenheim, J.,. . . Jasute, E. (2015). A global snapshot of computer science education in K-12 schools. In *ITiCSE working group reports* (pp. 65–83). New York: ACM. https://doi.org/10.1145/2858796.2858799.

Keller, J. M. (1987a). Strategies for stimulating the motivation to learn. *Performance and Instruction, 26*(9), 1–8. https://doi.org/10.1002/pfi.4160260802.

Keller, J. M. (1987b). The systematic process of motivational design. *Performance and Instruction, 26*(9–10), 1–8. https://doi.org/10.1002/pfi.4160260902.

Kölling, M. (2010). The Greenfoot programming environment. *ACM Transactions on Computing Education, 10*(4), 1–21. https://doi.org/10.1145/1868358.1868361.

Krugel, J., & Hubwieser, P. (2017). Computational thinking as springboard for learning object-oriented programming in an interactive MOOC. In *IEEE global engineering education conference (EDUCON'17)*.

Kurhila, J., & Vihavainen, A. (2015). A purposeful MOOC to alleviate insufficient CS education in Finnish schools. *ACM Transactions on Computing Education, 15*(2), 1–18. https://doi.org/10.1145/2716314.

Lethbridge, T. C. (2014). Teaching modeling using Umple: Principles for the development of an effective tool. In *IEEE 27th conference on software engineering education and training (CSEE&T)* (pp. 23–28).
Lewis, J. (2000). Myths about object-orientation and its pedagogy. In *SIGCSE '00, proceedings of the 31st SIGCSE technical symposium on computer science education* (pp. 245–249). New York: ACM. https://doi.org/10.1145/330908.331863.
Linck, B., Ohrndorf, L., Schubert, S., Stechert, P., Magenheim, J., Nelles, W.,... Schaper, N. (2013). Competence model for informatics modelling and system comprehension. In *IEEE global engineering education conference (EDUCON'13)* (pp. 85–93).
Liyanagunawardena, T. R., Lundqvist, K. O., Micallef, L., Williams, S. A. (2014). Teaching programming to beginners in a massive open online course. In : *OER14, building communities of open practice (OER"14)* (pp. 1–7).
Mishra, P., Yadav, A., & Deep-Play Research Group. (2013). Rethinking technology & creativity in the 21st century. *TechTrends, 57*(3), 10–14. https://doi.org/10.1007/s11528-013-0655-z.
Piccioni, M., Estler, C., Meyer, B. (2014). SPOC-supported introduction to programming. In Å. Cajander, M. Daniels, T. Clear, A. Pears (Eds.), *Innovation & technology in computer science education (ITiCSE'14)* (pp. 3–8). https://doi.org/10.1145/2591708.2591759.
Rosson, M. A. A. S. (1990). The cognitive consequences of object-oriented design. *Human Computer Interaction, 5*(4), 345–379.
Rumbaugh, J., & Blaha, M. (1991). *Object-oriented modeling and design*. London: Pearson Education Limited.
Schwill, A. (1994). Fundamental ideas of computer science. *EATCS Bulletin, 53*, 274–295.
Seehorn, D., Carey, S., Fuschetto, B., Lee, I., Moix, D., O'Grady-Cunniff, D., . . . Verno, A. (2011). *CSTA K-12 computer science standards: Revised 2011.*
Settle, A., Vihavainen, A., Miller, C. S. (2014). Research directions for teaching programming online. In *Proceedings of the international conference on frontiers in education computer science and computer engineering (FECS'14).*
Skoric, I., Pein, B., Orehovacki, T. (2016). Selecting the most appropriate web IDE for learning programming using AHP. In *39th international convention on information and communication technology, electronics and microelectronics (MIPRO'16)* (pp. 877–882). IEEE. https://doi.org/10.1109/MIPRO.2016.7522263.
Staubitz, T., Klement, H., Teusner, R., Renz, J., Meinel, C. (2016). CodeOcean – a versatile platform for practical programming exercises in online environments. In *IEEE global engineering education conference (EDUCON'16)* (pp. 314–323). https://doi.org/10.1109/EDUCON.2016.7474573.
Striewe, M., & Goedicke, M. (2013). JACK revisited: Scaling up in multiple dimensions. In *Lecture notes in computer science, 8th European conference, on technology enhanced learning (EC-TEL'13). Scaling up learning for sustained impact* (pp. 635–636). Berlin: Springer Berlin Heidelberg. https://doi.org/10.1007/978-3-642-40814-4_88.
Vihavainen, A., Luukkainen, M., & Kurhila, J. (2012). Multi-faceted support for MOOC in programming. In R. Connolly (Ed.), *ACM digital library, proceedings of the 13th annual conference on information technology education* (p. 171). New York: ACM. https://doi.org/10.1145/2380552.2380603.
Wing, J. (2006). Computational thinking. *Communications of the ACM, 49*(3), 33–35. https://doi.org/10.1145/1118178.1118215.

Chapter 6
Introducing Computational Thinking Through Spreadsheets

John Sanford

6.1 Background

To a large extent culture defines who we are. The human species is gregarious. Humans like to interact with other humans. These interactions follow acceptable protocol. Cultural change is generally a slow process baring cataclysmic occurrences. Computational thinking requires a cultural change. It involves a change in the way humans think of using words and numbers and relationships. Advances in computational education require the educators themselves to think that way, and this alone presents a considerable difficulty.

As prehistoric cultures progressed, they developed methods for recording information and keeping a record of significant occurrences and features of their lives. The earliest such records constituted an oral history. Eventually skills of reading, writing, and arithmetic were acquired by a select few in each community. These skills certainly did not include digital computational thinking because the technology did not exist. The digital computer with all its implications as a problem-solving partner is an extremely recent newcomer to the human cultural scene.

Ancient societies produced some astoundingly intricate and accurate mechanisms as evidenced by the Antikythera mechanism discovered in 1901 of the island of Antikythera. This mechanism, thought to have been constructed in 150 BCE, calculated positions of the sun and moon and provided other celestial data. The abacus, the Roman hodometer, water clocks, and similar counting devices certainly exhibited a computational nature. Nonetheless, people at that time would never have considered such instruments as presenting a new way of thinking. They were tools intended to assist humans who were employing existing methods of thinking.

J. Sanford (✉)
Thomas Jefferson University, Philadelphia, PA, USA
e-mail: SanfordJ@philau.edu

M. S. Khine (ed.), *Computational Thinking in the STEM Disciplines*,
https://doi.org/10.1007/978-3-319-93566-9_6

Digital computer developments of the twentieth century CE created profound differences. Some researchers and practitioners began to use digital computers interactively. The computer became a partner in many areas of research and analysis. This partnership really represented a new paradigm. This was a new way of thinking. People used digital machines to solve problems whose solutions were not available by other means then or now. Pursuit of this human-machine partnership added a new feature to the long established skills of reading, writing, and arithmetic. A somewhat popular though not exclusive term for this added skill is "computational thinking." Participants in the digital computational field are still struggling to clearly define the concept of computational thinking. Whatever it is, it should become as much a natural activity as writing a note or calculating change for transactions in a department store.

How should the pedagogical community approach the new paradigm? Forecasting the future is a popular pastime but one attended by questionable accuracy. Reviewing the milieu of existing science fiction may be as useful as any other forecasting method. Current developments in the field of artificial intelligence suggest that human-computer cooperation may become extensive. Such a conclusion is not really helpful. It rather clouds the issue of appropriate education choices.

The feeling of urgency may exist, but progress is slow. Many states, cities, and even nations are moving ahead with computer science curriculum. For just a few examples:

- In 2017 the state of Mississippi, USA, will roll out a pilot computer science program in 34 school districts (Wright (2016)).
- San Francisco established a pilot CS program for middle grades in the 2015–2016 school year (Twarek (2016)).
- In 2012 the small nation of Estonia introduced programming for its first graders (Olson (2012)).

The Internet reports quite a few pilot programs intended for instruction of young people in computer coding. Scratch is an online programming language developed at Massachusetts Institute of Technology's (MIT) Media Lab and is designed to be easy if not intuitive for young children. Children who wish to participate can join a Scratch community to design games and create stories. See https://scratch.mit.edu/.

Harvard University offers a Scratch Ed community for educators to join an online community and develop projects for children (http://scratched.gse.harvard.edu/).

Minecraft is a popular game programming site for children. Like Scratch, it has its own programming techniques and coding language of sorts (https://minecraft.net/).

As noted, these programs are pilot programs and participants self-select, which means they are not necessarily representative of the general population. Such programs are ad hoc attempts to add coding as a separate component to the participants' education. They are not a part of any standard curricular.

6.2 What Is Education?

We are entitled to ask, "What is the purpose of education?" Is education intended as an apprenticeship directed toward useful employment of the adult? Is education intended as a means of propagating the advantages of an existing culture? Is education intended to provide the student with knowledge and skills that will allow the individual to lead a happy and fulfilling life?

The last option is the one that classically applied to those attaining an education at the college level. Frequently the historical goal of college education in the western world was preparation of the youth for service in church and state. In 1860 the United States of America created "land-grant" colleges. The focus of these institutions shifted toward an agricultural and mechanical education. Land-grant colleges were intended to provide an education for all social classes. So perhaps the answer to the question is yes to all three options mentioned in the paragraph above.

Should there be a fundamental educational core for everyone with divergence as the child matures? Some countries have national tests to select students who will go on to higher academic levels and hence to more prestigious and higher-paying occupations. This approach tends to be less widespread today but is still practiced in some societies. Often family wealth is an important consideration.

Today we recognize that the education system should offer all students the same educational opportunities starting with the first grade or even preschool. Diversification may be applicable at higher levels. At least one educational goal should be "to provide the student with knowledge and skills that will allow the individual to lead a happy and fulfilling life."

With these considerations in mind, we must not allow considerations of computational thinking to diminish emphasis on the skills and knowledge that have been demonstrated in the past as valuable for a fulfilling life in an advanced society. Individuals must develop their own mental capacity for reading, writing, arithmetic, logic, and music. In fact there is evidence that logic and music are important progenitors for problem-solving ability. Students need to read and discuss their readings. Students need to have mastered fundamental arithmetic skills required for normal daily life just as they did before the digital age. We must not allow computational learning to become a siren song to lure the child away from mental agility. And yet we need to introduce computational thinking early in the education process so that it becomes a familiar capability on a par with reading, writing, arithmetic, and other skills.

Children should not become addicted to an electronic calculator or computer. It is easy to use Google for solution of simple mathematical problems that should be done on paper or in the student's head. Today cell phones possess personal assistants, Siri, Alexa, and Google. More personal assistant apps are available at app stores. Soon these artificial intelligence apps will become all too obliging. The digital assistance is not limited to mathematics. Google and other applications will turn text into audio and spoken word into text. Translation into foreign languages is readily available. In the near future artificial intelligent applications will provide analysis of paragraphs and even entire books. The future is a little frightening. Students who develop a dependency on these applications too early in life will certainly underdevelop their mental capacity.

So there is the problem. This treatise will recommend early introduction of digital solutions, particularly utilizing spreadsheet technology because it is easy to use and readily available. But that introduction should at an appropriate time.

6.3 The Approach to a Solution

There are some well-known guidelines for problem-solving though, as already noted, there are situations in which other approaches are necessary. Generally established approaches are facilitated rather than being replaced by computational thinking methods. The following methodologies are valuable:

- Stating the problem and inherent assumptions
- Systems analysis
- Defining constants and variables (state vector)
- Defining a mathematical model
- Selecting a digital computational mechanism where appropriate
- Utilizing graphical presentations where appropriate

Stating assumptions and listing parameters are a good practice to provide clarity of the task at hand. It is particularly essential in providing repeatability. Other researchers cannot follow or replicate your work if you have not provided all the initial conditions.

Systems analysis involves separation of a problem into separate parts that can be studied individually.

Defining a mathematical model may not apply in every case. Perhaps it should be called a computational model. This model is the framework that defines methodology which will be pursued enroot to the eventual solution.

If we are to use digital computational methods, we must select some application software or some computer coding language. There is indeed a wide assortment from which to choose. Software for use in computers includes more application languages than we have space to list, specialized application software, and even specialized digital machines. A very short partial list of specialized application software would include spreadsheets, Mathcad, Maple, project evaluation and control software, linear programming and goal programming, etc. Discussion on the particular value of spreadsheets will be presented shortly.

6.4 What Is Computer Science?

Computer scientists have championed the introduction of computer coding for all elementary and intermediate students. As previously noted, some school districts are introducing computer coding in early grades, sometimes very early grades. In fact, computer scientists have been championing this approach for some time although as suggested by Denning, computational thinking did not originate with computer scientists but rather with the various scientific fields, Denning (2017).

Computer coding teaches a valuable approach to logical thinking. Students writing computer code in whatever language must clearly specify variables and constants that will be part of the problem. They learn to deal with sequential logic and predicate logic, (if. . . then... else). The risk is that the student will spend more emphasis on language structure and less on problem structure. One popular approach for teaching coding to young people involves having them create games. Computer games are popular, so they will capture the child's interest and at the same time will provide valuable coding experience. But computer code does not directly display itself in the algebraic form typically employed in introductory texts on arithmetic or mathematics. Solutions using computer code may appear more as an adjunct than as natural partner to classical educational methods. Some students develop a natural proclivity for coding just as some naturally have exceptional ability with music and musical instruments. But other students find quite the opposite and for them learning computer code proves to be a difficult task.

Harvard University offers a popular course for computer science majors and nonmajors (CS50X (2017)). The website for the course suggests that it provides:

- "A broad and robust understanding of computer science and programming.
- How to think algorithmically and solve programming problems efficiently.
- Concepts like abstraction, algorithms, data structures, encapsulation, resource management, security, software engineering, and web development.
- Familiarity in a number of languages, including C, Python, SQL, and JavaScript plus CSS and HTML."

This must be good description of computer science or at least an introduction to it. Computer science as described here is very much involved with computer language and structures. But in order to deal with the problem-solving environment of business or profession, one should also have familiarity and experience with systems analysis, stating inherent assumptions, etc., as was discussed previously. The goal is to think of the digital process as part of the solution methodology and not as an add-on feature. Coding and programming are different from the usual methods encountered in precollege education, and they may well appear to be existing in a separate silo.

6.5 What Is Computational Thinking?

What is computational thinking? This is a good question to ask. It is not easy to answer. Much has already been written on the subject. One of the early and significant discussions was given in 2006 by Jeannette M. Wing of Carnegie Mellon School of Computer Science (Wing (2006)). See also Sanford (2013). As is often the case when attempting to delineate first occurrences, even earlier discussions of the topic are to be found, and no attempt to present an accurate history will be made here.

In his recent article on Computational Thinking in Science, Peter Denning commented that, "Scientists who used computers found themselves routinely designing new ways to advance science. They became computational designers as well as experimenters and theoreticians" (Denning (2017)). He describes the process as one of the solutions comprised of computational steps. But as he points out, these steps may or may not be reducible to analytic algorithms.

Sometimes problem-solvers may obtain solutions through digital computation and yet have no idea how the solutions were actually obtained. Such solutions might involve neural networks, genetic algorithms, crowd sourcing, or other methods known or yet to be discovered. If there is a distinguishing factor to computational thinking, it must be that it is a partnership of human mind and digital machinery. One goal for education, even early childhood education, should be to prepare future researchers to use this partnership. Students should think of digital assistance as just part of a natural process as common as using a pencil or a reference book. And this is computational thinking. Too bad that there is no single word, such as informatics, that would present a good description. In fact, informatics already has a well-known connotation that does not meet the requirement.

The student or researcher needs to think of the problem solution as information in the Claude Shannon sense (Shannon (1949)). In that sense one will operate on the known data to develop a solution. In some cases it is just natural that the path to this solution will involve digital computation. We want to be mindful that the path does not destroy information. The paths to problem solutions should, as much as possible, preserve methods, allow for expansion, clearly delineate assumptions, and preserve repeatability. Students should be introduced to this methodology early in their educational career and in such a way that it augments and does not inhibit essential mental development associated with reading, writing, mathematics, music, science, and logic.

6.6 What Is Important?

So, the pedagogical concern with computational thinking really refers to electronic digital computational thinking and to a partnership of human mind and digital machinery. This partnership should become second nature. The student and/or the researcher should think of using digital machinery as readily as one might think of using a pencil and paper. But, most of us think of digital computation as an add-on not as an integral part of problem-solving. Most young people today grow up using cell phones, social media, and even electronic calculators but do not readily turn to digital machinery for assistance in adjusting recipes for an afternoon picnic or budgeting vacation expenses. If computing machinery is used as an assistant in such activities, the solutions become dynamic. Information is readily available concerning the effect of varying different parameters. Most importantly, if digital assistance is used in small things, then it will naturally become an integral part of large things.

The best way for a classroom instructor to present digital computation as a partnership of human mind with digital machines would be to go through the textbook and provide teaching examples to accompany any topic and every topic where such examples could be applicable. Students should work at least one example. This presents digital solutions as strongly associated with the textbook material rather than some add-on feature.

It is not appropriate to present here material from a particular copyrighted textbook in order to develop computer examples that would couple with it. However, verbal descriptions often do not convey a concept as completely as a picture. So beginning with the next section, we present randomly selected examples of computer solutions that might fit with typical educational topics in elementary and intermediate classes. Keep in mind that these examples are only part of the picture. The other part is the textbook or lecture material that is supported by the computer examples. Tied together they constitute instruction in computational thinking.

Many people think that they are using computational thinking if they use an existing program to find answers that apply only to a fixed set of data. The fullest meaning of computational thinking applies to a way of thinking that prepares a mathematical model which is appropriate for digital machinery. The result is dynamic. It may be used to evaluate variations in result that would be expected from variations in input data. The examples presented in this chapter emphasize this.

There is a large collection of software available from which to choose for the examples just discussed. This includes search software, plotting software, and calculating software. Some applications will allow users to write equations “free form” much as they would with pencil and paper. This chapter suggests the use of spreadsheets. It presents support for this choice. However the field of software development is progressing rapidly, and other excellent choices may be available by the time you read these words.

6.7 Why Spreadsheets?

It is easy to say that spreadsheet software can be used in ways that resemble a pencil and paper approach with the added advantages that words are typewritten and mathematical calculations are done for you instead of separately on a calculator. It is easy to say that the computer solution on a spreadsheet can be developed and displayed in a way parallel to the familiar solution printed in a textbook and as such will be easy for students to visualize. But, "Seeing is believing!" The following pages present a succession of examples intended to demonstrate the applicability of this medium throughout the precollege educational experience. Discussion of more advanced applications can be found elsewhere.

These examples are in no way unique. They do not present any particularly original concepts. In fact their simplicity and commonality are the hallmark of the argument for their introduction as a complimentary feature. The examples here are only a few. Spreadsheet software can be used for just about any problem in precollege mathematics, physics, and science. Spreadsheets update all formulas every time a change is made rather than execute a sequence of instruction as indicated in computer programming. This means that the "Do Loop" is not applicable. Difference equations (or time delays) can generally be implemented by assigning time to successive columns or successive rows. All the other basic programming features such as sequential logic, predicate logic, and even table lookup are available in spreadsheet. I addition they contain a wealth of special features such as finding maximums, minimums, sorting, and others. There will be instances where the spreadsheet is not a valid replacement problem-specific software. Such instances might include specialized tutorial software, software to simulate chemical processes, software for advanced mathematical concepts, etc.

Spreadsheets are widely used for business applications, and knowledge of their capabilities is an asset for knowledge workers.

Descriptions presented in the following pages are definitely not intended as a tutorial. People who wish to learn more about development of spreadsheet methodologies can refer to any number of simple introductory books or even to free material on the Internet.

Remember the ultimate goal is for students to become comfortable with organizing problems in a particular way that will facilitate a computer solution. The student should be comfortable using the computer as an integral part of problem solutions.

Spreadsheets are very visual and as such almost self-teaching (well not quite). They enjoy a wide range of applications in business today and will almost certainly continue to do so in the future. Elementary school teachers may already be familiar with this application for recording grades, seating assignments, and countless other uses.

Microsoft is the leading purveyor of spreadsheet software. It is part of their office suite. And, several open-source office suites containing spreadsheet software are available. All offer similar functionality. In addition they have a macro language that may be used by older students to create stand-alone computer programs. Microsoft Excel uses Visual Basic. This is not the currently preferred language for coding instruction, but if a student has finished an introductory class in some other language, such as Python, Visual Basic will not appear difficult.

The following list itemizes some of the arguments for early introduction of spreadsheets:

1. They are almost as visual as using paper and pencil.
2. Students can begin to produce useful material with minimal instruction.
3. They have an extensive library of easily used features and functions.
4. Graphical presentations are easily produced.
5. It is highly unlikely that they will be superseded any time soon.

So when should students first encounter a spreadsheet?

They begin learning addition, multiplication, and division in the early grades. There is no doubt that students should still memorize addition, subtraction, multiplication, and division "tables" so that they are able to perform common numerical computations mentally. Although calculators are all too available, they have no place for simple calculations. It would be a mistake to use digital aids on any occasion where the problem is easily done without them. Like most other features of the human body, the brain will benefit from exercise.

At the proper time students can bring their arithmetic knowledge to a spreadsheet application. A small amount of overhead instruction is necessary but not arduous. How do you initiate a spreadsheet? How do you save it when you are through with it? How do you open a previously saved spreadsheet? Students will probably already be familiar with such processes because they will have been playing computer games. Also, how do you erase? How do you copy?

The next step is the art of creating functions or equations. At the start these will be only simple equations like $20 = 5 \times 4$. The computer will use an asterisk instead of the familiar multiply sign and a slash instead of the familiar divide sign. And that is enough for a start. The rest is developed as the need arises.

Figure 6.1 shows a sample. The student writes =5*4 but leaves the 20 for the computer to insert by itself. If the student leaves a space at the start, the spreadsheet shows the formula. If the student does not leave a space, the computer shows the correct answer. The student can experiment with this sort of calculation in order to become comfortable with it.

Can this be made more interesting? Suppose we develop an arithmetic game. True enough, many computer games are already available to teach arithmetic. But this game the student will create by herself.

	A	B	C
1		My Name	
2		space	no space
3		= 5*4	20
4		= 4 + 4 + 4	12
5		=4/2	2
6		=456/32	14.25
7			
8			

Fig. 6.1 When writing equations, type a space first, and see the formula (equation). Leave no space and see the computer's answer as shown in column C

6.8 Where Is Computational Thinking?

The above example uses a spreadsheet to accomplish the same thing that could be accomplished with pencil and paper. However it demonstrates the ability of a computer solution to extend the problem as is done where long division is easily added.

In early grades the student would not yet have been introduced to long division. Division of 456 by 32 could be effected with an electronic calculator, and students will probably have such an app on their cell phones. They should look for that on their cell phones. All of this sharing of computer and problem-solving inculcates what we call computational thinking.

Each and every one of the examples presented in this chapter inculcates computational thinking when it is presented along with standard textbook or lecture material. They exemplify the partnership of the human mind and electronic computing to solve problems. Without a computer such problems would be solved with paper and pencil and, in some cases, with the aid of an electronic calculator. The computer approach presents a dynamic solution where the effect of variations in parameters can be easily examined and where results can be easily presented as a chart if applicable.

6.9 An Arithmetic Game with a Spreadsheet

Figure 6.2 shows how the "game" will appear when we are done. Construction is quite simple. The student will type in the labels and then type the numbers in column B. After placing the first two numbers, the student can select those same numbers

	A	B	D	E
1		My Name		
2		Lets learn multiplication		
3		multiply by		
4		this number	3	
5		0	0.00	TRUE
6		1	3.00	TRUE
7		2	6.00	TRUE
8		3	9.00	TRUE
9		4	12.00	TRUE
10		5	15.00	TRUE
11		6	18.00	TRUE
12		7		FALSE
13		8		FALSE
14		9		FALSE
15		10		FALSE

Fig. 6.2 The multiplication game
This figure shows the spreadsheet as it will appear when completed. The student has partially filled in column D

together and copy them to the remainder of the column. The computer knows to increase each number according to the example of the first two.

Next the student enters the formula = B5*D4 into cell C5. The student copies this formula to the bottom of the column. The computer will automatically advance the number parameter, B5 to B6, and B7, and so on, progressively as the formula is copied. The dollar signs lock the parameters that they precede so that they will not change.

The student enters = C5 = D5 in cell E5 and copies this formula to the bottom.

Note that Fig. 6.2 shows column C as very narrow. This column will have the correct product value, and we want to hide it. The column width can be adjusted with the mouse so that it is very narrow. To play the game, the student types the product of 3 and 0 in column D, and the computer places TRUE or FALSE in column E.

Features learned in this spreadsheet are:

1. How to copy numbers to create a sequence
2. How to copy formulas
3. How to type formulas like = B5*D4 and = C5 = D5
4. Use of $ for absolute reference
5. How to change the width of a column

That is not too much! Student should be invited to create more games for addition and subtraction but not division. Division presents a problem because the computer uses decimals and not fractions. Decimals are probably a later introduction.

Perhaps it would be fun to have the numbers in column B appear in random order instead of sequentially. We would introduce two new features, the RAND() function and a simple sorting routine. First place = RAND() in cell A5 and copy it to cell A15. Next sort the range A5:E15 using column A. Figure 6.3 shows a typical result after this step. But when one looks at column A in Fig. 6.3, the numbers do not appear to be in ascending order as we expect them to be. At the completion of the

Fig. 6.3 The randomized multiplication game

	A	B	(	D	E
1		My Name			
2		Lets learn multiplication			
3		multiply by			
4		this number		3	
5	0.393	5		15.00	TRUE
6	0.4813	3		9.00	TRUE
7	0.8727	7		21.00	TRUE
8	0.9283	8		24.00	TRUE
9	0.1453	0		0.00	TRUE
10	0.546	1		3.00	TRUE
11	0.9339	4		12.00	TRUE
12	0.3148	6			FALSE
13	0.1954	10			FALSE
14	0.0131	9			FALSE
15	0.9829	2			FALSE

sort operation, the spreadsheet recalculates all functions on the sheet. So then the random numbers are all changed. No matter! The intent was to rearrange the numbers in column B, and this has been accomplished. The final step would be to make column A width very narrow so we would not see the random numbers.

Indeed, every time the student types an answer into a cell in column C, all the random numbers will change, but they will not be seen because the column width has been made small. There are "hide" and "reveal" instructions available, but that would add unnecessary complication. The student has played the game and typed in some answers. Every time a new number is entered, all the RAND() values change. But we don't care because after the range A5:E15 is sorted, it remains the way it is.

The additional features to be learned are:

1. The RAND() function
2. Sorting

It is important that the students be encouraged to follow this pattern and to create new games. Some student may use imagination and ingenuity to do new things. That would show that they are becoming comfortable with computer use.

6.10 A Game to Add Columns of Figures

Another game could add different numbers in a column. We start with three numbers in a column. The student picks them at random and types them into columns as seen in Fig. 6.4. At the top of each column, the student enters the formula to sum the numbers in that column. Figure 6.4 shows the result with the formula just as it is being entered into cell D2. When the student presses ENTER to finish the formula in D2, the letters will disappear, and the proper sum value will appear.

The game is to enter the correct sums for the columns into cells B6, C6, and D6. Then the "FALSE" responses in row 7 will change to "TRUE." Can the student remember what to type in cell B7? It would be = B2 = B6. Of course the game is not much of a game if the correct answer is visible in row 2. To hide these numbers, change row height. Another trick would be to change the text color to white so it will not be visible.

Fig. 6.4 Adding columns game

	A	B	C	D	E
1					
2		10	22	=SUM(D2:D5)	
3		3	8	10	
4		4	4	5	
5		3	10	2	
6					
7		FALSE	FALSE	FALSE	

To continue the student will have to change the numbers in the columns by hand. If RAND() is used, the numbers will change every time anything is done to the spreadsheet and nothing will work properly.

6.11 Word Problems

Computer use is a perfect adjunct when word problems are introduced in math classes. This is true no matter what software is used. Computational modeling of the problem allows reuse of the same template for other similar problems. This is particularly applicable to spreadsheets because when properly constructed, the sheet contains its own labeling and instructions. The spreadsheet provides a self-contained report and solution at the same time. And often a model will bring to mind new imaginative activities beyond the original scope of the initial problem.

Word problems introduced in the lower grades generally involve simple addition, subtraction, multiplication, and division. For example, a problem might say that Oscar has six apples and Juan has four apples. How many do they have together? If Carla joins them and they give her two apples, how many will Oscar and Juan have left? Such problems can be modeled on a spreadsheet which offers the advantage that it is neat and ordered and provides a pleasing presentation. These simple problems can be done in one's head with little or no formal organization. However they provide learning experience on the computer. Good form dictates that a short restatement of the problem be placed at the top of the sheet. All parameters of the problem should be placed on the sheet, usually at the top left. These constitute what are often referred to as the state vector or state variables. Actually some of them may be constants for the sake of the problem. The state variables do not include universal constants such as π that are available as functions on the computer. But they should include other constants such as the acceleration of gravity that will be approximated because it is actually not a constant and is not a simple calculation.

Figure 6.5 shows a simple practice problem. Here Carla has $7.50 and Oscar has $9.00. How much do they have together?

	A	B	C	D	E	F	G
1	Part one: Carla and Oscar are friends. Carla has $7.50 and Oscar has $9.00						
2	How much money do they both have together						
3	Carla	$7.50					
4	Oscar	$9.00		Carla + Oscar=		$16.50	
5	Juan	$8.24					
6							
7	Part two: Juan has $8.24. if he joins them how much will they all have together						
8							
9				All three =		$24.74	

Fig. 6.5 Presented for spreadsheet practice

When Juan joins them with \$8.24, how much do all three have together? The student would not create a new sheet for the second part but rather add Juan to the state vector and add part two to the existing sheet. The sheet will contain formulas such as = SUM(B3:B4) and = SUM(B3:B5). Students should not do the work in their head and just type the answer. And if the problem changes so that Oscar has \$8.00, the new solutions are obtained by changing the value in the state vector from \$9.00 to \$8.00.

As previously mentioned, the spreadsheet serves as a template for similar problems. The student should be guided to use the same sheet for a problem where Carla has only \$3.50 and Oscar has only \$5.00. New results are obtained by simply changing the initial condition values in the state vector.

Problems like these offer an introduction to spreadsheet capabilities and methods, but they do not suggest the significant advantages that are available through the computer. Eventually these advantages should be introduced. The next example demonstrates array capability and the easy display methods that are available in the spreadsheet.

In this problem, Carla and Oscar are friends. Carla has \$7.50 and Oscar has \$9.00. They decide to pool their resources and buy comic books together. They will share the books. Their favorite comics are "Tales for Zombies" which costs \$3.00 per copy and "Squeaky Crypt" which costs \$2.30 per copy. Because they have not bought any of these for some months, their collections are a bit behind. They will buy the current issues and back issues. So they might buy four "Tales of Zombies," and then they would have enough money for only one issue of "Squeaky Crypt." What are the various combinations of issues that they could buy with the money that they have?

Figure 6.6 shows the spreadsheet for this problem. The state vector has a value for Carla + Oscar. Together they have \$16.50. Even though this appears in the area for constants, it should be calculated with a formula. It is not a fundamental data item.

The outer boarders of the array show the number of each type of comics purchased. Students can type in the numbers 1, 2, 3, and so on. But they should remember how to type 1 and 2 and then select them and copy. The computer will know that it is to increase each number by one.

Formulas will be used for all the other numbers. The formula in cell C14 is =C13*\$B\$8. The purpose of each dollar sign is to lock the parameter that follows it. When copying this formula cross the row to cell F14, the \$B\$8 will never change and will always point to the cost of "Tales for Zombies."

The only tricky formulas are the ones in the center that produce the values 5.30, 8.30, and so on. These numbers show the total dollar cost. So if Carla and Oscar buy one of each of the comic books, they will spend \$5.30. The formula is =C \$14 + \$B15. Here again the dollar sign locks the parameter that follows it. This allows the student to write the one formula and copy it across and down to achieve the total array. The numbers that are possible without exceeding funds available are shown in bold. Conditional formatting is available to allow the computer to set bold format automatically, but this is an advanced feature. Also an IF function can be used

	A	B	C	D	E	F
1	Carla has $7.50 and Oscar has $9.00. They are going to pool					
2	their money and buy comic books. "Tales for Zombies" cost $3.00 per					
3	copy and "Squeaky Crypt" costs $2.30 per copy.				What are their	
4	purchase options?					
5						
6	Carla	$7.50				
7	Oscar	$9.00				
8	Tales	$3.00				
9	Squeaky	$2.30				
10	Carla+ Oscar	$16.50				
11			1, 2, 3. etc. are numbers of issues in that			
12			column or that row.			
13		Tales →	1	2	3	4
14	Squeaky ↓	$0.00	$3.00	$6.00	$9.00	$12.00
15	1	$2.30	**5.30**	**8.30**	**11.30**	**14.30**
16	2	$4.60	**7.60**	**10.60**	**13.60**	16.60
17	3	$6.90	**9.90**	**12.90**	**15.90**	18.90
18	4	$9.20	**12.20**	**15.20**	18.20	21.20
19	5	$11.50	**14.50**	17.50	20.50	23.50
20	6	$13.80	16.80	19.80	22.80	25.80

Fig. 6.6 Showing costs for combinations of comic books

to avoid displaying any numbers that exceed the available funds, but this also is an advanced concept. Most likely the student can set the bold format by hand after inspecting the numbers to see which ones are too large. Perhaps the student may be invited to think of how to add the IF feature after the basic solution is created.

Features learned in this problem:

1. The use of absolute reference, i.e., use of dollar signs in a formula
2. How to set text to bold
3. Maybe how to use an IF function

Students should be encouraged to see how easy it is to obtain new results if Oscar or Carla has a different amount of money or if the cost of a comic book changes. One might ask how the problem would be different if Carla and Oscar had a friend who wished to join in the project.

One might ask how to approach the problem if the selection is to be among three different comic books. If there are three comic books, the array must be three-dimensional. It is not likely that a young student will think of that. Such a problem would be beyond the elementary level. The resultant display would involve a set of two-dimensional arrays, one for each different number of issues of the third comic book. Computer programming code for such a problem would involve three loops. That would be simple, but the display would not differ from the spreadsheet display. It is not easy to display three dimensions on a two-dimensional sheet of paper.

Data is often displayed in arrays. For example, the temperature at noon of every day in a month would be an array. Other examples would be inches of rain each day for a month, student grades on a quiz, and names of students in a class. Sometimes people are interested in the numerical average of the data or the total value of all the data. Sometimes we want to sort the data into alphabetical order. There are ample opportunities for examples and problems.

6.12 Student Projects

The intention thus far has been to show how spreadsheets and computer technology can fit as part of more usual instruction rather than as separate stand-alone instruction. After students have developed fundamental facility with spreadsheets, they will enjoy projects that offer more self-expression or even group activities. One example of such a group project might be planning a lemonade stand to raise money for some charity. This is a good project, and there are many lemonade recipes on the Internet. There are even Internet sites that will help with planning.

The solution of any problem should start with the accepted scientific method. First identify all the information that might apply to the problem. What are the physical requirements? What locations are possible? What are advantages of each location? What criteria will be used to select the best location? What hardware such as table, chairs, pitchers, and the like may be required? Does required hardware impact site selection criteria?

People often neglect specification of selection criteria. This oversight may result in selections based more on whim rather than on attributes.

Of course much of the hardware can be borrowed from friends, relatives, or other sources, and these factors are listed as problem assumptions. But students must plan for the expendable items and must establish a price for a cup of lemonade. Modeling the material costs on a spreadsheet will be helpful. It will provide a clear display of all the data and calculations. Further it will allow examining various scenarios that will lead to setting a price and estimating the probable profit of the enterprise.

The first step in modeling the expendables is to obtain a good recipe, preferable from the Internet. Using the Internet rather than human help will encourage resourcefulness and independence. The next step is planning and designing the spreadsheet itself. The various spreadsheets developed by different groups may be similar but will probably differ in appearance. The first part of the sheet should contain a brief description, names, and a date. Then there needs to be places for the recipe, for all the raw materials, and for any constants such as how many cups will be available from a gallon, and so on. When purchases are made, how many units are contained in a package? For example, how many pounds of sugar are in one sack? All numbers need to be entered in separate cells all by themselves with no labels. However, dollar signs are actually part of a number.

	A	B	C	D	E	F	G	H
1	PRELIMINARY PLAN for lemonade stand, Carls & Oscar 4/5/2017							
2	Open Saturday and Sunday afternoons on two weekends.							
3	They will not consider any sales taxes because the money is for a charity.							
4	Borrow table and containers.							
5	One time purchase of cups because they are 1,000 per package.							
6	Refrigerate lemonade unsold on Saturday.							
7	Discard lemonade unsold on Sunday.							
8	One package of ice per day for customers on demand.							
9	Aater chilled overnight so minimal ice will be required.							
10	Estimate 4 gallons per day							
11	After the model is complete, select a sales price to achieve desired profit							
12	after considering what customers might be willing to pay.							
13								
14	RECIPE for ONE GALLON				CONSTANTS			
15	Lemons	(2 cups)	8		Lemons per pound		4	
16	Cups sugar		2		Cups of sugar per pounc		2	
17	Cup hot water		1		Ounces per gallon		128	
18	Pints cold water		6.5		ounces per cup (8 oz.)		7	
19					Cups per gallon		18.2857	
20								
21	FIXED COSTS		Units	Cost/Unit			# of Days	TOTALS
22	Cups	pkg.	1000	$24.49			1	$24.49
23	Ice	pound	15	$2.00			4	$8.00
24								$32.49
25								
26	VARIABLE COSTS							
27	Item	Sold by	Units	Cost/Unit	Cost/Gallon	Gallons/Day	# of Days	TOTALS
28	sugar	pound	2	$1.80	$1.80	4	4	$28.80
29	lemons	pound	1	$1.70	$3.40	4	4	$54.40
30	water	free			0	4	4	$0.00
31					$5.20			$83.20
32	Total Expense							$115.69
33								
34	SELLING PRICE CALCULATION							
35	Estimated Sales		units	Price	$/gallon	gallons/day	# of days	Total
36	Cups		1	$0.75	$13.71	4	4	$219.43
37	Profit							$103.74

Fig. 6.7 Showing possible lemonade sale configuration

Figure 6.7 shows one possible configuration. Teachers should give some guidelines as a start for the spreadsheet:

1. Begin with a very brief description and include names and the date.
2. Organize in columns with column headings for units, cost per unit, number of days, etc.
3. Think of fixed costs and variable costs separately.

4. Variable costs are costs that depend on how many cups you sell, like lemons and sugar.
5. For raw materials, think of how you buy them, per pound, per unit, etc. and their cost.
6. Set up the spreadsheet so that there are certain numbers entered as numbers such as cost of sugar, cost of ice, number of days, etc.
7. Other numbers should be calculated using a formula that refers to the fixed numbers. Then if you change the cost of lemons, all the other numbers will change (recalculate).

6.13 Equations

Consider further use of spreadsheets and computer technology as part of normal in-class instruction in mathematics. Equation (6.1) is a simple example that would have a simple solution:

$$\mathbf{12 = 3X + 3} \tag{6.1}$$

Yes, the solution is X = 3, but it might be instructive to see the locus of all points defined by 3X + 3 = Y. Figure 6.8 shows this locus. It is easy to produce, and the method is well worth the time it would take a student to learn it. Two columns define

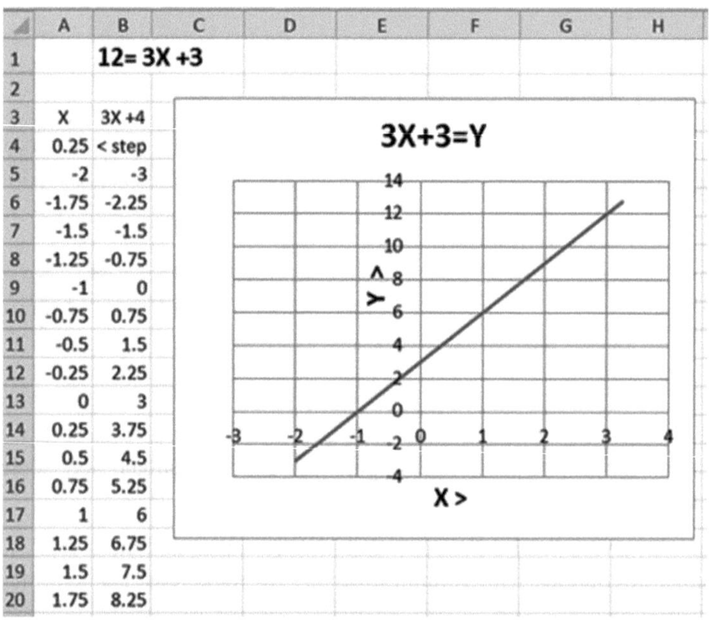

	A	B
1		12= 3X +3
2		
3	X	3X +4
4	0.25	< step
5	-2	-3
6	-1.75	-2.25
7	-1.5	-1.5
8	-1.25	-0.75
9	-1	0
10	-0.75	0.75
11	-0.5	1.5
12	-0.25	2.25
13	0	3
14	0.25	3.75
15	0.5	4.5
16	0.75	5.25
17	1	6
18	1.25	6.75
19	1.5	7.5
20	1.75	8.25

Fig. 6.8 Showing locus of all Y values for Y = 3X + 3

the chart shown in Fig. 6.8. One is a column of X values and the other a column of Y values. Create the columns by selecting a starting value for X and placing it in a cell, in this case cell A5. The value of X should increase by some fixed amount for each cell. This amount of increase can be called the step. Place the step value in cell A4. Now the formula in cell A6 is = A5 + A4. Copy this down the column for a reasonable way.

The next column has the computation of =3X + 3 obtained by referring to the X value from the corresponding row in the X column. In other words, cell B5 contains =3*A5 + 3. Copy this formula down the column, and create the chart (graph). All the spreadsheet software contains graphing capabilities although they differ slightly. It is important to use the X-Y plot feature.

Now the student can see that Y = 12 when X = 3. This is so simple that it might appear trivial, but the technique has broad application. For example, consider Eq. (6.2):

$$\mathbf{3X^2 + 2X - 3 = 0} \tag{6.2}$$

There is, of course, a quadratic formula that may be used to solve this equation for the value of X.

Examples are offered on the Internet. Some schools may have advanced math application software such as Mathcad, Maple, or similar applications that will provide easy methods for finding the solution and even for graphing the function. However, the spreadsheet is included with office suites, and it is essentially free if you own the suite. So we can use it. Figure 6.9 shows the result.

If the student is advanced sufficiently in the use of spreadsheets, values for the equation coefficients, a, b, and c, may be placed at the head of the sheet and referenced in the various formulas where they are required. For a typical example, see Eq. (6.3):

$$= \$B\$1{*}A6\hat{}2 + \$B\$2{*}A6 + \$B\$3 \qquad \text{formula in cell B6 of Fig.6.9} \tag{6.3}$$

The solutions are read where the graph intersects the x-axis. Using coefficients in cells B1, B2, and B3 provides the student with the capability of changing the C value and watching the entire curve move up or down. The student can see when the curve has two solutions, only one solution, and when it has no solution (not valid).

Figure 6.9 is a template. It stands ready to deal with any quadratic problem. The only modifications required for a new problem are changing the coefficients and changing the step and start values.

Graphical solutions are also applicable to third-, fourth-, and higher-order equations. As previously mentioned software that will provide such solution is readily available. In fact at least one Internet site is available that will provide roots of a third-order equation. There is still some advantage to using a graphical solution that one creates for oneself. For one thing behavior of the function away from the zero crossing is available. The student can examine sensitivity of the roots to coefficient changes. At what value of the constant term will there be two solutions and only one solution?

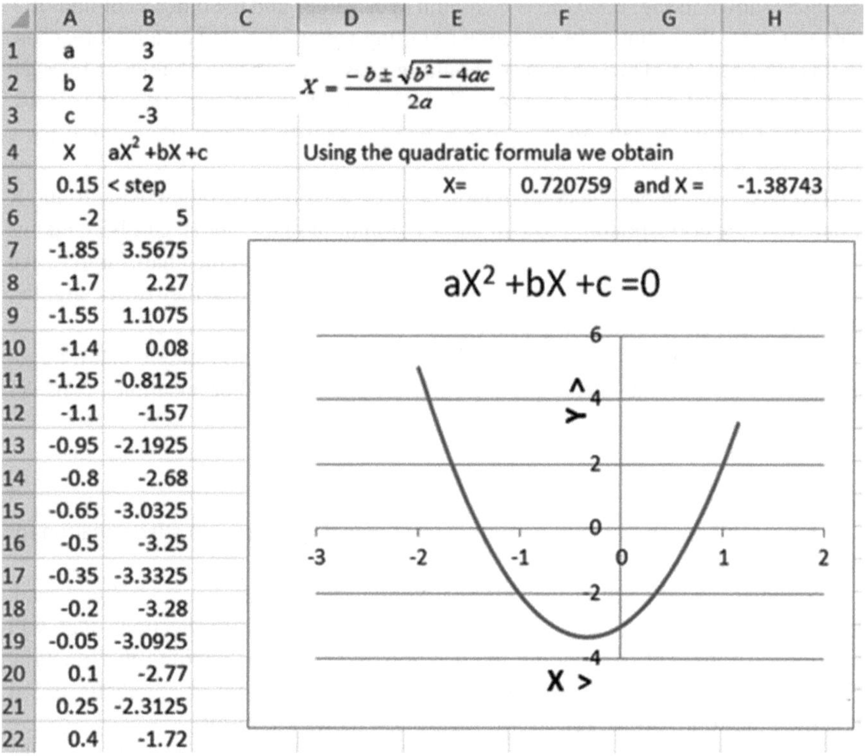

	A	B	C	D	E	F	G	H
1	a	3						
2	b	2		$X = \frac{-b \pm \sqrt{b^2 - 4ac}}{2a}$				
3	c	-3						
4	X	aX^2 +bX +c		Using the quadratic formula we obtain				
5	0.15	< step			X=	0.720759	and X =	-1.38743
6	-2	5						
7	-1.85	3.5675						
8	-1.7	2.27						
9	-1.55	1.1075						
10	-1.4	0.08						
11	-1.25	-0.8125						
12	-1.1	-1.57						
13	-0.95	-2.1925						
14	-0.8	-2.68						
15	-0.65	-3.0325						
16	-0.5	-3.25						
17	-0.35	-3.3325						
18	-0.2	-3.28						
19	-0.05	-3.0925						
20	0.1	-2.77						
21	0.25	-2.3125						
22	0.4	-1.72						

Fig. 6.9 Showing solution of a quadratic equation

As can be seen in Fig. 6.10, the coefficients are listed at the top of the spreadsheet so that they can be readily changed. One can obtain a more accurate determination of the routs by expanding the graph in the area of zero crossings. This requires only that the starting point and step size be changed. A very small step size will expand the area of interest. Figure 6.11 shows this. And, the spreadsheet of Fig. 6.10 is a template for use with any third-order polynomial.

6.14 Simulating a Problem in High School General Science

Describing the trajectory of an object thrown up into the air is a common general science problem. It is so common that multiple solutions are available on the Internet. A slight variation would be a rifle bullet trajectory. If the rifle is held at some angle, there will be horizontal and vertical velocities. The vertical component is given by multiplying muzzle velocity by the sin of the angle. For the horizontal component, one uses the cos instead of the sin.

	A	B
1	a	3
2	b	3
3	c	-4
4	d	-2
5	X	$aX^3 + bX^2 + cX + 3$
6	0.15	< step
7	-2	-6
8	-1.85	-3.32738
9	-1.7	-1.269
10	-1.55	0.235875
11	-1.4	1.248
12	-1.25	1.828125
13	-1.1	2.037
14	-0.95	1.935375
15	-0.8	1.584
16	-0.65	1.043625

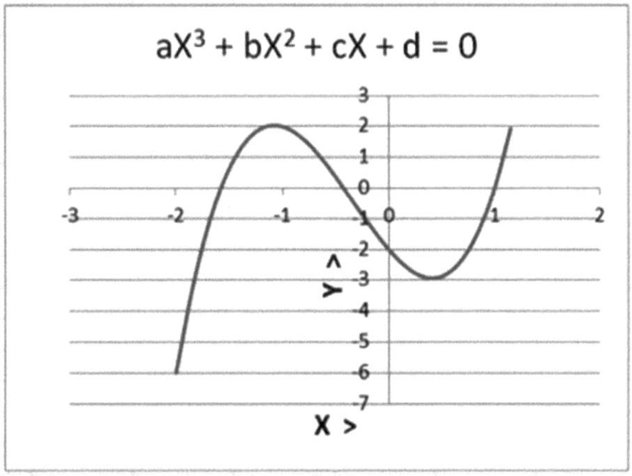

Fig. 6.10 Showing graphical solution of third-order equation

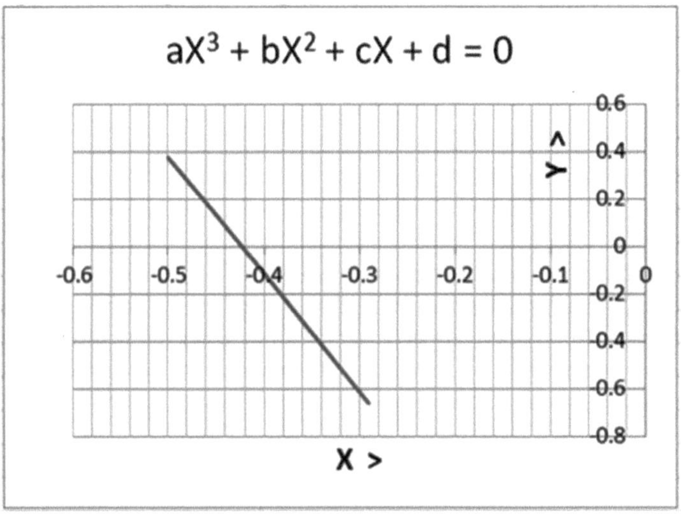

Fig. 6.11 Showing expanded graph in area of the root $X = -0.423$

Problems like this provide good learning experiences because they have many parts that must be investigated in succession. A spreadsheet allows these components to be displayed in an orderly fashion.

The maximum height of the trajectory occurs when the velocity resulting from gravity equals the upward component of muzzle velocity. The two velocity vectors act in opposite directions. So the time spent traveling up, T_U, is obtained as the upward component of muzzle velocity divided by the acceleration of gravity. This is shown in Eq. (6.4). It is the same time that the bullet will require to fall back down to the height of the rifle. The height above the rifle that the bullet will reach in T_U seconds is given by Eq. (6.5):

$$T_U = V_U/g \tag{6.4}$$

$$D_U = \frac{1}{2}gT_U^2 = V_U T_U \tag{6.5}$$

Total distance from apex to the ground, D_T, is equal to $D_U + D_R$, where D_R is the distance from the rifle to the ground. The time for the bullet to travel from the apex to the ground is given in Eq. (6.6). So the total time of flight, up and down, is $T_D + T_U$:

$$T_D = \sqrt{\frac{2D_T}{g}} \tag{6.6}$$

These equations are derived through use of the calculus, but they are typically presented without derivation in elementary general science textbooks. Some pedagogical assistance is required because the student did not use the calculus to develop the equations and so will not necessarily recognize that the sequence of calculations is important. One must first find the total distance from ground to the apex and then employ Eq. (6.6). Then the time in flight, T, is the sum of T_U and T_D as shown in Eq. (6.7):

$$\mathrm{T} = \mathrm{T_U} + \mathrm{T_D} \tag{6.7}$$

Once the total time of flight is available, the horizontal range can be found as the product of flight time and horizontal component of muzzle velocity, assuming zero air drag.

Figure 6.12 shows atypical spreadsheet solution. This solution utilizes an Excel add-in called "Solver" to select an angle that maximizes the horizontal range.

The Solver add-in is not really necessary. The student can use different values for the angle and optimize heuristically. Students should be encouraged to change initial conditions such as such muzzle velocity and angle to see what effect these have on trajectory.

Plotting the trajectory would be fun if the students are sufficiently familiar with spreadsheets. By the time they reach general science they should be.

As always, the first steps in solving a problem are a clear statement of the problem and a clear statement of parameters (state vector) and influencing factors.

The riffle bullet trajectory problem demonstrates this orderly development. Each of the calculated values is shown in the column for calculated values. Each is calculated using a function that relates to problem constants or other calculated values. The spreadsheet provides an easy method of describing the succession of steps and allows them to be labeled and visually displayed. The spreadsheet provides not only a solution but essentially a report of the solution all packaged together.

All of the values must be related. All formulas must reference the constants in the constants column. Most constants are not laws of nature but are initial conditions that can be changed. Changing any one of them must result in a new solution. Perhaps the

Fig. 6.12 Showing the bullet trajectory problem

	A		B	C	D
1	NAME Fred		DATE 3/1/2017		
2	Bullet trajectory using gravity at sea level				
3	air drag neglected				
4			Symbol	constants	calculated
5	angle to the horizontal		A	44.976	
6	acceleration of gravity	ft/s^2	g	32.174	
7	muzzle velocity ft/s		V	390	
8	Height of rifle	ft.	D_R	4	
9	feet in a mile		M	5280	
10	upward muzzle velocity		V_U		275.66
11	Horizontal muzzle velocity		V_H		275.89
12	Time traveling up		T_U		8.57
13	Time traveling down		T_D		8.58
14	Time to hit the ground		T		17.15
15	Max height above rifle	ft.	D_U		1180.86
16	Max total height	ft.	D_T		1184.86
17	Horizontal distance	ft.	D_H		4731.42
18	Horizontal distance	miles			0.90
19	Formulas used in the model are shown below.				
20	$V_H = \cos(A) \times V$				
21			$D_U = \frac{1}{2} g T_U^2$	$T_D = \sqrt{\frac{2D_T}{g}}$	
22	$V_U = \sin(A) \times V$				

most significant risk inherent in spreadsheet solutions is that the problem-solver may perform mental calculations and insert them into the solution process. In that case changing the initial conditions or constants would not produce a correct solution for the new constants.

Students will and should display originality in design so all spreadsheets will not all look like Fig. 6.12. Presenting students with a template would defeat the fundamental intention of creating a comfortable relationship with computational thinking. There is little doubt that solutions for this exact same problem will be found on the Internet. Students should resist looking at them before obtaining their own solution unless they are stumped.

6.15 Trigonometry

Trigonometry is generally considered to be a later high school subject. Students would typically use calculators in order to obtain required level of accuracy for even simple problems. Built-in functions are available in spreadsheets for both plane and spherical trigonometry and do not require loading special libraries. Simple diagrams can be created right on the spreadsheet using a feature that allows insertion of shapes.

Students still need to know the fundamental rules that are applicable to trigonometry, such as that the sum of all interior angles of a triangle will equal 180° in Euclidian trigonometry. Students will need to know what functions to employ for the

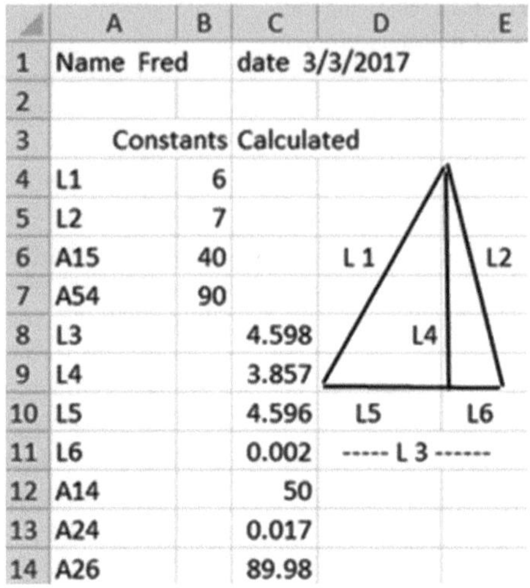

	A	B	C	D	E
1	Name	Fred	date	3/3/2017	
2					
3		Constants	Calculated		
4	L1	6			
5	L2	7			
6	A15	40			
7	A54	90			
8	L3		4.598		
9	L4		3.857		
10	L5		4.596		
11	L6		0.002		
12	A14		50		
13	A24		0.017		
14	A26		89.98		

Fig. 6.13 A typical trigonometric problem

solution to a given problem. The advantage of the spreadsheet over the calculator is that problem visibility is as always present. Changes may easily be made to accommodate new problems that are similar to the given one. Figure 6.13 presents a typical problem.

6.16 Simultaneous Algebraic Equations

Introduction of simultaneous equations is a late high school topic. If software similar to Mathcad, Maple, etc. is available, it will present a medium that is easier to use than the spreadsheet. But convenience and expense incline toward use of a spreadsheet. It is there and it is free with any office suite. The spreadsheet will perform matrix operations, addition, subtraction, multiplication, and inversion. Knowledge of matrix operations is valuable for mathematics and for computer science.

Spreadsheets use functions such as MMULT and MINVERSE to accomplish multiplication and inversion for division. These are array operations, and so the input parameters for them are arrays. The easiest way to specify an input array is to select it with the mouse, but one can specify it by identifying the two diagonal cells. Since the response will be an array, one must select the area that the response array will occupy and then type a function such as = MMULT(. . . .

The method for solving simultaneous equations involves the matrix equations shown in (6.8) and (6.9)

where Y, A, and X are matrices. A^{-1} is the inverse of A.

$$\{Y\} = [A] \bullet \{X\} \tag{6.8}$$

$$[A]^{-1} \bullet \{Y\} = \{X\} \tag{6.9}$$

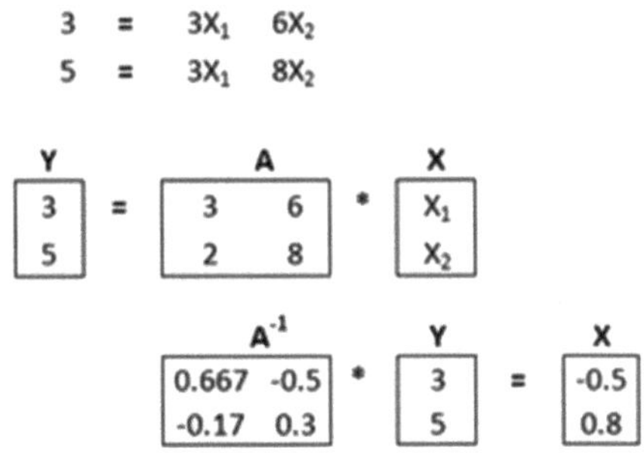

Fig. 6.14 Showing matrix operations

The X matrix is a column matrix containing $X_1, X_2, \ldots X_n$. Actually it is blank in the spreadsheet the labels, X_1, X_2, etc., are understood. After performing the operation shown in (6.9), X will contain the solution for all values of $X_1, X_2, \ldots X_n$ dictated by the simultaneous equations. The Y matrix is also a column matrix containing all the Y values that correspond to the rows of the simultaneous equations. The matrix, A, contains the coefficients of all the X values in the simultaneous equations. Figure 6.14 shows this as it might appear in a spreadsheet. The letters Y, A, etc., above the matrices are not necessary. They are added for clarity as are X_1 and X_2 in the initial X matrix. In fact, the entire Y matrix in the lower equation is added for clarity. The original Y matrix can be pre-multiplied by A^{-1} to yield the answers for X values.

One interesting feature is that the results shown on the spreadsheet are dynamic. If one changes a value in the A matrix or the initial Y matrix, the solution changes to the correct values for the new input. If a changed parameter produces a set of equations for which there is no solution, Excel puts "#NUM!" in the cells. Other spreadsheets may use a different notification.

The simultaneous equations may be many in number, limited only by practical computational considerations. It is valuable for students to know how to solve two simultaneous equations by hand. Certainly more than two simultaneous equations can be solved by hand, but the process does become tedious for large numbers of equations.

6.17 Conclusion

This chapter has presented some concepts related to computational thinking.

1. Computational thinking is nothing more than developing a normal tendency for individuals to organize and present problems for solution with the aid of digital computing machinery.
2. It is a partnership of human mind and digital machinery.
3. Digital machinery should be as readily used as paper and pencil.
4. Computational thinking should be incorporated in early childhood education.

5. Computational thinking should be included throughout the education years.
6. It involves a methodology for problem-solving and not just computer programming.
7. Spreadsheet software is suggested as a medium, but there are other options.

Each of the examples shown in the chapter presents a computerized method of problem-solving. When instructors develop similar problems to augment textbook material, they are inculcating computational thinking.

Problem-solving methods such as systems analysis and use of mathematical models are touched on in the chapter. These are not really specific to computational thinking any more than they are components of problem-solving, or computer science, or other disciplines. The chapter has not stressed their existence.

In this world "we must run as fast as we can, just to stay in place" (Carroll). By the time the grade school child of today reaches college age, the human interface with computers will almost certainly be though artificial intelligence and robotics. Does that mean that this concept of computational thinking will fade in significance? No! If humans have a problem, then they must describe the problem even for artificial intelligence. A significant feature present in the examples contained in this chapter is that the first step is describing the problem.

References

CS50X. (2017). Harvard University https://www.edx.org/course/introduction-computer-science-harvardx-cs50x January 13, 2017.

Denning, P. J. (2017). Computational thinking in science, American Scientist January 2017, publication of Sigma Xi, January–February 2017 edition pages 13–17.

Olson, P. (2012). *Why Estonia has started teaching its first-graders to code*. Forbes.com http://www.forbes.com/sites/parmyolson/2012/09/06/why-estonia-has-started-teaching-its-first-graders-to-code/#7de3a61b5790. Sighted 1/13/2017.

Sanford, J. F. (2013). Core concepts of computational thinking. *International Journal of Teaching and Case Studies, 4*(1), 1–12.

Shannon, C. E. (1949). The mathematical theory of communication. ISBN 0-252-72546-8.

Twarek, B. (2016). San Francisco Unified School District, STEM Dept. http://www.csinsf.org/pilot.html. Sighted 1/31/2017.

Wing, M. (2006). Computational thinking. *Communications of the ACM, 49*(3), 33–35.

Wright, C. M. (2016) (CS4MS) pilot program in 34 Mississippi school districts. http://www.mde.k12.ms.us/TD/news/2016/04/28/mde-targets-expansion-of-computer-science-education-in-pilot-programs-teacher-training. Sighted 1/31/2017.

Other Related Papers by This Author

Sanford, J. F., & Naidu, J. T. (2016). Computational thinking concepts for grade school. *Contemporary Issues in Education Research, 9*(1), 23.

Sanford, J. F., & Naidu, J. T. (2017). Mathematical modeling and computational thinking. *Contemporary Issues in Education Research, 10*(2), 159–168.

Part II
Computational Thinking and Teacher Education

Chapter 7
Preparing Pre-service Teachers to Promote Computational Thinking in School Classrooms

Charoula Angeli and Kamini Jaipal-Jamani

7.1 Introduction

The importance of teaching computational thinking across the K–12 curriculum has been strongly argued for by many educational scholars and researchers (Wing 2006; Grover and Pea 2013; Guzdial 2008). At the same time, there is evidence that teacher education departments lack the knowledge and skill to teach pre-service teachers about computational thinking (Yadav et al. 2011, 2014). While in general, the research in this area is scarce; some preliminary evidence exists showing that the introduction of computational thinking modules in the teaching of existing teacher education courses can have positive results on pre-service teachers' understanding of computational thinking concepts (Yadav et al. 2014). Thus, the issue of how to promote computational thinking in pre-service teacher education is timely, and any new studies undertaken for this research purpose are much needed and are fully warranted (Gretter and Yadav 2016).

The present study responds directly to calls for more research into how to include the teaching of computational thinking in existing teacher education courses (Yadav et al. 2011, 2014; Hodhod et al. 2016; National Research Council 2011). In particular, the study herein proposes the use of scaffolded programming scripts as one instructional method for teaching computational thinking to novice pre-service teachers. The study is undertaken within the context of educational robotics activities with LEGO WeDo, a robotics construction kit for education (Kim and Coxon 2016; Geist 2016), for teaching a science lesson about gears and their functions.

C. Angeli (✉)
Department of Education, University of Cyprus, Nicosia, Cyprus
e-mail: cangeli@ucy.ac.cy

K. Jaipal-Jamani
Faculty of Education, Brock University, St. Catharines, ON, Canada

M. S. Khine (ed.), *Computational Thinking in the STEM Disciplines*,
https://doi.org/10.1007/978-3-319-93566-9_7

7.2 A Definition of Computational Thinking

While the concept of computational thinking in education can be traced back to the work of Seymour Papert, who strongly advocated the idea of children developing algorithmic thinking through the Logo programming language (Papert 1980), Wing's (2006) article has rekindled the interest in promoting computational thinking in education. Collective efforts aiming at developing a definition for computational thinking include the two National Academy of Sciences workshops (National Research Council 2010, 2011) and the initiative undertaken by Royal Society (2012) and also workshops organized by the Computer Science Teachers Association (CSTA) and the International Society for Technology in Education (ISTE).

The 2010 National Research Council's report differentiated computational thinking from computer literacy, computer programming, and computer applications and broadened the term to include core concepts from the discipline of computer science, such as abstraction, decomposition, pattern generalizations, visualization, problem-solving, and algorithmic thinking.

Similarly, the Royal Society (2012) defined computational thinking as "the process of recognizing aspects of computation in the world that surrounds us, and applying tools and techniques from computer science to understand and reason about both natural and artificial systems and processes" (p. 29).

CSTA and ISTE developed an operational definition of computational thinking as a problem-solving process that includes, but is not limited to, the following elements: (a) formulating problems in a way that enables us to use a computer and other tools to help solve them; (b) logically organizing and analyzing data; (c) representing data through abstractions, such as models and simulations; (d) automating solutions through algorithmic thinking (i.e., a series of ordered steps); (e) identifying, analyzing, and implementing possible solutions with the goal of achieving the most efficient and effective combination of steps and resources; and (f) generalizing and transferring this problem-solving process to a wide variety of problems.

Despite the fact that currently there is not one unanimous definition of computational thinking, after a systematic examination of what is currently known in the literature, Grover and Pea (2013) and Selby (2014) concluded that researchers have come to accept that computational thinking is a thought process that utilizes the elements of abstraction, generalization, decomposition, algorithmic thinking, and debugging (detection and correction of errors). Abstraction is the skill of removing characteristics or attributes from an object or an entity in order to reduce it to a set of fundamental characteristics (Wing 2011). While abstraction reduces complexity by hiding irrelevant detail, generalization reduces complexity by replacing multiple entities which perform similar functions with a single construct (Thalheim 2000). For example, programming languages provide generalization through variables and parameterization. Abstraction and generalization are often used together as abstracts are generalized through parameterization to provide greater utility. Decomposition is the skill of breaking complex problems into simpler ones (Wing 2008; National Research Council 2010). Algorithmic thinking is a problem-solving skill related to devising a step-by-step solution to a problem and differs from coding (i.e., the technical skills required to be able to write code in a programming language) (Selby 2014). Additionally, algorithmic notions of sequencing (i.e., planning an algorithm, which involves putting actions in the correct sequence) and algorithmic notions of flow of control (i.e., the order in which individual instructions or steps in an algorithm are evaluated) are also considered important elements of computational thinking (Lu and Fletcher 2009). Debugging is the skill to recognize when actions do not correspond to instructions and the skill to fix errors (Bers et al. 2014).

Similarly, Perrenet et al. (2005) and Perrenet and Kaasenbrood (2006) proposed a hierarchy, known as the PGK hierarchy, of four levels for developing computational thinking skills:

4. The *problem* level is the fourth and highest level in the hierarchy. At this level, one thinks of problems as objects and can develop solutions to problems as black boxes. Solutions are elegant and stripped of unnecessary details.
3. The *object* level is the third level in the hierarchy. At this level, one can think computationally in ways that do not depend on upon the specifics of a programming environment. Mainly, one can think regarding functions (generalizations) that rely on inputs. Thus, while abstraction at the fourth level involves thinking about solutions to problems as black boxes, at the third level common patterns are recognized, and abstractions are generalized through parameterization for wider use.
2. The *program* level is the second level in the hierarchy. At this level, one understands algorithms as sets of instructions written in a specific programming language. Algorithmic thinking is a problem-solving skill related to devising a step-by-step solution to a problem. Sequencing (i.e., planning an algorithm, which involves putting actions in the correct sequence) and flow of control (i.e., the order in which individual instructions or steps in an algorithm are evaluated) are considered essential elements of thinking computationally at this level.
1. The *execution* level constitutes the lowest level in the hierarchy. At this level, one understands algorithms as computer programs that run on specific machines and receive values or signals as inputs. While this level is the lowest in the hierarchy, it is an important one, because here one develops and practices the skill of debugging, i.e., the skill to recognize and fix errors when actions/outputs do not correspond to instructions.

Accordingly, in this study, pre-service teachers' computational thinking was examined using the four-level hierarchy of computational thinking as proposed by Perrenet and his colleagues and focused on the teaching of computational thinking through the use of scaffolded programming scripts within the context of educational robotics activities using LEGO WeDo.

7.3 The Teaching of Computational Thinking

A systematic review of the literature on the teaching of computational thinking in pre-service education revealed a dearth of research articles published in referred scholarly education journals. Nonetheless, a large number of related studies about the teaching of computational thinking, not necessarily within the context of pre-service education, have been published in referred computer science conference proceedings. Based on the results of this analysis, the teaching of computational thinking, thus far, has been pursued using five different approaches: (a) unplugged activities, (b) building-block programming, (c) tangible programming, (d) digital game creation through computer programming, and (e) educational robotics. It is worth mentioning that the results of this literature review are in agreement with Selby (2012, 2014) and Lye and Koh (2014), who stated that computer programming has been a favorite approach to adopt for teaching computational thinking, as it is the thing to do in order to automate a solution to a problem. In particular, four of the five approaches identified herein involved computer programming.

7.3.1 Unplugged Activities

This approach deploys unplugged activities to teach children computational thinking – that is, activities without the use of a computer. Prottsman (2014) reported on the development of the *Thinkersmith* curriculum in 2011, which introduced a stand-alone set of unplugged activities for K–8 specifically designed to provide students with strong computer science foundations without using computers. Lessons in this curriculum, such as Binary Baubles, used materials found in games and crafts to teach authentic computer science concepts. In 2013, *Code.org* expanded on what *Thinkersmith* created and offered a 20-hour unplugged curriculum for grades K–8. After the wide adoption of this curriculum by teachers worldwide, in 2015 *Code.org* developed further the existing 20-hour unplugged curriculum, which now includes more than 55 lessons. *CS Unplugged* is another unplugged computer science approach proposed by Bell et al. (2015). *CS Unplugged* is a collection of activities that teach computational thinking through engaging games and puzzles that use cards, string, crayons, and lots of physical movement. Students learn about binary numbers and algorithms without using computer programming.

7.3.2 Building-Block Programming Approach

Researchers who adopted the building-block programming approach (e.g., Fessakis et al. 2013; Resnick et al. 2009a, 2009b; Portelance et al. 2016) deployed Logo or Logo-like programming languages, such as Scratch, to teach young students various computational thinking skills, such as abstraction, generalization, and algorithmic thinking. In these building-block programming environments, one creates computer programs by simply snapping graphical blocks together into stacks, which represent sequences of instructions. In addition, these environments are usually "low floor" (easy to learn how to use) and "high ceiling" (afford the development of complex and sophisticated programs) and allow children to engage in rich programming activities with them.

7.3.3 Tangible Programming

Other researchers used tangible programming to teach computational thinking (Wang et al. 2014; Kazakoff et al. 2013). Tangible programming has been proposed in order to make programming more direct and less abstract for young children. It is a technique that combines computer programming and tangible interaction using physical objects to interact with the computer. There are several tangible programming tools appropriate to be used with children, such as T-Maze (Wang et al. 2014), Tern (Horn and Jacob 2007), Toque (Tarkan et al. 2010), and Twinkle (Silver and Rosenbaum 2010).

7.3.4 Digital Game Design and Creation

A fourth approach for teaching computational thinking to children is digital game design and creation (Kumar 2014; Tsalapatas et al. 2012; Denner et al. 2014). According to Dalal et al. (2009), rapid computer game creation is an innovative pedagogical approach for teaching computational thinking, because it allows the creation of games quickly without formal knowledge of programming. Students can easily create objects with visual representations and assign properties to them. Similarly, Tsalapatas et al. (2012) reported on *cMinds Learning Suite*, a learning intervention that used game-based visual programming toward building computational thinking skills. Results from different studies that adopted *cMinds Learning Suite* in different European countries showed high learner motivation in engaging in computational thinking activities using the tool and improved problem-solving skills.

7.3.5 Educational Robotics

Lastly, during the last decade, the research community has embraced educational robotics with genuine enthusiasm as an approach for teaching computational thinking to students (Stoeckelmayr et al. 2011; Janka 2008; Bers 2010; Bers et al. 2014;

Kazakoff and Bers 2012; Alimisis and Kynigos 2009; Benitti 2012; Bredenfeld et al. 2010; Johnson 2003). Benitti (2012) indicated that robotics have the potential to be effective teaching tools; learning gains for the students are not guaranteed just by the mere use of robotics. A major factor that directly influences student learning is how well the teacher knows how to use the technology as well as how to use the technology to teach a specific subject matter (Thomaz and Cakmak 2009). Vollstedt et al. (2007) also found that the comfort level of the teacher to use computers and program with robotics to be a major reason hindering the implementation of teaching with robotics in classrooms.

Recently, very few researchers have focused on teaching the teachers how to teach with robotics to enhance student learning. For example, Bers et al. (2013) investigated the effect of a 3-day robotics workshop on 32 pre-primary educators. They found that there were statistically significant gains in participants' knowledge of technology, pedagogy, and robotics content. Similarly, Burrows et al. (2012) designed a 2-week professional development (PD) to teach simple programming to pre-collegiate teachers using a hands-on, trial and error method. Pre-collegiate teachers used LEGO Mindstorms robotics kits to learn computer science through engineering design problems. According to the results, there was an increase in knowledge of and confidence to integrate robotics into classroom instruction and lesson plans.

Based on the results of the preceding studies, targeted robotics interventions that were designed in ways that provided hands-on activities that involved problem-solving, engineering design, programming, as well as modeling about how these could be applied in teaching promoted pre-service teachers' computational thinking skills. Our study builds on these previous studies and involves the use of LEGO WeDo to engage pre-service teachers in meaningful learning activities to learn about the science concept of gears. We also extend the literature on the teaching of computational thinking within the context of pre-service teacher education by introducing a scaffolded programming strategy, developed from computational theoretical principles, to develop computational thinking.

7.4 Research Questions

To this end, the present study sought to answer the following:

(a) How did learning with scaffolded programming scripts in the context of robotics activities influence pre-service teachers' computational thinking?
(b) How did learning with scaffolded programming scripts in the context of robotics activities influence pre-service teachers' understanding of gears and their functions?

7.5 Method

7.5.1 *Participants*

Twenty-one third-year pre-service teacher education students enrolled in a science education course at a Canadian university participated in the study after ethics approvals were received and all participants signed consent forms. All research participants had basic computing skills and completed, before their participation in this study, an educational technology course on teaching with technology in elementary and secondary education. The participants had no previous experience with neither computational thinking nor the use of educational robotics in classroom teaching.

7.5.2 *The LEGO WeDo Education Construction Set*

LEGO WeDo, which is designed for students ages 7 and up, enables novices to construct robots and easily program the robots using the LEGO WeDo programming language. It is an ideal tool to use in teacher education courses for promoting pre-service teachers' computational thinking skills, because, later, the teachers themselves can also use it with their students. Also, LEGO WeDo constitutes a "low floor" (i.e., easy to learn how to use) and "high ceiling" (i.e., affords the development of vibrant computer programs) (Resnick et al. 2009a, 2009b) robotics environment allowing novices to engage in interesting computational thinking tasks with it.

7.5.3 *Data Collection Instruments*

Research data were collected using (a) four programming tasks and (b) a knowledge test about gears. All research instruments were developed by the authors and checked for validity by experts in computational thinking.

Four programming tasks, related to three robots, namely, "The Dancing Birds," "The Smart Spinner," and "The Roaring Lion," were used for assessing pre-service teachers' computational thinking. For each programming task, students were given a textual description of behavior that the robot had to exhibit, and they had to first write down pseudocode for a suitable algorithm (see Appendix B for an example). Then, they were asked to code the algorithm using the LEGO WeDo programming language, run the computer program, and debug it, if necessary. The intention with each one of the four programming tasks was for the students to think and provide a general solution for the task at hand that could be utilized in various other cases. After debugging, students were asked to write down the corrected/revised computer program, if any. This way the researchers were able to compare the computer programs students wrote before debugging, with those they wrote/changed after debugging. The researchers used the PGK hierarchy (as explained in Sect. 7.2) for assessing students' computational thinking. Also, the correctness of the computer programs was scored on a scale from 0 to 10 points. The criteria used were as follows: (a) the LEGO WeDo programming instructions were put in the correct sequence (4 points), (b) the flow of control was correctly decided (3 points), and (c) programming errors were identified, removed, and fixed (3 points). A total score was used for recording performance on each programming task. Students were allowed up to 15 min for each programming task.

The knowledge test, administered both as a pre- and a posttest, consisted of five questions about gears and their functions. A cumulative score was used for rating performance on the test. The highest possible score on the test was 15 points. Each question was scored on a scale from 1 to 3 points: 1 for incorrect, 2 for partially correct, and 3 for correct. Fifteen minutes were allowed for administering the knowledge test.

7.5.4 *Scaffolded Programming Scripts*

Three scaffolded programming scripts, in paper form, were used for teaching pre-service teachers computational thinking. The first programming script was about the robot "The Dancing Birds" and consisted of seven sections. The second programming script was about the robot "The Smart Spinner" and consisted of five sections, and the third script was about the robot "The Roaring Lion" and also consisted of five sections. In essence, the three robots were selected, because their construction covered aspects of gears, and, thus, they were used as learning materials to teach pre-service teachers about gears. Accordingly, for each robotic construction, a programming script was designed and developed to teach students how to program each robot to demonstrate specific behaviors or actions.

Each section in a programming script consisted of four programming subtasks sequenced from simple to more complex with continuous fading support for the learner (see Appendix A for a complete example). Scaffolding was provided in the form of programming tasks ranging from worked-out programming tasks to semi-completed programming tasks to new programming tasks. Worked-out programming tasks provided full instructional scaffolding for the learner; semi-completed programming tasks provided partial scaffolding and new programming tasks no scaffolding. To better explain how these ideas were used for the design of the programming scripts, the authors here describe the layout of the first section of "The Dancing Birds" programming script.

The section consisted of four programming tasks as follows. The first task provided a completed computer program (a worked-out programming task) written in the LEGO WeDo programming language. The students were asked to use LEGO WeDo to generate the program, run the program, and write in the space provided what they observed. The purpose of the first programming task was for the pre-service teachers to realize that specific computer instructions corresponded to specific robotic actions and that sequencing was one essential aspect of programming. For the second task, students were given a completed computer program again. They were asked to run the program and change it if needed so that the robot exhibited a particular behavior. The second task introduced students to debugging, that is, to recognize that specific instructions did not correspond to actions and to remove and fix errors. The third programming task asked students to complete a semi-completed program using the LEGO WeDo programming language. Finally, for the fourth task, students were asked to write pseudocode for an algorithm for the robot to perform a particular action. Then, students were asked to write LEGO WeDo code for the pseudocode, run the program, change it if needed, and write the new/revised code in the space provided. The purpose here was for the students to think abstractly first about providing a general solution to a problem, expressed in a high-level language, before coding it in a particular programming language.

7.5.5 Research Procedure

Research data were collected in two 3-hour class meetings facilitated by an instructional technology educator and a science education faculty who collaborated on this research study. During the first phase, students were kindly asked to complete the self-efficacy questionnaire and the pretest on gears. Then, the science educator introduced them to the science terminology of gears and their functions using LEGO WeDo robotic models. After that, the instructional technology professor presented computational thinking as an essential skill that everyone should have in the twenty-first century, using a series of real-life examples, such as fixing a leaky roof or pipe, looking up a name in an alphabetically sorted list, and planning a long road trip. The idea here was for the students to realize the importance of computational thinking in carrying out everyday tasks. Then, students were asked to work in dyads to build the robots using LEGO WeDo for the purpose of further investigating concepts and phenomena about gears. Each student in each dyad had his or her LEGO WeDo construction set. Thus, while the researchers/instructors encouraged collaboration and exchange of ideas between the students, at the same time, they provided all students with their LEGO WeDo kit, so they could all have equal time and engagement with the robotics constructions and programming. The students were first introduced to the elements of the LEGO WeDo kit before building their first robot ("Dancing Birds"). Subsequently, the first programming script was handed out to them. In the beginning, the instructional technology educator explained the programming tasks and discussed each computer command (coding) systematically. The instructional technology professor discussed notions of what is an algorithm, sequencing of instructions, debugging, and flow of control in the form of loops. While this process was followed systematically for the first two sections of the "Dancing Birds" programming script, soon after that, students were able to assume full agency of their activities with the programming environment of LEGO WeDo and engage in self-directed learning. At the end of the first programming script, students were asked to work individually on a new programming task that was used for assessment purposes.

Seven days later, the second 3-h research phase followed. Students were welcomed back and reminded of what was accomplished during the first meeting. Additionally, for repeated measure purposes, students were asked to work on the same programming task as the one they worked on at the end of the first phase. Then, they were asked to construct the "Smart Spinner" and program the robot using the scaffolded programming script that was given to them. At the end of this activity, students were asked to work individually on a new programming task related to the "Smart Spinner" that was used for final assessment purposes. Subsequently, the same procedures were followed for the "Roaring Lion."

7.6 Results

7.6.1 *Pre-service Teachers' Computational Thinking*

In order to answer the first research question "How did learning with scaffolded programming scripts in the context of robotics activities influence pre-service teachers' computational thinking?," the computer programs students wrote for the four programming tasks (i.e., one for the "Dancing Birds" script in the first research phase that was also repeated in the second research phase, one for the "Smart Spinner," and one for the "Roaring Lion") were analyzed using the PGK hierarchy. The correctness of the programs was also determined. To remind the reader, for each programming task, students had first to write pseudocode for an algorithm followed by a computer program using the LEGO WeDo programming language, run the program, debug it, and change it if needed.

According to the results, for the first programming task (i.e., the one related to the "Dancing Birds" robot), two (9.52%) students only wrote pseudocode, as shown in Fig. 7.1, before writing actual computer code using the LEGO WeDo programming

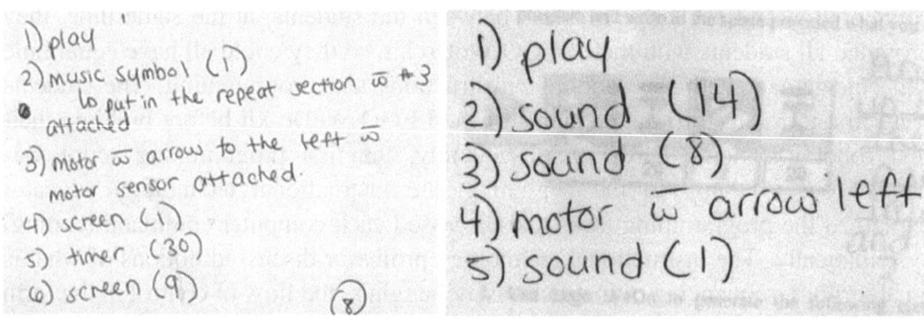

Fig. 7.1 Two examples of expressing an algorithm in words before writing computer code

Fig. 7.2 Two examples of writing computer code directly

language. The remaining 19 (90.48%) rushed into writing computer code directly, as shown in Fig. 7.2, bypassing the pseudocode phase. For the first programming task, all students debugged their computer programs, and four (19.05%) of them wrote revised programs. For the second (which was a repetition of the first task), third, and fourth programming tasks, all students wrote computer code directly. Also, all of them debugged their programs. Three (14.29%) students wrote revised programs for the third programming task, and two (9.52%) of them wrote revised programs for the fourth programming task.

These results showed that pre-service teachers were able to demonstrate computational thinking at the two lowest levels of the PGK hierarchy only, that is, at the execution level (Level 1) and program level (Level 2), showing no evidence of computational thinking at Levels 3 and 4. Thus, learning with scaffolded programming scripts in this study helped novice pre-service teachers to develop debugging skills and coding skills in regard to the specifics of a specific programming language. In essence, pre-service teachers were able to understand computer programs as sets of instructions written in the LEGO WeDo programming language and were able to run their programs on specific inputs, such as signals that could be identified by the LEGO WeDo sensors. There was no evidence of thinking computationally at higher-order levels, independent from the specifics of the programming environment at hand.

In regard to the correctness of the computer programs, three repeated measures analyses were performed between the first and second administration of the first programming task, the second and third programming tasks, and the third and fourth programming tasks. The mean performances were found to be 8.68 ($SD = 0.94$), 9.10 (second administration of the first task, $SD = 0.72$), 9.45 ($SD = 0.51$), and 9.65 ($SD = 0.33$), for the first, second, third, and fourth programming task, respectively. Repeated measures analyses revealed statistically insignificant differences between the first and second administration of the first programming task, $F(1, 19) = 3.33$, $p = 0.08$. The results also showed statistically significant differences between the second and third programming tasks and the third and fourth programming tasks, $F(1, 19) = 30.03$, $p < 0.05$, and $F(1, 19) = 12.67$, $p < 0.05$, respectively.

7.6.2 Conceptual Understanding About Gears

In regard to the second question "How did learning with scaffolded programming scripts in the context of robotics activities influence pre-service teachers' conceptual understanding about gears?," descriptive statistics showed an average performance on the pretest knowledge test on gears of 9.62 ($SD = 1.88$) and an average performance on the posttest knowledge test of 12.48 ($SD = 1.63$). A paired sample t-test was subsequently performed showing statistically significant differences between pre-service teachers' pre- and posttest knowledge scores, $t(20) = 5.84$, $p < 0.01$, indicating the effectiveness of the intervention on students' conceptual understanding of gears and how they work.

7.7 Discussion

Overall, the results of this study are very encouraging if one considers the fact that the research participants had no prior experiences with computational thinking. In particular, the results of the research revealed the significant development of novice pre-service teachers' computational thinking skills at the program level (Level 2) and execution level (Level 1) of the PGK hierarchy. While there was no evidence of development at the two higher levels of the PGK hierarchy, these findings are significant, as they showed that novice pre-service teachers with no prior experience with computer programming and debugging developed these two skills in a relatively short amount of training time. The results of the repeated measures analyses, which showed a gradual improvement of students' programming skills over the 6-hour intervention with the three scaffolded programming scripts, also supported these conclusions.

It is also noted that the significant results of the study regarding the development of computational thinking skills at the two lowest levels of the PGK hierarchy may also be related to the choice made regarding using LEGO WeDo. LEGO WeDo, as a "low floor" "high ceiling" environment, enabled students to learn the specifics of the programming language quickly without imposing a high mental load on their cognitive resources. Also, as the LEGO WeDo programming environment does not afford the use of functions, it was not possible for the researchers to explain thinking in more abstract forms using LEGO WeDo computer programs. This constraint related to the affordances of the LEGO WeDo programming language had a direct impact on how well the researchers in this study were able to demonstrate computational thinking at the two highest levels of the PGK hierarchy. In reality, and based on the learning materials used in this study (i.e., the design of the scaffolded programming scripts), the teaching of computational thinking was restricted to Level 1 and Level 2 of the PGK hierarchy. Hence, the results are consistent with those aspects of computational thinking the researchers were able to teach with LEGO WeDo.

Furthermore, the results showed that even when students were specifically asked to write pseudocode in textual form for their algorithms, two students only did so and only for the first programming script related to the "Dancing Birds," while the majority of them coded directly using the LEGO WeDo programming language from the start. These findings strongly showed that writing algorithms in narrative form was a difficult skill for the pre-service teachers to develop within the time constraints of the study.

The results are also significant because they indicate that scaffolded programming scripts can promote novice pre-service teachers' computational thinking in the context of existing teacher education courses, without the need to restructure teacher education programs. In this study, this was achieved through the means of a close collaboration between an instructional technology faculty member and a science education faculty member. It is worth noting here that the preparation for teaching this 6-hour module on computational thinking was lengthy and intense because the researchers needed to make several learning design decisions about how to best couple the teaching of computational thinking using educational robotics activities with the teaching of gears and their functions. According to the findings, students not only developed some aspects of computational thinking, but they also further improved their conceptual understanding about gears. This result is directly related to the decision made to use the three robots of "The Dancing Birds," "The Smart Spinner," and "The Roaring Lion" to explicitly demonstrate how the constructions of these robots were based on principles of how gears work. Also, within this context, the results showed significant gains in pre-service teachers' self-efficacy to teach with robotics indicating that students developed confidence in teaching with robotics in their classrooms.

Regarding future studies, researchers can assume more advanced designs moving away from one-group-only designs, as it was the case in this study, to two- or more-group designs. Also, the sample of the studies needs to increase. These improvements allow for more rigorous accounts on the effectiveness of the interventions enabling researchers to develop a better understanding of the teaching of computational thinking in teacher education courses and beyond.

Also, while the development of novice pre-service teachers' computational thinking was promoted in this study through the use of scaffolded programming scripts, as the primary instructional method used, the field can benefit from future studies that will address the design, development, and assessment of various other instructional methods. As the results showed, the scaffolded programming scripts afforded the development of computational thinking at the two lowest levels of the PGK hierarchy. An issue that arises directly from these findings is about what side of the PGK hierarchy computational thinking teaching strategies should be tried to tackle first. For example, is it better to first teach students how to create abstractions and generalizations of problems (two highest levels of the hierarchy) before teaching them how to develop algorithms and write and debug computer code (two lowest levels), or the other way around? This issue is important to investigate further with important implications for learning design and utilization of programming environments. While we foresee pros and cons for both sides of the debate, we firmly believe that the literature about the teaching of computational thinking can greatly benefit from future studies pursuing this line of research.

Appendices

Appendix A: Scaffolded programming script for the robot "The Dancing Birds"

Part I

1. Use LEGO WeDo to generate the following computer program. Run the program and write in the space provided what you observe.

2. Use LEGO WeDo to generate the following computer program. Run the program and change it (if needed), so that the robot makes a noise first followed by a spin to the right. Write in the space provided the new code, if any.

3. Use LEGO WeDo to generate and complete the following computer program so that the robot does the following in this order: (a) spins to the right, (b) makes a sound, and (c) makes a different sound than before. Then, write in the space provided the completed code.

4. Without using LEGO WeDo, write (with a pen) in the space below the code for a computer program so that the robot does the following in this order: (a) spins to the right, (b) makes a sound, (c) makes a different sound, and (d) spins to the left.

5. Now, use LEGO WeDo to run the program you wrote in 4 above. If needed, change it so that the robot does whatever is specified in 4 above. Then, write the new code in the space provided below.

Part II

1. Use LEGO WeDo to generate the following computer program. Run the program and write in the space provided what you observe.

2. Use LEGO WeDo to generate the following computer program. Run the program and change it (if needed), so that the robot does the following in this order: (a) spins to the right, (b) waits for 2 s, (c) spins to the left, (d) waits for 2 s, and (e) stops spinning. Write in the space provided the new code.

3. Use LEGO WeDo to generate and complete the following computer program so that the robot does the following in this order: (a) spins to the right, (b) waits for 3 s, (c) spins to the left, (d) waits for 3 s, (d) makes a sound, and (e) makes a different sound. Then, write in the space provided the new code.

4. Without using LEGO WeDo, write (with a pen) in the space below the code for a computer program so that the robot does the following in this order: (a) makes a sound, (b) makes a different sound, (c) spins to the right, (d) makes a different sound than the previous one.

5. Now, use LEGO WeDo to run the program you wrote in 4 above. If needed, change it so that the robot does whatever is specified in 4 above. Then, write the new code in the space provided below.

Part III

1. Use LEGO WeDo to generate the following computer program. Run the program and write in the space provided what you observe.

2. Use LEGO WeDo to generate the following computer program. Run the program and change it (if needed), so that the robot shows the picture of a beautiful beach.

3. Use LEGO WeDo to generate and complete the following computer program so that the robot does the following in this order: (a) shows the picture of a wood, (b) waits for 3 s, (c) shows the picture of the bottom of the sea, and (d) waits for 1 s.

4. Without using LEGO WeDo, write (with a pen) in the space below the code for a computer program so that the robot does the following in this order: (a) spins to the right, (b) shows a picture with clouds, and (c) spins to the left.

5. Now, use LEGO WeDo to run the program you wrote in 4 above. If needed, change it so that the robot does whatever is specified in 4 above. Then, write the new code in the space provided below.

Part IV

1. Use LEGO WeDo to generate the following computer program. Run the program and write in the space provided what you observe.

2. Use LEGO WeDo to generate the following computer program. Run the program and change it (if needed), so that the robot does the following in this order: (a) spins to the right, (b) shows the picture of a cave, and (c) spins to the left.

3. Use LEGO WeDo to generate and complete the following computer program so that the robot does the following in this order: (a) waits for 2 s, (b) spins to the left, (c) shows a picture on the screen, (d) waits for 2 s, and (e) shows a new picture on the screen.

4. Without using LEGO WeDo, write (with a pen) in the space below the code for a computer program so that the robot does the following in this order: (a) spins to the left, (b) waits for 10 s, and (c) stops spinning, and (d) shows a picture with clouds.

5. Now, use LEGO WeDo to run the program you wrote in 4 above. If needed, change it so that the robot does whatever is specified in 4 above. Then, write the new code in the space provided below.

Part V

1. Use LEGO WeDo to generate the following computer program. Run the program and write in the space provided what you observe.

2. Use LEGO WeDo to generate the following computer program. Run the program and change it (if needed), so that the robot does the following in this order: (a) spins to the right, (b) waits for 9 s, (c) makes a sound, (d) spins to the left, (e) makes a sound, and (f) spins to the right.

3. Use LEGO WeDo to generate and complete the following computer program so that the robot does the following in this order: (a) shows a picture, (b) spins to the left, and (c) makes the sound of a whistle five times.

4. Without using LEGO WeDo, write (with a pen) in the space below the code for a computer program so that the robot does the following in this order: (a) spins to the left, (b) waits for 2 s, (c) makes the sound of a whistle six times, and (d) spins to the right.

5. Now, use LEGO WeDo to run the program you wrote in 4 above. If needed, change it so that the robot does whatever is specified in 4 above. Then, write the new code in the space provided below.

Part VI

1. Use LEGO WeDo to generate the following computer program. Run the program and write in the space provided what you observe.

2. Use LEGO WeDo to generate the following computer program. Run the program and change it (if needed), so that the robot makes the sound that you see below for ever.

3. Use LEGO WeDo to generate and complete the following computer program so that the robot does the following in this order: (a) makes the sound of a whistle, (b) makes a kissing sound, and (c) repeats this sequence forever.

4. Without using LEGO WeDo, write (with a pen) in the space below the code for a computer program so that the robot makes a sequence of four different sounds and repeats this sequence forever.

5. Now, use LEGO WeDo to run the program you wrote in 4 above. If needed, change it so that the robot does whatever is specified in 4 above. Then, write the new code in the space provided below.

Part VII

1. Use LEGO WeDo to generate the following computer program. Run the program and write in the space provided what you observe.

2. Use LEGO WeDo to generate the following computer program. Run the program and change it (if needed), so that the robot (a) makes a sound and (b) makes a new sound that is repeated forever.

3. Use LEGO WeDo to generate and complete the following computer program so that the robot does the following in this order: (a) makes a sound, (b) shows a picture, and (c) repeats this sequence forever.

4. Without using LEGO WeDo, write (with a pen) in the space below the code for a computer program so that the robot (a) waits for 1 s, (b) makes a sequence of three different sounds, and (c) repeats this sequence forever.

5. Now, use LEGO WeDo to run the program you wrote in 4 above. If needed, change it so that the robot does whatever is specified in 4 above. Then, write the new code in the space provided below.

Appendix B

Without using LEGO WeDo, write (with a pen) in the space below the code for a computer program so that the robot does the following in this order:

1. Spins to the left
2. Makes the sound of a bird
3. Makes the sound of a bird
4. Makes the sound of a bird
5. Spins to the right
6. Shows the picture of a sky
7. Waits for 3 s
8. Shows the picture of a cave

References

Alimisis, D., & Kynigos, C. (2009). Constructionism and robotics in education. In D. Alimisis (Ed.), *Teacher education on robotic-enhanced constructivist pedagogical methods* (pp. 11–26). Athens: ASPETE.

Bell, T. C., Witten, I. H., Fellows, M. R., Adams, R., & McKenzie, J. (2015). *CS Unplugged: An Enrichment and extension programme for primary-aged students*. Retrieved from http://csunplugged.org/wp-content/uploads/2015/03/CSUnplugged_OS_2015_v3.1.pdf

Benitti, F. B. V. (2012). Exploring the educational potential of robotics in schools: A systematic review. *Computers & Education, 58*(3), 978–988.

Bers, M. U. (2010). The TangibleK robotics program: Applied computational thinking for young children. *Early Childhood Research & Practice, 12*(2), 1–20.

Bers, M., Seddighin, S., & Sullivan, A. (2013). Ready for robotics: Bringing together the T and E of STEM in early childhood teacher education. *Journal of Technology and Teacher Education, 21* (3), 355–377.

Bers, M. U., Flannery, L., Kazakoff, E. R., & Sullivan, A. (2014). Computational thinking and tinkering: Exploration of an early childhood robotics curriculum. *Computers & Education, 72*, 145–157.

Bredenfeld, A., Hofmann, A., & Steinbauer, G. (2010). Robotics in education initiatives in Europe-status, shortcomings and open questions. In *Workshop proceedings of international conference on simulation, modeling and programming for autonomous robots* (pp. 568–574). Darmstadt: Germany.

Burrows, A. C., Borowczak, M., Slater, T. F., & Haynes, J. C. (2012). *Teaching computer science & engineering through robotics: Science & art form*. Problems of Education in the 21st Century, 47.

Dalal, N., Dalal, P., Kak, S., Antonenko, P., & Stansberry, S. (2009). Rapid digital game creation for broadening participation in computing and fostering crucial thinking skills. *International Journal of Social and Humanistic Computing, 1*(2), 123–137.

Denner, J., Werner, L., Campe, S., & Ortiz, E. (2014). Pair programming: Under what conditions is it advantageous for middle school students? *Journal of Research on Technology in Education, 46*(3), 277–296.

Fessakis, G., Dimitracopoulou, A., & Palaiodimos, A. (2013). Graphical interaction analysis impact on groups collaborating through blogs. *Journal of Educational Technology & Society, 16*(1), 243–253.

Geist, E. (2016). Robots, programming and coding, Oh My! *Childhood Education, 92*(4), 298–304.

Gretter, S., & Yadav, A. (2016). Computational thinking and media & information literacy: An integrated approach to teaching twenty-first-century skills. *TechTrends*, 1–7. https://doi.org/10.1007/s11528-016-0098-4.

Grover, S., & Pea, R. (2013). Computational thinking in K–12: A review of the state of the field. *Educational Researcher, 42*(1), 38–43.

Guzdial, M. (2008). Education: Paving the way for computational thinking. *Communications of the ACM, 51*(8), 25–27.

Hodhod, R., Khan, S., Kurt-Peker, Y., & Ray, L. (2016). Training teachers to integrate computational thinking into K-12 teaching. In *Proceedings of the 47th ACM Technical Symposium on Computing Science Education* (pp. 156–157). New York: ACM.

Horn, M. S., & Jacob, R. J. (2007). Designing tangible programming languages for classroom use. In *Proceedings of the 1st international conference on Tangible and embedded interaction* (pp. 159–162). New York: ACM.

Janka, P. (2008). Using a programmable toy at preschool age: Why and how. In *Teaching with robotics: didactic approaches and experiences. Workshop of International Conference on Simulation, Modeling and Programming Autonomous Robots (SIMPAR 2008).*

Johnson, J. (2003). Children, robotics, and education. *Artificial Life and Robotics, 7*(1–2), 16–21.

Kazakoff, E., & Bers, M. (2012). Programming in a robotics context in the pre-primary classroom: The impact on sequencing skills. *Journal of Educational Multimedia and Hypermedia, 21*(4), 371–391.

Kazakoff, E. R., Sullivan, A., & Bers, M. U. (2013). The effect of a classroom-based intensive robotics and programming workshop on sequencing ability in early childhood. *Early Childhood Education Journal, 41*(4), 245–255.

Kim, K. H., & Coxon, S. V. (2016). In M. K. Demetrikopoulos & J. L. Pecore (Eds.), *Fostering creativity using robotics among students in STEM fields to reverse the creativity crisis* (pp. 351–365). Amsterdam: SensePublishers.

Kumar, D. (2014). Digital playgrounds for early computing education. *ACM Inroads, 5*(1), 20–21.

Lu, J. J., & Fletcher, G. H. (2009). Thinking about computational thinking. *ACM SIGCSE Bulletin, 41*(1), 260–264. Chattanooga, TN: ACM. Retrieved 24 June, 2015 from URL http://portal.acm.org/citation.cfm?id=1508959&dl=ACM&coll=portal

Lye, S. Y., & Koh, J. H. L. (2014). Review on teaching and learning of computational thinking through programming: What is next for K-12? *Computers in Human Behavior, 41*, 51–61.

National Research Council (NRC). (2010). *Committee for the workshops on computational thinking 2010. Report of a workshop on the scope and nature of computational thinking*. Washington, DC: The National Academies Press.

National Research Council. (2011). *Committee for the workshops on computational thinking: Report of a workshop of pedagogical aspects of computational thinking*. Washington, DC: National Academies Press.

Papert, S. (1980). *Mindstorms: Children, computers, and powerful ideas*. New York: Basic Books.

Perrenet, J., Groote, J. F., & Kaasenbrood, E. (2005). Exploring students' understanding of the concept of the algorithm: Levels of abstraction. *ACM SIGCSE Bulletin, 37*(3), 64–68 ACM.

Perrenet, J., & Kaasenbrood, E. (2006). Levels of abstraction in students' understanding of the concept of the algorithm: The qualitative perspective. *ACM SIGCSE Bulletin, 38*(3), 270–274 ACM.

Portelance, D. J., Strawhacker, A. L., & Bers, M. U. (2016). Constructing the ScratchJr programming language in the early childhood classroom. *International Journal of Technology and Design Education, 26*(4), 489–504.

Prottsman, K. (2014). Computer science for the elementary classroom. *ACM Inroads, 5*(4), 60–63.

Resnick, M., Maloney, J., Monroy-Hernández, A., Rusk, N., Eastmond, E., Brennan, K., & Kafai, Y. (2009a). Scratch: Programming for all. *Communications of the ACM, 52*(11), 60–67.

Resnick, M., Maloney, J., Monroy-Hernández, A., Rusk, N., Eastmond, E., Brennan, K., Millner, A., Rosenbaum, E., Silver, J., Silverman, B., & Kafai, Y. (2009b). Scratch: Programming for all. *Communications of the ACM, 52*(11), 60–67.

Selby, C. C. (2012, November). Promoting computational thinking with programming. In *Proceedings of the 7th workshop in primary and secondary computing education* (pp. 74–77). New York: ACM.

Selby, C. C. (2014). *How can the teaching of programming be used to enhance computational thinking skills?* Unpublished doctoral dissertation. Southampton: University of Southampton.

Silver, J. S., & Rosenbaum, E. (2010). Twinkle: programming with color. In *Proceedings of the fourth international conference on Tangible, embedded, and embodied interaction* (pp. 383–384). Cambridge, MA: ACM.

Stoeckelmayr, K., Tesar, M., & Hofmann, A. (2011). Pre-primary children programming robots: A first attempt. In *Proceedings of 2nd international conference on robotics in education* (pp. 185–192). Vienna: INNOC – Austrian Society for Innovative Computer Sciences.

Tarkan, S., Sazawal, V., Druin, A., Golub, E., Bonsignore, E. M., Walsh, G., & Atrash, Z. (2010). Toque: designing a cooking-based programming language for and with children. In *Proceedings of the SIGCHI conference on human factors in computing systems* (pp. 2417–2426). New York: ACM.

Thalheim, B. (2000). *Entity-relationship modeling: Foundations of database technology.* New York: Springer.

The Royal Society. (2012). *Shut down or restart? The way forward for computing in UK schools.* London: The Royal Society.

Thomaz, A. L., & Cakmak, M. (2009). Learning about objects with human teachers. In *Proceedings of the 4th ACM/IEEE international conference on human robot interaction* (pp. 15–22). San Diego: ACM.

Tsalapatas, H., Heidmann, O., Alimisi, R., & Houstis, E. (2012). Game-based programming towards developing algorithmic thinking skills in primary education. *Scientific Bulletin of the "Petru Maior" University of Targu Mures, 9*(1), 56.

Vollstedt, A. M., Robinson, M., & Wang, E. (2007). Using robotics to enhance science, technology, engineering, and mathematics curricula. In *Proceedings of american society for engineering education pacific southwest annual conference,* Honolulu: Hawaii.

Wang, D., Wang, T., & Liu, Z. (2014). A tangible programming tool for children to cultivate computational thinking. *The Scientific World Journal, 2014,* 1. https://doi.org/10.1155/2014/428080.

Wing, J. M. (2006). Computational thinking. *Communications of the ACM, 49*(3), 33–35.

Wing, J. M. (2008). Computational thinking and thinking about computing. *Philosophical Transactions of the Royal Society A, 366*(1881), 3717–3725.

Wing, J. M. (2011, March). Computational thinking *VL/HCC*, 3.

Yadav, A., Zhou, N., Mayfield, C., Hambrusch, S., & Korb, J. T. (2011). Introducing computational thinking in education courses. In *Proceedings of the 42nd ACM technical symposium on computer science education* (pp. 465–470). New York: ACM.

Yadav, A., Mayfield, C., Zhou, N., Hambrusch, S., & Korb, J. T. (2014). Computational thinking in elementary and secondary teacher education. *ACM Transactions on Computing Education (TOCE), 14*(1), 5.

Chapter 8
Computational Thinking in K-12: In-service Teacher Perceptions of Computational Thinking

Phil Sands, Aman Yadav, and Jon Good

8.1 Introduction

Much of what we know about computational thinking comes from early research in educational practices using computers (Papert 1980; Pea and Kurland 1984) and from common conceptions of how computer scientists think about problems designed to be solved by computers (Denning 2009). Wing (2006) formalized computational thinking in an influential article discussing the ways computer scientists think about problems and how skills associated with computing are broadly applicable in other disciplines. Wing sparked a discussion about how educators should prepare students for careers influenced by computing and where core computational thinking concepts could be integrated into K-12 curricula (Barr and Stephenson 2011; Grover and Pea 2013; Yadav et al. 2014). Almost a decade later, teaching computational thinking skills to students has permeated at all levels of elementary and secondary schools. This integration is being done through the generation of new curricula within computer science education programs – the AP computer science principles course is one notable example – as well as in other content areas, such as mathematics and science (Weintrop et al. 2016). With this increased interest, however, comes key questions about how in-service teachers conceptualize computational thinking, especially teachers who are not trained in computer science. Namely, how do these teachers understand computational concepts as they work to apply them in their classrooms? Further, what steps do we need to take to help in-service teachers integrate computational thinking into their curriculum?

Most of the attention on embedding computational thinking during the past decade has focused on preservice teachers (Yadav et al. 2011, 2014). While this

P. Sands · A. Yadav (✉) · J. Good
College of Education, Michigan State University, East Lansing, MI, USA
e-mail: ayadav@msu.edu

M. S. Khine (ed.), *Computational Thinking in the STEM Disciplines*,
https://doi.org/10.1007/978-3-319-93566-9_8

information can help guide in-service teachers' professional development, we have yet to identify the unique challenges that exist in introducing computational thinking to non-computing teachers. A better understanding of in-service teachers' conceptions of computational thinking can guide design of teacher professional development programs. In a recent survey, we examined how K-12 in-service teachers perceive computational thinking within elementary and secondary classrooms. We present results from the survey and provide recommendations for developing professional development programs around computational thinking practices. We also discuss specific areas within the computational thinking model that lend themselves to the nature of applied problem-solving in K-12 classrooms.

8.2 Background

In considering computational thinking and its application to student preparation, Wing (2008) pointed to the links between CT and the wide variety of disciplinary skills traditionally taught in K-12 classrooms. These connections focus on the ubiquitous nature of computing and the nature of abstraction as it pertains to STEM career pathways. In addition, Wing stressed that computational thinking was not the same as the practice of programming; rather, she argued that the skills used in programming are useful for problem-solving in multiple contexts. Denning (2009) argued for the use of computational thinking ideas as the "third leg of science," a component of the inquiry process as much as it is a separate and distinct discipline. While Wing and Denning differed in how computational thinking was framed, they both agreed on the benefits for students from learning computer science. Regardless of which perspective one takes, it is apparent that the connections between computing and K-12 curricula are deep enough to justify the interest in further embedding these ideas in classrooms.

Since Wing (2006) introduced computational thinking, there have been several attempts to expand on what ideas encapsulate CT. Wing proposed that computational skills include abstraction, problem decomposition, pattern recognition, algorithmic thinking, and logical thinking. In attempting to draw connections between these skills and an educational model in Bloom's taxonomy, Selby (2015) organized a variation of these ideas by perceived difficulty: evaluation, algorithm design, generalization, abstraction of functionality, abstraction of data, and decomposition. Barr and Stephenson (2011) proposed nine major computational thinking concepts and abilities to be used within K-12 classrooms across core content areas. These include data collection, data analysis, data representation, problem decomposition, abstraction, algorithms and procedures, automation, parallelization, and simulation. This set is echoed in the work of Grover and Pea (2013), who offered that CT was comprised of abstractions and pattern generalizations, systematic processing of information, symbol systems and representations, algorithmic notions of flow of control, structured problem decomposition, iterative, recursive, and parallel thinking, conditional logic, efficiency and performance constraints, and debugging and systematic error detection. A more complex set of skills were described by the National Research Council (2010) including:

> reformulation of difficult problems by reduction and transformation; approximate solutions; parallel processing; checking and model checking as generalizations of dimensional analysis; problem abstraction and decomposition; problem representation; modularization; error prevention, testing, debugging, recovery and correction; damage containment; simulation; heuristic reasoning; planning, learning, and scheduling in the presence of uncertainty; search strategies; analysis of the computational complexity of algorithms and processes; and balancing computational costs against other design criteria. (p. 3)

Given the wide variety of skills that can be connected to computational thinking, the lack of a clearly defined subset of skills may confuse educators trying to implement these practices.

Computational thinking skills have also appeared in recent updates to K-12 curriculum frameworks, such as Next Generation Science Standards (NGSS) as well as other curricula designed to teach introductory computing skills. The Next Generation Science Standards (NGSS) include the use of CT as an important practice to develop scientific understanding (NGSS Lead States 2013). The College Board created a new Advanced Placement computing course focusing on six key computational thinking practices, with the goal of attracting a more diverse group of students to computer science (2014). Similarly, Google introduced the CS First initiative to provide traditional computer science activities and lessons focused on computational thinking primarily for use by out-of-school organizations.

Considering that the onus for implementing these programs is on educators with limited experience in computing, a concern is the risk of conflating computational thinking with computer science or mathematics. There is also a potential for those implementing computational thinking ideas to imply that both CT and CS require the use of programming in all contexts (Fletcher and Lu 2009). In order to address this issue, it has been suggested that educators encourage the use of computational thinking skills at an early age, concentrating more on the innate thought processes that are associated with computing as opposed to specific computing tools. By doing so, educators can reduce the barriers for entry for students taking computing courses later in their academic careers (Margolis et al. 2010). This group includes not just students that develop further interest in computer science but also students interested in other fields engaging with computing in some form.

In spite of the potentially overwhelming set of skills that can be included in definitions of computational thinking, it is possible to implement most of the core ideas in primary and secondary classrooms without overemphasizing technical abilities. Examples can include digital storytelling, simple data collection, and the encouragement of scientific investigation (Lee et al. 2014). Considering that teachers may be using these skills in primary school classrooms already (Mannila et al. 2014), this suggests a need to help move teachers from implicit to explicit practices grounded in an understanding of why computational practices are relevant to student development.

8.3 Need

Computational thinking practices have the potential to develop student interest in how computing plays a role in other disciplines, specifically STEM. In order to see the benefits of student exposure to these computing concepts, we need to train both preservice and in-service teachers in computational thinking practices regardless of academic discipline. Across the United States, academic standards have been rewritten to include computational thinking as a core principle of curriculum implementation. Examples of this include the Next Generation Science Standards which include computational thinking concepts (NGSS 2013), Indiana's K-8 science standards (Indiana Department of Education 2017), and Texas' Essential Knowledge and Skills for elementary education (Texas State Board of Education 2012). Designing teacher professional development program should focus on augmenting teachers existing competencies while relying on established best practices, in order to align courses with the major components of computational thinking. As an important step in this process, we need to understand in-service teachers' current perceptions of computational thinking (Prieto-Rodriguez and Berretta 2014). In identifying areas of need, the transition can then be made to connecting professional development with classroom integration of CT. This study examined in-service teachers' conceptions of computational thinking and was guided by the following research questions:

1. How do in-service teachers conceptualize computational thinking as it would manifest in classroom practice?
2. How does teachers' subject area influence their computational thinking conceptualizations?
3. How does teachers' grade level taught influence their computational thinking conceptualizations?

8.4 Methods

Participants Seventy-four elementary and secondary teachers from a Midwestern state participated in the study. Of these teachers, 65 were female and 9 were male. Teachers taught at a variety of levels in the K-12 spectrum but could be divided roughly into primary school (N = 45) and secondary school (N = 29) levels. For the purposes of this study, we included grades K-6 as primary school teachers and grades 7–12 as secondary school teachers. Lastly, we considered those teachers that taught primarily STEM subjects (N = 29) versus those that were in non-STEM subjects (N = 55). STEM subjects included mathematics, science, computers, or technology.

Survey The survey included ten Likert scale questions based on prior work examining preservice teachers' perceptions of computational thinking (Yadav et al. 2011, 2014). The survey items began with the phrase "Computational thinking involves..." followed by a short stem that either belonged or did not belong to the broader perception of computational thinking. Teachers responded to the items on a Likert scale with five potential response values. These included "strongly agree," "agree," "disagree," "strongly disagree," and "don't know." Table 8.1a includes the list of survey items, and Table 8.1b includes how we characterized whether the item aligned with literature's conceptions of computational thinking. It should be noted in this table that the concept of "coding/programming" was not categorized due to disagreement over whether programming is an essential element of teaching CT in classrooms (Denning 2009; Wing 2006; Brennan and Resnick 2012). The internal reliability of these items was assessed using Cronbach's alpha ($\alpha = 0.92$). In addition, the survey included items to collect demographic information regarding teachers' gender, grade level taught, and subjects taught.

The survey was distributed at the Michigan Association for Computer Users in Learning (MACUL) conference. Participants were recruited at an exhibition booth for university K-12 outreach programming.

Table 8.1a Items included in the teacher survey

Computational thinking involves...
... solving problems
... using heuristics/algorithms
... logical thinking
... thinking like a computer
... coding/programming
... doing mathematics
... using computers (e.g., office tools)
... knowing how to use a computer
... using technology in your teaching
... playing online games

Table 8.1b How researchers categorized items from the teacher survey

Computational thinking involves...	Computational thinking does not involve...
... solving problems	... doing mathematics
... using heuristics/algorithms	... using computers (e.g., office tools)
... logical thinking	... knowing how to use a computer
... thinking like a computer	... using technology in your teaching
	... playing online games
It is unclear whether or not computational thinking involves...	
... coding/programming	

8.5 Data Analysis

Likert response was given a numerical value from 1 to 4 ("strongly agree," 1; "agree," 2; "disagree," 3; "strongly disagree," 4), and missing responses and those marked as "don't know" were excluded from these calculations. We used descriptive analysis for each of the survey items to view patterns in teachers' conceptions of computational thinking. In addition, Mann-Whitney U test was used to analyze the influence of teachers' subject area and grade level taught on their conceptions of computational thinking. Mann-Whitney U test, a nonparametric alternative test to the independent t-test, was used due to the ordinal nature of the data. The data was analyzed using the R statistical package.

8.6 Results

Majority of the teachers in our study were most confident that computational thinking involved logical thinking (100%), doing mathematics (100%), and solving problems (99%). To a lesser degree, majority of the teachers also agreed that computational thinking involved using heuristics or algorithms (93%), using computers (86%), using technology in teaching (82%), and knowing how to use a computer (76%). Teachers' conceptions of computational thinking are shown in Fig. 8.1, and the descriptive statistics are presented in Table 8.2.

8.6.1 STEM vs Non-STEM Teachers

STEM refers to teaching and learning in the fields of science, mathematics, engineering, and technology (Gonzalez and Kuenzi 2012). For the purpose of this study, teachers that specified their primary area as one of the natural sciences or engineering (e.g., computer science, physics, chemistry, etc.) were included within STEM. This group was categorized as "STEM" teachers, and those outside of these disciplines was categorized as "non-STEM" teachers. For this study, most of the primary school teachers were removed from the STEM analysis because these educators commonly teach all domains. Only those primary educators that specified a domain specialization were considered in this analysis. Table 8.3 shows the breakdown by grade level and STEM specialization.

As shown in Fig. 8.2, results showed that STEM teachers had the greatest confidence that computational thinking involved doing mathematics (100%), logical thinking (100%), solving problems (100%), using computers (96%), and using heuristics or algorithms (96%). The non-STEM teachers showed similar beliefs that computational thinking involved doing mathematics (100%), logical thinking (100%), solving problems (100%), and using heuristics or algorithms (93%). While

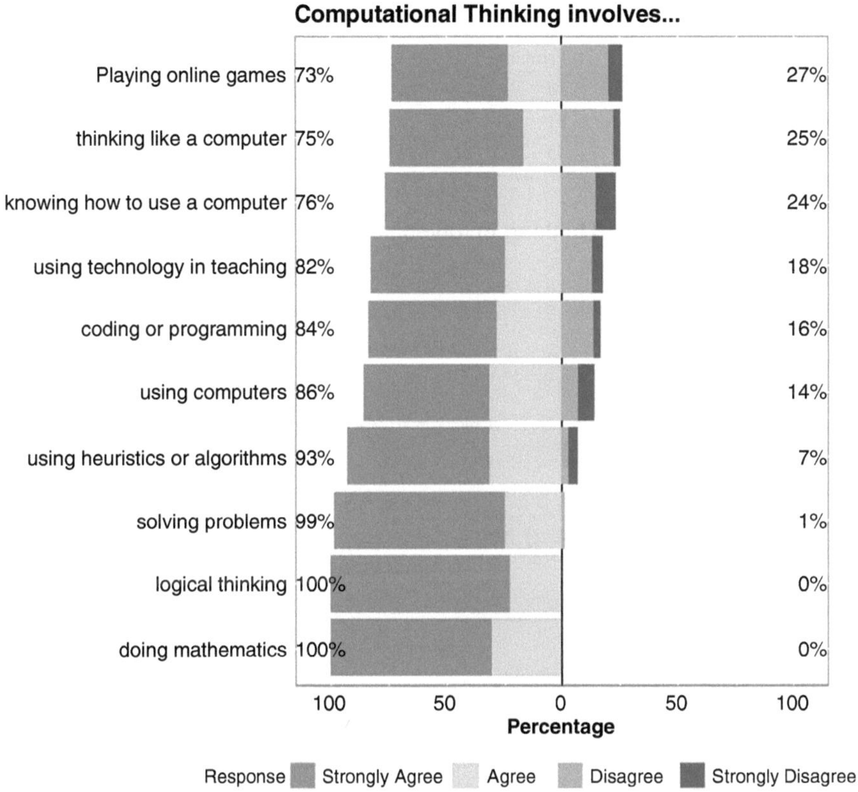

Fig. 8.1 Teachers' conceptions of computational thinking

Table 8.2 Descriptive statistics on teachers' conceptions of computational thinking

Computational thinking involves...	Mean	Standard deviation
... doing mathematics	1.31	0.46
... using computers (e.g., office tools)	1.67	0.90
... solving problems	1.28	0.48
... using heuristics/algorithms	1.5	0.76
... logical thinking	1.23	0.42
... thinking like a computer	1.70	0.92
... knowing how to use a computer	1.84	0.99
... using technology in your teaching	1.65	0.88
... playing online games	1.83	0.97
... coding/programming	1.64	0.83

Note: The scale was from 1 (strongly agree) to 4 (strongly disagrees)

Table 8.3 Primary and secondary teachers considering STEM vs non-STEM teaching credentials

	Primary	Secondary	
STEM	14	15	29
Non-STEM	31	14	45
	45	29	

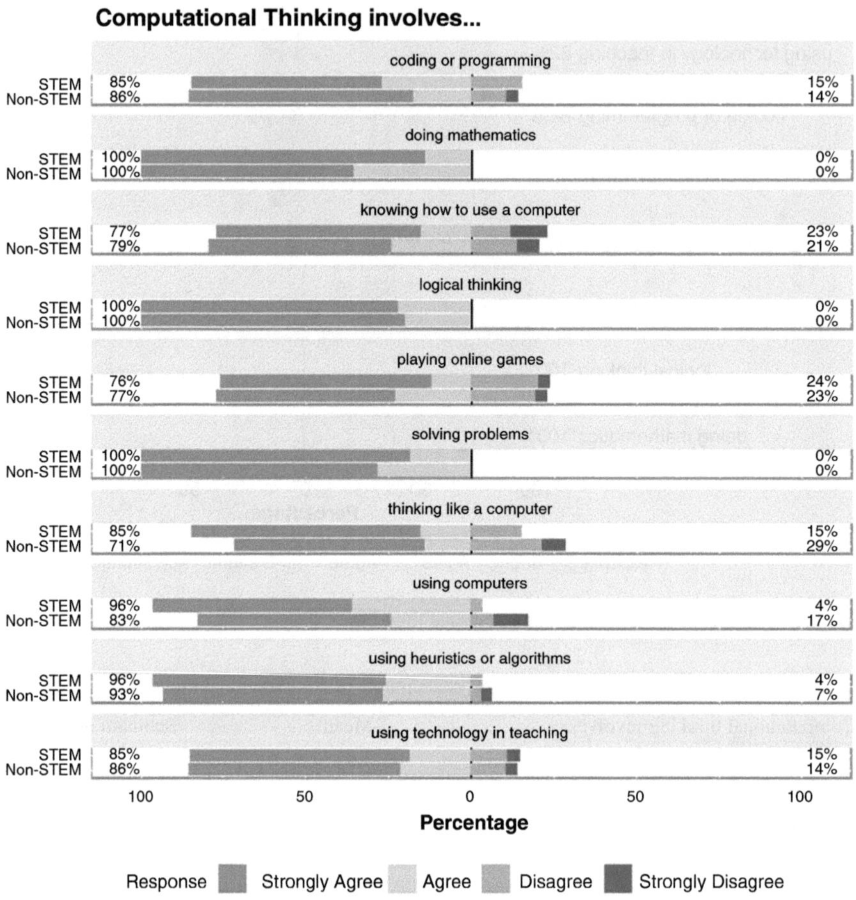

Fig. 8.2 STEM vs. non-STEM teachers and perceptions of computational thinking

there were similar responses between the STEM and non-STEM teachers on almost all of the items, two notable exceptions were “thinking like a computer” and “using computers.” This showed that non-STEM teachers were less likely to view those as computational thinking. It should be noted that “using computers” was described on the survey instrument as being akin to using office tools and other applications.

Table 8.4 Mann-Whitney U test comparing STEM vs Non-STEM teachers

Computational thinking involves...	U statistic	p-value
... doing mathematics	526	0.06557
... using computers	437.5	0.5706
... solving problems	473.5	0.3973
... using heuristics or algorithms	423.5	0.7236
... logical thinking	396	0.8475
... thinking like a computer	420	0.2644
... knowing how to use a computer	389	0.8276
... using technology in teaching	385	0.8967
... playing online games	349	0.6169
... coding or programming	333	0.5387

Mann-Whitney U results exhibited there was no significant difference between STEM and non-STEM teachers on how they conceptualized computational thinking (see Table 8.4 for the Mann-Whitney U statistics for each of the computational thinking items).

8.6.2 *Primary vs Secondary School Teachers*

Over the last decade, the high awareness of STEM curricula has led to more elementary teachers exploring ways to engage their students in technology (DeJarnette 2012); hence, we examined whether there were differences in how they conceptualized computational thinking when compared to secondary teachers. As shown in Fig. 8.3, results demonstrated that secondary teachers believed that computational thinking involved doing mathematics (100%), logical thinking (100%), solving problems (100%), and using heuristics or algorithms (100%). Similarly, primary teachers also viewed computational thinking as involving doing mathematics (100%), logical thinking (100%), and solving problems (98%). However, there were some differences between the two groups as secondary teachers disagreed at a higher rate whether computational thinking involved "knowing how to use a computer," "playing online games," and "using technology in teaching." In addition, they had uniform sentiment that "using heuristics or algorithms" belonged to computational thinking, while primary teachers showed some disagreement. Other items showed some differences, but none that were visually significant enough to note.

Mann-Whitney U results suggested no significant difference between primary and secondary teachers on how they conceptualized computational thinking (see Table 8.5 for the Mann-Whitney U statistics for each of the computational thinking items).

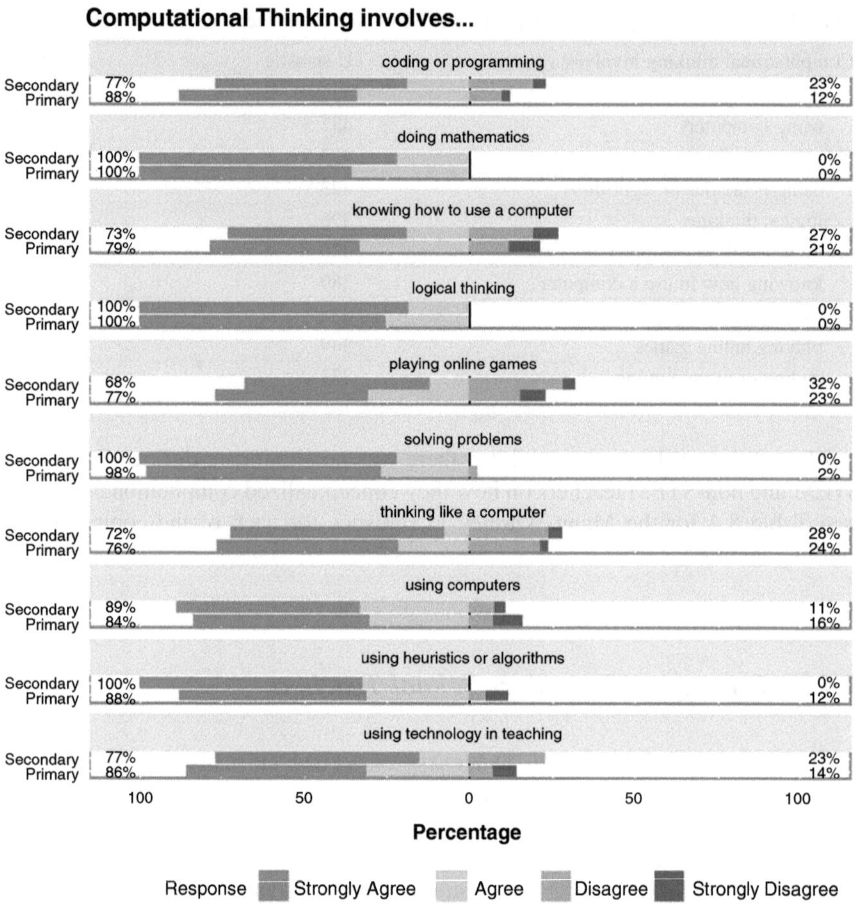

Fig. 8.3 Primary vs. secondary teachers and perceptions of computational thinking

Table 8.5 Mann-Whitney U test comparing primary and secondary teachers' perceptions

Computational thinking involves...	U statistic	p-value
... doing mathematics	688.5	0.24
... using computers	607	0.72
... solving problems	651	0.51
... using heuristics or algorithms	673.5	0.24
... logical thinking	621.5	0.50
... thinking like a computer	550.5	0.71
... knowing how to use a computer	571.5	0.73
... using technology in teaching	565	0.79
... playing online games	508.5	0.76
... coding or programming	525.5	0.92

8.7 Discussion

Overall, results suggested that while teachers conceptualized computational thinking in alignment with the literature, they also had some incorrect ideas about what computational thinking entailed. We also found that there were no differences on teachers' conceptions of computational thinking based upon either the content area (STEM vs. non-STEM) or grade level (primary vs. secondary). Computational thinking involves a set of skills that describe many of the same abilities inherent to programming and problem-solving with computers (Denning 2009). The responses given by the teachers in our study suggested that many educators have very little knowledge about what these skills are and lack awareness of how these skills can be implemented in their classrooms. The results suggest that there is much work to be done before in-service teachers are able to implement computational thinking in their classrooms.

Based on the literature, we classified what computational thinking entails (see Table 8.1b). Our results exhibited that teachers had the greatest confidence that CT involved "logical thinking" and "solving problems," which align with how computational thinking has been conceptualized recently (Denning 2017). On the other hand, teachers also viewed CT as "doing mathematics," which does not align with the common conception of computational thinking. Overall, we found that majority of the teachers strongly agreed with all the components of computational thinking outlined in the survey items and in many cases that teachers incorrectly agreed with concepts that we did not view as computational thinking. With these conceptions of computational thinking, a teacher simply using digital tools, such as Microsoft Office, might think that he/she is engaging his/her students in computational thinking. On the other hand, it is also possible that teachers might think that CT involves too many conceptual tasks to integrate.

Our results support the need to develop non-computing teachers' understanding of computational thinking if it is to permeate within K-12. Teachers, regardless of whether they taught a STEM subject or not, have similar ideas about computational thinking and sometimes hold incorrect conceptions. Given the prevalence of incorrect views related to computational thinking suggests that while CT maybe a buzzword in computing education, many teachers are not being introduced to the core components of computational thinking. While researchers have argued for the need to embed computational thinking within teacher education (Yadav et al. 2017), our results suggest the need to also train in-service teachers. This training needs to be content-specific on how to integrate computational thinking ideas into existing curriculum. Specifically, teachers need to be introduced to computational thinking in a way that meets their existing learning goals and fits within their pedagogical practices. Rather than adapting approaches designed for preservice teachers, we instead propose implementing a distinct strategy for integrating CT ideas aimed at teachers already working in K-12 classrooms.

In-service teacher professional programs need to provide support for content integration, allowing educators to utilize their existing body of knowledge while also meeting their needs with regard to time constraints and availability. Existing research into teacher professional development has found the difficulties of providing long-term gains in the classroom based on limited exposure to applied concepts through isolated workshop sessions (Harris and Sass 2011; Desimone 2009). Thus, in order to successfully train teachers to integrate computational thinking into K-12 classrooms, we need to develop ongoing and continuous professional development programs that help teachers develop a thorough understanding about what it means to think computationally and then engage their students in computing ideas (Yadav et al. 2017).

Professional development needs to draw upon teachers' expertise in their content knowledge, pedagogical knowledge, and pedagogical content knowledge. The Reading Apprenticeship model (Greenleaf et al. 2011) provides a framework to support teachers' learning of computational thinking concepts and develop students' understanding of how computation can be applied in specific subject areas. Specifically, professional development should point out clear connections and how computational thinking can meet subject area learning goals rather than just being an instructional add-on in the K-12 curriculum (Greenleaf et al.). Given the large number of demands teachers face and the time constraints of the classroom, we also need to address how to deliver the content to teachers. Schools of education should collaborate with departments of computer science to lead state-approved professional development certification programs in computing education. These low-cost flexible programs could be delivered online, to allow teachers to learn virtually and be a member of an online community of practice to discuss how computational thinking can be embedded to meet their subject-specific learning goals. As suggested by Yadav et al. (2017), we believe that an online community of practice would allow teachers to effectively integrate computational thinking to meet their curriculum needs.

Our findings have important implications for how professional development programs should be structured to ensure that teachers effectively integrate computational thinking in their classrooms. Results suggest that professional development needs to differentiate between the use of computing tools and the concepts and practices inherent to computational thinking. It might be beneficial to expose teachers to computational thinking without the use of computers, such as using the CS Unplugged curriculum (Bell et al. 2009). Focusing on unplugged activities might help teachers grasp how computational thinking and the use of computers in the classroom differ from one another. We believe that given Wing's (2006) description of computational thinking overlapped with aspects of problem-solving components, such as abstraction, problem decomposition, pattern recognition, and algorithmic thinking, a focus on problem-solving skills offers a low floor to get teachers interested in computational thinking. By using problem-solving as the focus, we feel that more teachers will be motivated to embed subcomponents of computational thinking in their regular academic subjects (Yadav et al. 2016).

This study had a few limitations, which has implications for generalizability of the findings. First, we acknowledge that the survey was based on a small number of teachers and may not have accurately represented teacher knowledge of computational thinking across the United States. The impact of this small group is also enhanced due to the large number of elementary teachers in our sample that were not included in our evaluation of STEM and non-STEM teachers. Additionally, given that participants in our study were volunteers might lead to self-selection bias, which limits generalizability of the results. It is also possible that the since teachers completed the survey at a conference focused on technology in education, they were more focused on computational thinking as involving use of technology/digital tools. At the same time, given that teachers interested in technology struggled with identifying computational thinking ideas suggests we have an uphill climb before CT becomes another core subject similar to reading, writing, and arithmetic as called for by Wing (2006).

In summary, we recognize the need to prepare students for twenty-first-century careers makes it essential for K-12 teachers to be prepared to integrate computational thinking concepts. This requires a multipronged approach to prepare teachers at the preservice and in-service level to become computationally literate.

References

Barr, V., & Stephenson, C. (2011). Bringing computational thinking to K-12: What is involved and what is the role of the computer science education community? *ACM Inroads, 2*(1), 48–54.

Bell, T., Alexander, J., Freeman, I., & Grimley, M. (2009). Computer science unplugged: School students doing real computing without computers. *The New Zealand Journal of Applied Computing and Information Technology, 13*(1), 20–29.

Brennan, K., & Resnick, M. (2012). New frameworks for studying and assessing the development of computational thinking. In *Proceedings of the 2012 annual meeting of the american educational research association,* Vancouver, Canada (pp. 1–25).

The College Board. (2014). AP Computer Science Principles: 2016–2017. Retrieved from https://advancesinap.collegeboard.org/stem/computer-science-principles/course-details

DeJarnette, N. (2012). America's children: Providing early exposure to STEM (science, technology, engineering and math) initiatives. *Education, 133*(1), 77–84.

Denning, P. J. (2017). Remaining trouble spots with computational thinking. *Communications of the ACM, 60*(6), 33–39. https://doi.org/10.1145/2998438.

Denning, P. J. (2009). Beyond computational thinking. *Communications of the ACM, 52*(6), 28–30.

Desimone, L. M. (2009). Improving impact studies of teachers' professional development: Toward better conceptualizations and measures. *Educational Researcher, 38*(3), 181–199.

Fletcher, G. H., & Lu, J. J. (2009). Education: Human computing skills: Rethinking the K-12 experience. Association for computing machinery. *Communications of the ACM, 52*(2), 23.

Gonzalez, H. B., & Kuenzi, J. J. (2012). *Science, technology, engineering, and mathematics (STEM) education: A primer*. Congressional research service. Retrieved from https://fas.org/sgp/crs/misc/R42642.pdf

Greenleaf, C. L., Litman, C., Hanson, T. L., Rosen, R., Boscardin, C. K., Herman, J., Schneider, S. A., Madden, S., & Jones, B. (2011). Integrating literacy and science in biology: Teaching and learning impacts of reading apprenticeship professional development. *American Educational Research Journal, 48*(3), 647–717.

Grover, S., & Pea, R. (2013). Computational thinking in K–12 a review of the state of the field. *Educational Researcher, 42*(1), 38–43.

Harris, D. N., & Sass, T. R. (2011). Teacher training, teacher quality and student achievement. *Journal of Public Economics, 95*(7), 798–812.

Indiana Department of Education (2017). *Indiana academic standards: Science & computer science*. Retrieved from http://www.doe.in.gov/standards/science-computer-science.

Lee, I., Martin, F., & Apone, K. (2014). Integrating computational thinking across the K–8 curriculum. *ACM Inroads, 5*(4), 64–71.

Mannila, L., Dagiene, V., Demo, B., Grgurina, N., Mirolo, C., Rolandsson, L., & Settle, A. (2014). Computational thinking in K-9 education. In *Proceedings of the working group reports of the 2014 on innovation & technology in computer science education conference* (pp. 1–29). New York: ACM.

Margolis, J., Estrella, R., Goode, J., Holme, J. J., & Nao, K. (2010). *Stuck in the shallow end: Education, race, and computing*. Cambridge: MIT Press.

National Research Council. (2010). *Report of a workshop on the scope and nature of computational thinking*. Washington, DC: National Academies Press.

NGSS Lead States (Ed.). (2013). *Next generation science standards: for states, by states*. Washington, DC: National Academies Press. Retrieved from http://ezproxy.msu.edu/login?url=http://search.ebscohost.com/login.aspx?direct=true&scope=site&db=e000xna&AN=867791.

Papert, S. (1980). *Mindstorms: Children, computers, and powerful ideas*. New York: Basic Books, Inc.

Pea, R. D., & Kurland, D. M. (1984). On the cognitive effects of learning computer programming. *New Ideas in Psychology, 2*(2), 137–168.

Prieto-Rodriguez, E., & Berretta, R. (2014). Digital technology teachers' perceptions of computer science: It is not all about programming. In *2014 I.E. Frontiers in Education Conference (FIE) Proceedings* (pp. 1–5).

Selby, C. C. (2015). Relationships: Computational thinking, pedagogy of programming, and bloom's taxonomy. In *Proceedings of the workshop in primary and secondary computing education* (pp. 80–87). New York: ACM.

Texas State Board of Education (2012). *Chapter 111. Texas essential knowledge and skills for mathematics subchapter a. Elementary.* Retrieved from http://ritter.tea.state.tx.us/rules/tac/chapter111/ch111a.html

Weintrop, D., Beheshti, E., Horn, M., Orton, K., Jona, K., Trouille, L., & Wilensky, U. (2016). Defining computational thinking for mathematics and science classrooms. *Journal of Science Education and Technology, 25*(1), 127–147.

Wing, J. M. (2006). Computational thinking. *Communications of the ACM, 49*(3), 33–35.

Wing, J. M. (2008). Computational thinking and thinking about computing. *Philosophical Transactions of the Royal Society A: Mathematical, Physical and Engineering Sciences, 366*(1881), 3717–3725.

Yadav, A., Gretter, S., Hambrusch, S., & Sands, P. (2017). Expanding computer science education in schools: Understanding teacher experiences and challenges. *Computer Science Education, 26*, 235–254. https://doi.org/10.1080/08993408.2016.1257418.

Yadav, A., Hong, H., & Stephenson, C. (2016). Computational thinking for all: Pedagogical approaches to embedding a 21st century problem solving in K-12 classrooms. *TechTrends, 60*, 565–568. https://doi.org/10.1007/s11528-016-0087-7.

Yadav, A., Mayfield, C., Zhou, N., Hambrusch, S., & Korb, J. T. (2014). Computational thinking in elementary and secondary teacher education. *ACM Transactions on Computing Education (TOCE), 14*(1), 5.

Yadav, A., Zhou, N., Mayfield, C., Hambrusch, S., & Korb, J. T. (2011, March). Introducing computational thinking in education courses. In *Proceedings of the 42nd ACM technical symposium on Computer science education* (pp. 465–470). New York: ACM.

Chapter 9
A Computational Thinking Curriculum and Teacher Professional Development in South Korea

Soohwan Kim and Hae Young Kim

9.1 Introduction

Changes in industry caused by the development of information and communication technology (ICT) have affected all areas of society, including educational environments. In this age of the "Fourth Industrial Revolution" (Schwab 2015), new technologies continue to integrate the physical, digital and biological worlds, thereby changing the economy, the demands of the workplace, and the educational needs of our youth. In Korea as elsewhere, educational policymakers have come to see that sofeware and computing education are no longer options but basic competencies that students must learn in order to succed in twenty-first century. Many nations (e.g., Korea, United Kingdom, United States, China, Japan, France) have prioritized computational thinking education, that is, education that teaches coding, algorithm, or digital literacy for K-12 students (Bocconi et al. 2016; ISTE and CSTA 2011; NRC 2010; Naace 2014; KICE 2015). As Corporate Vice-President of Microsoft Research Jeanette Wing noted, "Everyone can benefit from thinking computationally. My grand vision is that computational thinking will be a fundamental skill—just like reading, writing, and arithmetic—used by everyone by the middle of the 21st Century." (Wing 2008). The changes in computing education in Korea from 2005 to 2015 reflect the government's recognition of the importance of computing education. Accordingly, it has continuously pursued measures designed to give students the ability to apply technology in their daily lives and use it to make a better future for society (Code.org 2018; KICE 2015; MoE 2015; MoE et al. 2016; P21 2011).

The current 2015 computing education policy was intended to "strengthen SW [software] education in schools for computing education" (MoE 2014) by:

S. Kim (✉)
Chongshin University, Seoul, South Korea

H. Y. Kim
Independent Research Consultant, Seoul, South Korea

M. S. Khine (ed.), *Computational Thinking in the STEM Disciplines*,
https://doi.org/10.1007/978-3-319-93566-9_9

(1) establishing the foundation for SW education, (2) identifying and supporting SW talent, and (3) building the support system for SW education.

Such plans call on teacher educators to train pre-service and in-service teachers in computational thinking teaching and learning strategies to prepare them for effective implementation in the classroom. Understanding teacher perceptions can help teacher educators in computing education respond to problems with the new curriculum as well as teacher concerns.

9.2 The Evolution fo Computing Education in Korea

Up to 2005, Korean computing education consisted mainly of ICT education with a focus on using software applications and the internet (e.g., PowerPoint, Word processing, Excel). In 2005, the Ministry of Education attempted the first introduction of computer science into the curriculum of computing education, though the trial was not successful and the curriculum was not executed (Lee and Choi 2015). In 2009, the computing education curriculum was altered to include computational thinking and in 2015, the revised curriculum was finalized (see Fig. 9.1).

All primary school students are required to begin learning about computational thinking by providing at least 17 classroom hours of various activities involving educational programming languages or the Unplugged materials. Middle school students are supposed to spend at least 34 classroom hours exploring and learning about computational thinking practices with educational programming language and physical devices (e.g. sensor board, Pico board) (Ministry of Education 2015). All Korean high school students may take computing education as a normal elective subject, so any student who wants to learn computing education can select it.

The mandatory nature of these changes makes them quite meaningful. Because Korea has a national educational curriculum, all students now have equal access to the new computing education curriculum, the goals of which are to improve student competencies like information literacy, computational thinking, and collaborative problem-solving capability (see Fig. 9.2). Table 9.1 shows differences in the content

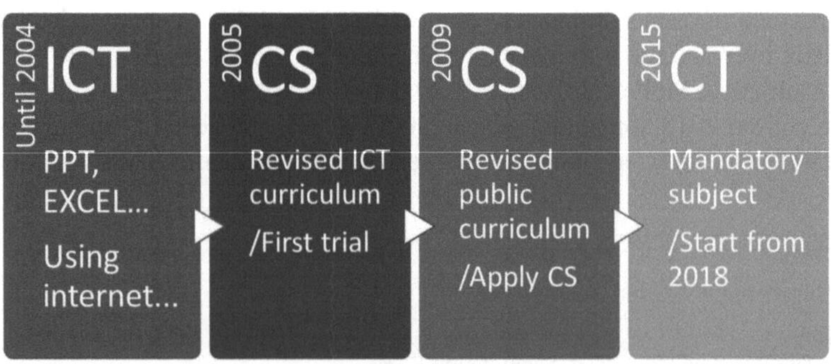

Fig. 9.1 Changes in Korean computing education, 2004–2015

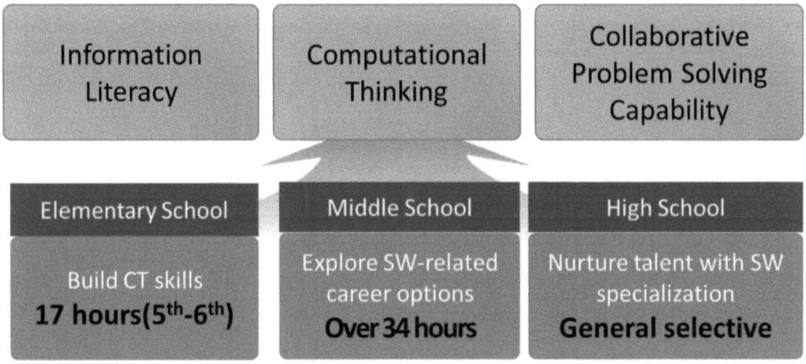

Fig. 9.2 Breakdown of 2015 revised computing education curriculum

Table 9.1 Changes in the content of the Korean computing education curriculum before and after *2015*

Level	Before	After	Contents
Elementary school (2019~)	ICT unit in practical arts (12 hours)	SW education in practical arts (over 17 hours)	Problem-solving process, algorithm, programming concepts Information ethics
Middle school (2018~)	Informatics (elective)	'Informatics' Over 34 hours (compulsory subject)	Problem-solving based on CT Developing algorithms and programming
High school (2018~)	'Informatics' (advanced elective)	'Informatics' (general elective)	Designing algorithms and programs with various fields

of the computing education curriculum for elementary, middle, and high school students before and after the 2015 curriculum change.

9.3 Current Status of Computing Education in Korea

In 2017, the Korean government is assigning and operating 1200 research or leading schools for computing education from elementary to high schools to prepare for the 2018 school year. In 2016, 160,000 students throughout Korea participated in an online coding party that operated on MS Kodu and the Korean Entry platform, which are block based coding tools (MoE et al. 2016).

The Korean Office of Education has taken many steps to help teachers with the 2015 curriculum. It continues to offer large-scale teacher training courses and has a plan to hire 500 new secondary school teachers with computer majors by the year 2020. In Korea, primary school teachers must teach all subjects, and all pre-service teachers must take a course training them to teach this new computing education.

All K-12 teachers are using to teach two educational programming languages: Scratch and Entry. Entry is a Korean style block programming language that has

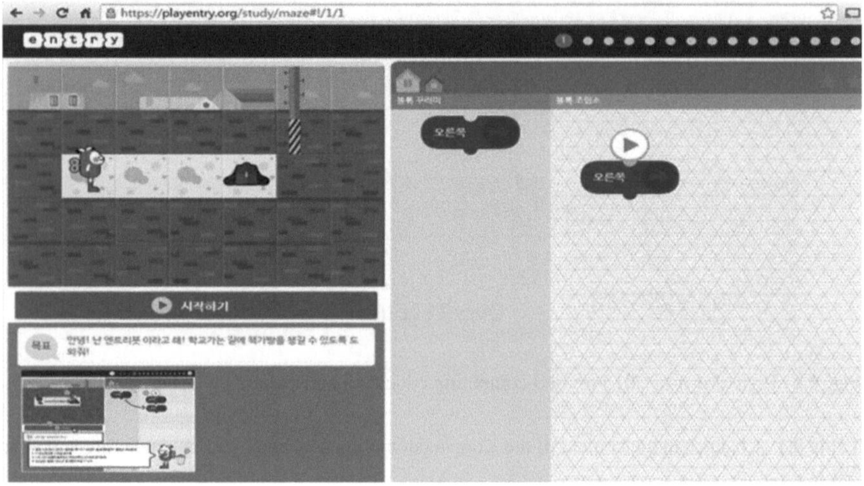

Fig. 9.3 Interface of entry (http://playentry.org)

been strongly influenced by Scratch, however Entry is based on a Learning Management System that enables teachers to make online classes for their students (Fig. 9.3).

Four government agencies have partnered to with educational associations, teacher groups, and private companies to ensure the successful implementation of the 2015 computing education curriculum:

- Four government agencies: Korean Ministry of Science and ICT; Korean Foundation for the Advancement of Science & Creativity; Korean Ministry of Education; Korean Education and Research Information Service
- Four academic associations: Korean Information Science Education Federation; Korean Association Of Information Education; Korean Association of Computer Education; Korean Institute of Information Scientists and Engineers
- Two teacher groups: Association of Teachers for Computing (primary school); Informatics and Computer Teachers of Secondary School

9.4 Steages of Teacher Concern about Computing Education

The effectiveness of educational reform depends on how well teachers practice the reform in their classroom. The 2015 South Korean computing education reform requires teachers to teach computing according to a national curriculum and standards. Yet, the success of this reform requires that teachers fully understand the revised computing education curriculum, including intentions, importance, goals, and methods needed to achieve those goals and achieve full implementation with

Table 9.2 Seven stages teacher concern

Area	Stage	Description
Self	0: Unconcerned	The individual indicates little concern about or involvement with the innovation.
	1: Informational	The individual indicates a general awareness of the innovation and interest in learning more details about it. The individual does not seem to be worried about himself or herself in relation to the innovation.
	2: Personal	The individual is uncertain about the demands of the innovation, his or her adequacy to meet those demands, and/or his or her role with the innovation.
Task	3: Management	The individual focuses on the processes and tasks of using the innovation and the best use of information and resources. Issues related to efficiency, organizing, managing, and scheduling dominate.
Impact	4: Consequence	The individual focuses on the innovation's impact on students in his or her immediate sphere of influence.
	5: Collaboration	The individual focuses on coordinating and cooperating with others regarding use of the innovation.
	6: Refocusing	The individual focuses on exploring ways to reap more universal benefits from the innovation, including the possibility of making major changes to it or replacing it with a more powerful alternative.

high fidelity and accountability. Clearly, the teachers' interest, motivation, and concerns regarding the computing education are integral to this implementation. Listening to teachers and identifying their needs in this endeavor is essential to finding avenues for appropriate educational supports rather than simply enforcing a top-down mandate for teachers to teach computing education across the country.

A first step in identifying the concerns, interests and thoughts of teachers is paying attention to their affective or emotional areas. Fuller (1969) proposed a Concern-Based Model of Teacher Development to explain how the concerns of pre-service (or novice) teachers may change over time, as they become professional teachers. Fuller categorized such concerns into three developmental stages: (1) concern with self as an early concern, (2) concern with tasks, and (3) concern with impact on student learning. Hall et al. (1973) extended Fuller's concern-based model to explore teachers' concerns and changes in concern when educational programs and reforms are initiated and implemented. They identified seven stages in teachers' concerns, starting with 0:: unconcerned, informational, personal, management, consequence, collaboration, and refocusing.

Table 9.2 provides a detailed explanation of each of these stages. The first three stages involve concerns with self and stage 3 is concerned with tasks. Stages 4 through 6 involve the impact of educational reforms on student learning. Based on this research, George et al. (2006) developed an assessment instrument called the Stages of Concern Questionnaire to examine these stages. The instrument includes 35 items with possible responses on an 8-point Likert scale with a Cronbach alpha of between .64 to .84, indicating good reliability.

George et al. (2006) proposed a hypothetical development of stages of concern like Fig. 9.4. They categorized teacher groups according to the extent to which one

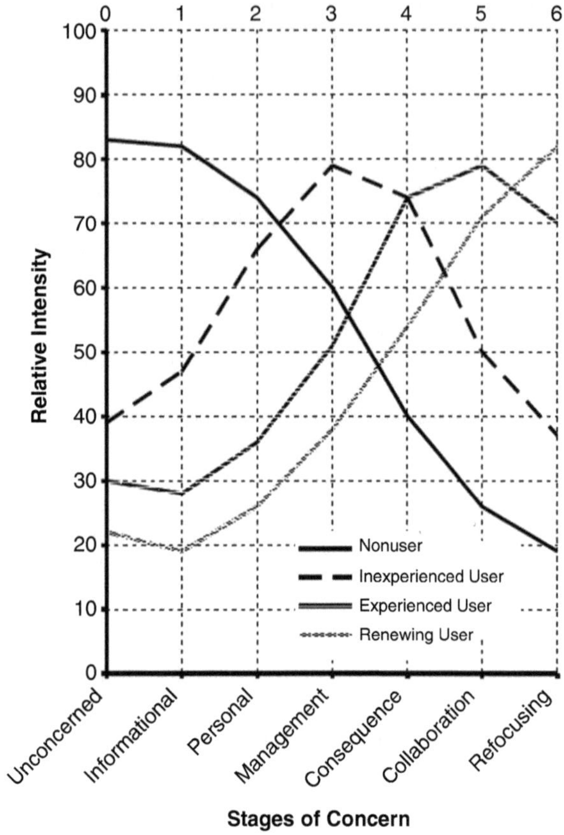

Fig. 9.4 Hypothesized development of stages of concern. (Adopted from George et al. 2006, p. 36)

uses and practices educatnoal reform as nonuser, inexperienced user, experienced user, and renewing user. Nonuser means a group of teachers who do not use or practice educational reform in their classrooms. Since they do not participate in the reform, nonusers may not be interested in the reform yet or may not have time to think about it due to their busy schedule. George et al. labeled this type of concern is unconcerned of stage 0. However, for some of nonuser teachers, they might be interest in the reform and want to know about it. George et al. (2006) called this as informational concern, which is stage 1. Typically, nonuser group have relatively high intensity of concern at stage 0-unconcerned and stage 1-informational.

Inexperienced users mean teachers who have just started practicing an educational reform but do not know yet effective ways to implement it. Therefore, inexperienced teachers have concerns in regard to managing a reform and conflicts with the other tasks. George et al. labeled this concern as management concern, which is stage 3. As teachers practice reform, they become experienced teachers and are more likely interested in student learning and effective teaching methods by teacher collaboration, which were labeled as Consequence concern of Stage 4 and Collaboration concern of Stage 5 repsectively. When the process of a reform reaches to a mature and stable period, teachers are able to reflect the contents of reform and an impact on student learning. As a result of their reflection, teachers have more concerns regarding how to revise the current reform for better results, which was labeled as Refocusing concern of Stage 6.

Kim and Kim (2016) conducted a survey to assess Korean teachers' stages of concern regarding computing education using the current version of the Stages of Concern Questionnaire (George et al. 2006). They collected data from 92 teachers from elementary, middle and high schools in Korea in early 2016. The analysis of the data revealed the following insights:

(1) The average Korean teachers presented a non-user profile of stages of concern, which showed the highest intensity of concern at Stage 0-Unconcerned and 1-Informational and lowest intensity of concern at Stage 4-Consequences and 5-Collaboration (2) Gender differences in the stages of concern existed in the data. Except for the fact that Stages 0 and 1 were the highest concerns in both genders, male teachers showed high intensity of concern at Stage 5-Collaboration and 6-Refocusing whereas female teachers showed high concerns at Stage 3-Management and Stage 2-Personal.

(3) The level of perceived software proficiency affected teachers' stages of concern. Teachers with high software proficiency showed significantly higher concerns at Stage 4-Consequence, Stage5-Collaboration, and Stage 6-Refocusing than the teachers with low software proficiency. Teachers with moderate software proficiency showed high concern at Stage 1-Information and Stage 2-Personal. Teachers with low software proficiency showed high concerns at Stage 1-Information and Stage 3-Mangement.

(4) Software education training experiences also affected teachers' stages of concern. Teachers who had software education training showed statistically higher concerns at Stage 4-Consequence, Stage 5-Collaboration, Stage 6-Refocusing than teachers with no training experience.

(5) The years of computing teaching experience affected teachers' stage of concern.

Figure 9.5 displays how the intensity of teacher concern changed in relation to years of computing education experiences. Teachers with less than 1 year of experience in computing education showed high intensity at Stages 0 and 1 and low intensity at Stages 4, 5, and 6. On the other hand, the teachers with more than 4 years of computing education experience show high intensity of concern at Stage 6-Refocusing and Stage 5-Collaboration.

Studies on teacher concerns have shown that teachers' concerns about computer education tend to differ based on factors such as software proficiency, software education training experience, and software teaching experiences. Since the implementation of the revised Korean computing education curriculum is now at a preparation stage, teachers concerns tend to be mostly at Stage 0-unconcerned and 1-Informational. For teachers who have high software proficiency, however, software education experience gained in their classroom and received software education training correspond to different concerns from teachers with low software proficiency, and no teaching and no training experience. By looking at teachers' concerns about the process of a specific educational reform (before-early-middle-later implementation), we can identify the needs of teachers at each stage and

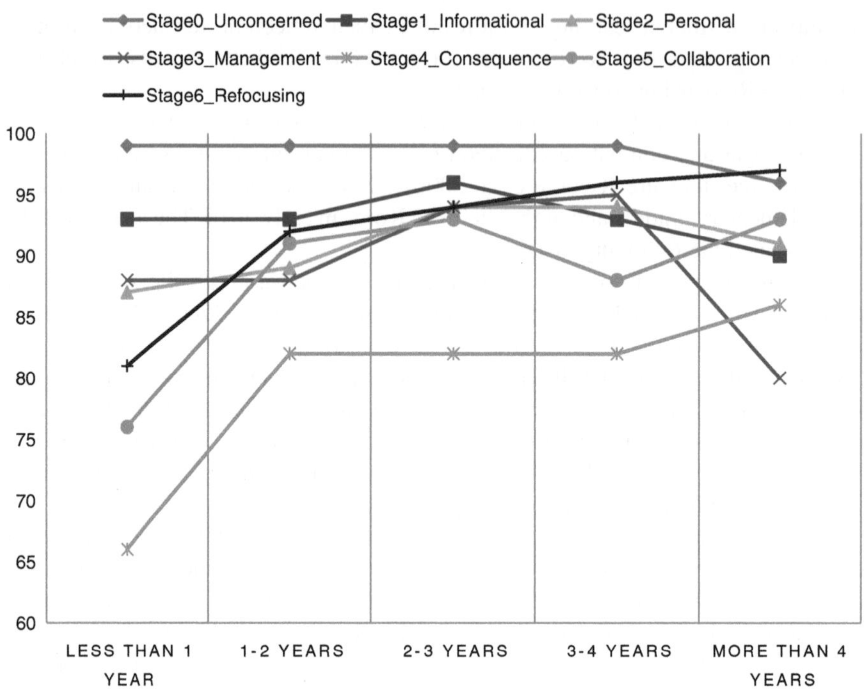

Fig. 9.5 Changes in stages of concern by computing education experience

provide appropriate supports based on individual needs. Thus, examining teacher's concerns and needs may ultimately contribute to the effective and efficient implementation of computing education reforms for K-12 students.

9.5 Analysis of Teacher Needs

We investigated the perceptions of 30 teachers from 15 elementary and 12 secondary schools in Seoul. We conducted seven focus group interviews and one group discussion using the qualitative method known as phenomenological research (Kim et al. 2017). Our analysis of teacher perceptions about current computing education resulted six distinct problems:

1. Lack of a teacher who majored in a computer-related field or computing education,
2. Little understanding by principals and non-computer scienc major teachers of the need for computing education,
3. The need to empower problem-solving capability and programming skill of compuer teachers,
4. Insufficient materials on teaching strategies and content for computing education,

5. Lack of lesson hours for K-12 students and a physical infrastructure for computing education,
6. Inconsistent computing education policies.

In addition, the teachers we interviewed suggested some specific policy changes to increase the future effectiveness of computing education.

1. Make a special department for informatics education in the Ministry of Education
2. Establish legal provisions mandating minimum informatics class times and integrated subjects
3. Plan and execute consistent long-term planning for informatics education
4. Provide more physical space for teacher self-study
5. Create regional computing education centers (e.g., university, library,museum), with a variety of resources
6. Construct an integrated support system of regional societies, universities, laboratories, and companies.
7. Create a training course on computing education for school principals and governors.
8. Consider both universal education and vocational education.
9. Support a free semester system and circles made by students themselves.
10. Hire supervisors and researchers with an appropriate educational philosophy and passion for computing education for education offices.
11. Develop an online system through which teachers can share teaching and learning materials and educational examples.
12. Develop computing education teacher training courses for different levels of expertise.
13. Establish an incentive system for excellent computing education teachers and support teacher groups.

9.6 Proposed Innovations for More Effective Computing Education

9.6.1 Teacher Training Course

Given the responses of the teachers we interviewed, we propose the establishment of a teacher training course for computing education. The objectives of the course include: expand computing education through guidance that involves essential content for schools, establish networks to support the leading schools, and understand the educational content of computing education. Table 9.3 shows the contents of a proposed basic teacher training course for 2015 computing education curricula.

Table 9.4 shows the content of an advanced one-day teacher training course, with teachers grouped by region and elementary or secondary school position. The course

Table 9.3 Time and content of teacher training basic courses

Time	Contents
1 h	Explanation of computing educational policy and running leading schools for computing education
1 h	Computational thinking and computing education
1 h	Educational contents of computing education (e.g. unplugged, EPL, physical computing)

Table 9.4 Detailed content of teacher training course

	Subject	Detailed contents
1	Explore computational thinking factors	Google computational thinking factors Creative computing factors Factors of CAS of U.K. Factors of Korean SW education *Practice: Make a table of contents of computational thinking factors*
2	Explore lesson plans for computational thinking	Google CT lesson plan CS first of MS Textbook of CAS Sample lesson plans of CSTA Korean textbooks for CT Report of teaching and learning model for CT *Practice: Make a worksheet for CT lesson*
3	Computing education curriculum	Revised computing education curriculum Explore hierarchy of educational contents Explore levels of educational contents *Practice: Make a worksheet for hierarchy of contents*
4–5	Make a lesson plan	Abstraction and algorithm Programming *Practice: Make teaching and learning plan*
6	Sharing and feedback	Present teaching and learning plan Peer feedback

content consists of activities involving problem-solving. In one, teachers will be asked to select six issues based on the educational environment of secondary school and then analyze and share some of the major factors, such as objectives, learning factors, teaching and learning considerations, educational environments. Each step will have problem-solving strategies that call for abstraction and the use of the automation method (e.g. extract core factor, analyze the current state, decomposition, modeling, algorithms, programming).

Table 9.5 in particular shows an advanced teacher training course to take place over 2 days. Based on a revised computing education curriculum, the course consists of reconstructing methods for learning content to improve students' computational thinking. This course is based on the 2015 revised curriculum in order to show how to reconstruct the curriculum for the needs of our current educational environments.

Table 9.5 Advanced teacher training course (secondary school)

Day	Contents	Description
Day 1	Reconstruction method of computing curriculum for computational thinking (2h)	Understand of the 2015 revised curriculum. Understand teaching and learning materials and educational environment for real life problem-solving using computational thinking.
	Practice about reconstruction of the curriculum (5 h)	Development and share lesson plan based on computational thinking with team.
Day 2	Practice of problem- solving and programming based on computational thinking (7 h)	Block based programming or text based programming for improving problem-solving in real life stuation. Physical computing programming

All teachers have to conduct a final project: to make lesson plans and materials based on computational thinking for their students.

9.6.2 Innovation in the Pre-Service Teacher's College Curriculum

In Korea, one needs a special certification to be a teacher for K-12. Most elementary school teachers have graduated from one of eleven special colleges of education; most secondary school teachers have graduated from a related educational department in a university.

Figure 9.6 presents some proposed changes to the curricula of primary pre-service teachers so as to emphasize computational thinking education. It has four parts: SW literacy, using SW, making SW, SW life (Jeong 2016).

The Korean government also has announced the plan of SW Core Universities for cultivating future human resources who can apply the computational thinking to solve real life problems (MSI 2018). The SW Core University will strengthen students competitiveness in SW fields and expand the value of SW by offering a revised curriculum that concentrates on the practical demands of the SW industry.

The SW Core Universities have to conduct the four following parts:

– Revise SW education curricula according to the needs of industry.
– Operate mandatory programs for all SW major students.
– Provide SW basic education for non-SW major students.
– Establish entrance standards for SW special students.
– Empower networking among SW core universities.

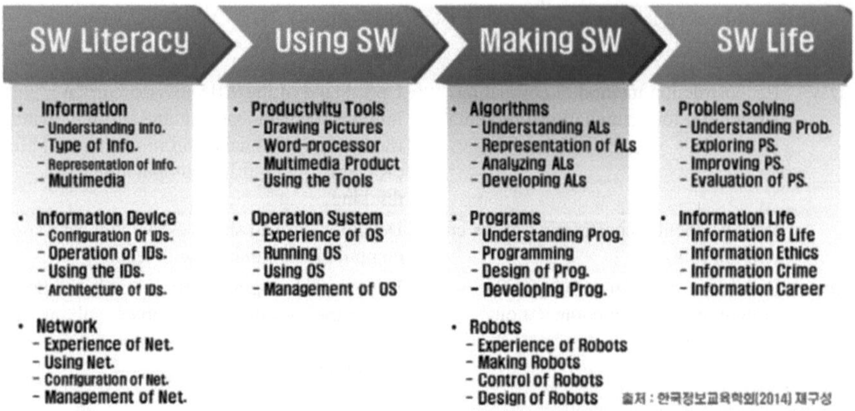

Fig. 9.6 Contents of framework for computing education for elementary pre-service teachers. Adapted from "Needs Analysis of Software Education Curriculum at National Universities of Education for the 2015 Revised National Curriculum" by Y. Jeong, 2016, *Journal of The Korea Association of Information Education, 20*(1), p. 85. Copyright 2016 by KAIE

9.7 Discussion and Conclusion

Among the most fundamental of our findings was the pressing need for the establishment of an appropriate physical infrastructure for computing education. The responses of teachers constantly spoke to this problem of infrastructure, and they identified it as the most important condition for successful computer education. Differences between schools in the quality of their computer facilities is a problem in many developing countries or countries with a large gap between rich and poor. Because some schools do not have the appropriate physical infrastructure, Unplugged method that it the teaching and learning method without a computer is an alternative of computing education, however coding process is necessary.

Therefore, it is necessary to make an environment that guarantees high educational quality and low investment cost. The One Laptop per Child project of Negroponte or the 'Hole in the wall' effort by Sugata are good alternatives to traditional computer labs. However these projects can only succeed if implemented by human resources with a real passion for their students.

Ultimately, successful computing education requires effective operational infrastructures and human resources. In education, the teacher is the primary human resource, so government must support them. There are two main methods for preparing teachers for the challenge of instituting the 2015 computing education curriculum in Korea. First, we must change the college curriculum for pre-service teachers in training so that it contains both pedagogical content knowledge and pedagogical methods for computing education. Second, we must plan and make available multiple levels of teacher training courses. The basic course should include the use of educational programming tools and various devices like Arduino, Sensor board, or Small Robot. Advanced courses should contain teaching and learning strategies for improving students' computational thinking. In addition, a convergence course should consist of a STEM project integrating other subjects. In the last year, KOFAC have been running basic and advanced courses for teacher training in Korea, some as trials for convergence training courses.

Another method is to support a self-organized teacher group and provide a place for them to study teaching and learning strategies. This concept is in line with making a free place for students, as in the future school by Isido (Ishido 2014). In fact, self-organized teacher groups are arising, thus government should strengthen their capacity voluntarily. Computing education teachers must create such groups and together attempt various challenges such as micro-teaching, learning about new devices, and studying the computing education curriculum. Because computing education requires a computer and various technological devices, a location with an appropriate infrastructure is a prerequisite. Little by little, the Korean government is seeking to make more of such places available. Both local governments and local offices of education should provide support for such teacher groups and locations as well.

Finally, training courses in computing education must be made available for governors and principals of schools. According to recent research(Choi et al. 2016), the perception of the necessity of computing education is rising among school principals, and this attitude has a positive effect on the entire system. Directors of schools who are convinced of the importance of computing education help to ensure that practice rooms and support for computing teachers remain available. The authority of principals is absolute in Korea, so the milieu they create within their context can have a major impact on the larger school environment and policymakers. We believe it is highly likely that principals who take these training courses will be much stronger supporters of computing education—and a facilitater of the proper professional development for teachers in their schools. The teacher is a primary factor in effective computing education but not the only one. As Toyama (2015) noted, successful computing education depends on the passion of *all* stakeholders, including leaders (governors or directors), teachers (performers), students, and parents (beneficiaries).

References

Bocconi, S., Chioccariello, A., Dettori, G., Ferrari, A., & Engelhardt, K. (2016). *Developing computational thinking in compulsory education-implications for policy and practice (No. JRC104188).* Seville: European Commission, Joint Research Centre.

Choi, S., Kim, H., Sung, J., Seo, Y., & Choi, Y. (2016). *A study of activating Gyeonggido SW education.* Korea: Gyeonggido institute of education.

Code.org. (2018, June 1). *Nine policy ideas to make computer science fundamental to K-12 education.* Retrieved June 1, 2018 from https://code.org/files/Making_CS_Fundamental.pdf

Fuller, F. F. (1969). Concerns of teachers: A developmental conceptualization. *American Educational Research Journal, 6*(2), 207–226.

George, A. A., Hall, G. E., & Stiegelbauer, S. M. (2006). *Measuring implementation in schools: The stages of concern questionnaire*. Austin: SEDL Retrieved June 1, 2018, from http://www.sedl.org/pubs/catalog/items/cbam17.html.

Hall, G. E., Wallace, R. C., & Dossett, W. A. (1973). *A developmental conceptualization of the adoption process within educational institutions*. Austin: Research and Development Center for Teacher Education, The University of Texas.

Ishido, N. (2014). *Children's creativity switch! Children's fort to play & learn – CANVAS's practice*. Tokyo: Film Art, Inc.

ISTE & CSTA. (2011). *Computational thinking in K-12 Education teacher resources* (2nd ed.). ISTE.

Jeong, Y. (2016). Needs analysis of software education curriculum at National Universities of Education for the 2015 revised natioanl curriculum. *Journal of The Korea Association of Information Education, 20*(1), 83–92.

Kim, H. Y., & Kim, S. H. (2016). Stages of concern of Korean teachers about software education and the relationship with teacher characteristics. *Journal of the Korean Association of Information Education, 20*(4), 387–400.

Kim, S., Kim, H., Kim, S., & Song, S. (2017). *A study of innovative method for future informatics education in digital innovation age*. Report 2016–56. Korea.

Korea Institute for Curriculum and Evaluation (KICE). (2015). *A study for the improvement of Korean Computer Information Literacy Education based on Result of ICILS 2013*. Research Resource ORM 2015–58. Korea.

Lee, T. U., & Choi, H. J. (2015). *Informatics education*. HANBIT Academy: Seoul.

Ministry of Education. (2014). *Activation plan for SW education in K-12*. Korea.

Ministry of Education. (2015). *2015 Revision Curriculum –Middle school information* (1). Korea.

Ministry of Education (MoE), Ministry of Science ICT and Future Planning (MSIFP), & KERIS. (2016). *Resources of support group for 2016 SW education leading schools*. Korea.

Ministry of Science and ICT (MSI). (2018, March). SW University. Retrieved May 20, 2018, from http://www.software.kr/um/um02/um0207/um020701.do

Naace, Havering Education service, and Computing at School (Nacce). (2014). *Switched on computing year1*. London: Rising Star UK Ltd.

National Research Council (NRC). (2010). *Report of a workshop on the scope and nature of computational thinking*. Washington, DC: National Academies Press.

Partnership for 21st Century Skills (P21). (2011, March). *Framewokr for 21st century learning*, Retrived June 1, 2018 from http://www.p21.org/storage/documents/1.__p21_framework_2-pager.pdf

Schwab, K. (2015, December 12). *The fourth industrial revolution*. Retrieved June 1, 2018, from https://www.foreignaffairs.com/articles/2015-12-12/fourth-industrial-revolution

Toyama, K. (2015). *Geek Heresy*. New York: PublicAffairs.

Wing, J. M. (2008). Computational thinking and thinking about computing. *Philosophical Transactions of The Royal Society, 366*, 3717–3725.

Part III
Computational Thinking in Schools and Higher Education

Chapter 10
Exploring the Scope and the Conceptualization of Computational Thinking at the K-12 Classroom Level Curriculum

Georgios Fessakis, Vasilis Komis, Elisavet Mavroudi, and Stavroula Prantsoudi

10.1 Introduction

While it is a rather common belief that the integration of computer science (CS) in modern general education is required for pedagogical and socioeconomic reasons, approaches to this attempt vary significantly. Briefly speaking, one can distinguish three main models – (*1*) the "CS as a separate subject" model, which emphasizes the fundamental concepts and skills of computer science, mainly programming. This model, in many cases, results in the diffusion of the misconception that "CS = programming" (Denning 2009). (*2*) The information communication technologies (ICT) model, which aims at the development of digital literacy (synonymous of ICT literacy). In this case, the emphasis is placed on the familiarization with software applications and the fluent use of digital technology. In the best case, this model also concerns the use of ICT as a cognitive tool (e.g., use of microworlds, simulations, and other learning software) for the construction of knowledge by students in various school subjects, combined with the constructivist pedagogy. The ICT model does not ensure that students develop the required CS competences. This is because it percepts computers as tools in science and, at best, computation as a method of science. These perceptions do not cover the current conceptions of computation as a process of nature and of computing as the study of natural and artificial information

G. Fessakis (✉)
Faculty of Humanities, University of the Aegean, Rhodes, Greece
e-mail: gfesakis@rhodes.aegean.gr

V. Komis
Department of Educational Sciences, University of Patras, Patras, Greece

E. Mavroudi · S. Prantsoudi
Greek Ministry of Education, Athens, Greece

M. S. Khine (ed.), *Computational Thinking in the STEM Disciplines*,
https://doi.org/10.1007/978-3-319-93566-9_10

processes (Denning and Martell 2015). (3) The third model constitutes a combination of CS and ICT models using curricula that bridge digital literacy and CS topics, in coordination with the curricula of other school subjects, with the intention of ICT-enhanced teaching and learning. The bridging of ICT and CS standards and curricula constitutes a pragmatic solution for modern public educational systems (Brinda et al. 2009). This model, if wisely implemented, could support the development of computational literacy (diSessa 2000) for all literate citizens, which concerns the use of CS concepts and methods as tools for general problem-solving as well as a medium to study other disciplines. Furthermore, it could cover the introduction of the fundamental and distinct principles and practices of CS instead of reducing CS to a few categories of its key practices like programming or application of computational thinking (Denning 2009).

Each of the three, CS integration in education, models has significantly distinct consequences and requirements for their implementation in a state- or nationwide educational system scale. For example, the preparation of teachers, the curriculum, and the teaching/learning material require a different approach in each model. Despite the efforts of several international organizations (such as the Association for Computer Machinery/Computer Science Teachers Association (ACM/CSTA) and the International Society for Technology Education (ISTE)) to propose model curricula, standards, teachers' training, learning material, research support, and policies, the position of CS and ICT still varies significantly among state and national educational systems. What most countries seem to be mainly delivering is ICT literacy, thus reproducing the misconception that "CS = digital devices and productivity software applications" and, consequently, limiting students' opportunities to develop their competence for computational resources utilization.

As CS is currently considered the fourth great scientific domain along with the traditional physical, life, and social sciences (Rosenbloom 2004), the integration of CS in general education is an imperative issue that should attract the interest of the education community and educational policy-makers. Recently, the role of CS in general education came in the center of the interest in many countries, mainly because of the recognition of computing as a core part of Science, Technology, Engineering, and Mathematics (STEM) Education (Henderson et al. 2007). The conducting of a meaningful and comprehensive discussion on the integration of CS in K-12 education, with the participation of all stakeholders, is currently of great importance. The CS education community's recently increased interest in the concept of CT is particularly helpful to initiate and fuel this dialogue. With the publication of the article entitled "Computational Thinking" (Wing 2006), Janette Wing set out her vision of recognizing CT as a fundamental competency that all literate citizens should develop through compulsory education, to complement the three other core skills, that is, reading, writing, and mathematics. Since Wing (2006) raised the issue of CT, as a conceptual tool to approach the role of CS in general education, extended discussion on the scope of the term got launched.

No sooner had this discussion come to a close than various initiatives, as well as full curricula – geared toward the development of CT in education – emerged. An important factor, which – at policy level – highlights this focus on CT was that, since

2009, the National Science Foundations' (NSF) Computer Information, Science and Engineering (CISE) Research Infrastructure, recognizing the importance of CT and the essential role it can play in education and society, decided that CT should be a required element in all proposals submitted for sponsorship under the program CISE Pathways to Revitalized Undergraduate Computing Education (CPATH). The early attempts of implementing CT integration in education have already been providing feedback information thus making a critical review of the concept, one that takes into consideration the sporadic theoretical criticism (Denning 2009; Easterbrook 2014) seem well-timed and fruitful. The present work constitutes an attempt to contribute to the clarification of the CT term within the context of K-12 education. A central issue of this study/chapter is to outline the conceptualization of CT as it is explicitly or implicitly presented through the curricula of K-12 education. Furthermore, the paper takes a cautious look at the expected conceptualization of CT by the teachers. The instructional events and learning activities that a teacher employs in his/her everyday practice and which formulate the classroom curriculum constitute a projection of his/her interpretation and pedagogical translation of the programmatic curriculum (Deng 2009; Fesakis and Serafeim 2009). Teachers' preparation is very significant for the development of the CT full pedagogical potential (Yadav et al. 2011, 2014), or, as Cuny (2011) points out, the biggest challenge does not lie in the curriculum but in the effective teacher preparation and support. In this paper, we attempt to explore the teachers' CT content theory using content analysis on the learning activities designs which constitute the product of the teachers' pedagogical translation of the programmatic curriculum. Finally, the study tests Denning's (2009) claim that CT is not a unique and distinctive characteristic of CS, neither adequate to express the whole field of CS. The rest of the chapter will unfold in four sections as follows: Section 10.2 presents the historical evolution of the concept along with some criticism it has been receiving. The research rationale and methodology are discussed in Section 10.3, while the research findings are presented and initially interpreted in Section 10.4. The paper closes with summarizing and concluding remarks in Section 10.5.

10.2 Theoretical Framework

10.2.1 Historical Evolution of the CT Concept

Jeannette Wing (2006) recoined the term CT in her homonymous article where she formulated the following definition:

> Computational thinking involves solving problems, designing systems, and understanding human behavior, by drawing on the concepts fundamental to computer science. (Wing 2006)

Wing's arguments about the importance of CT for general education are developed along two axes: (1) On the one hand, CT constitutes a set of skills, techniques, methods, and attitudes of reaching solutions to a wide range of problems.

Abstraction and analysis constitute the prominent tools in addressing the complexity of the problems. CT appears to equip us with methods and models that make the representation and solution of complex problems feasible, a perspective that would otherwise be impossible. (2) On the other hand, CT confronts us with the challenge of machine intelligence asking for what people can do better than machines and what machines are better than humans at. Attempts to answer the question of what types of problems are solvable by a computer promote science and technology toward exploring new frontiers.

As Grover and Pea (2013) remark, long before Wing's article rekindled the idea and triggered actions that would back up its implementation, the value of learning computer programming as a general competence had been highlighted by two other computer science pioneers, Alan Perlis and Seymour Papert. More specifically, in 1962, Alan Perlis put forth the idea that all college students should understand computation theory and learn computer programming as a medium to study a wide variety of other topics, more effectively (Guzdial 2008). He thus proposed the teaching of computer programming to all students. A few years later, in 1967, according to the LOGO Foundation, Papert (1991) created LOGO programming language specially designed as an alternative approach to mathematics teaching and algorithmic thinking. In the years that followed, Papert made popular the idea of programming as a means of developing algorithmic thinking, promoting his robotic turtle and LOGO programming language in the context of K-12. Computer science has for the first time become popular and in fact, at a time when the widespread and the possibilities offered by technology nowadays, seemed like science fiction scenarios (personal computers appeared in the late 1970s and spread during the 1980s). Papert was also the one who first coined the term "computational thinking" for education, in 1996, within his work with LOGO programming in MIT (Papert 1996). Furthermore, in 2000, Andrea diSessa introduces the definition of computational literacy, to describe how computers can turn into powerful catalysts for change in education and how everyone – apart from consumer – can become a creator of dynamic and interactive representations as cognitive tools (diSessa 2000). DiSessa supports the value of using computers as a medium for exploring other disciplines (Abelson and diSessa 1980) so the two terms, "computational literacy" and "computational thinking," although different, tend to be used as synonyms (Grover and Pea 2013).

More recently, Isbell and Stein argued that the curricula of computing courses should be revisited to include core competences in modeling, scales and limits, simulation, abstraction, automation, and interpretation of data, also known as the computationalist mind-set (Isbell et al. 2009). The term has later been expanded to include reasoning at multiple levels of abstraction, the use of mathematics to develop algorithms, and understanding the dimension of scale.

Wing's action toward the direction of integrating CT in basic education brought the scientific community up to serious questions. Questions related to which exact aspects of CS could contribute to the solution of problems across the spectrum of human research, as well as which people have the appropriate training to effectively support such a venture (Barr and Stephenson 2011). Furthermore, Guzdial (2008) raises the issue of what non-computing students understand about computing, as well as issues related to the tools and the structure/organization of the courses so that they become more challenging and easily accessible to all students. To examine the "nature of computational thinking" and what it implies in the academic and educational field, along with the pedagogical aspects of CT (National Research Council 2010), the National Academy of Sciences organized a first workshop, the closing of which left many questions unanswered. Thus, a second workshop followed, which leads to the revision of the original definition of CT by Wing herself and to the formulation, hence, of a second definition, according to which:

> Computational thinking is the thought processes involved in formulating problems and their solutions so that the solutions are represented in a form that can be effectively carried out by an information-processing agent. (Wing 2011)

At the same time, CSTA in collaboration with ISTE organized workshops aiming at creating an "operational definition," that is, a list of the key concepts and skills related to CT, along with examples of how these could be incorporated into different subjects. The two organizations mentioned above display extensive experience in the development of standards, teaching material, professional development programs for teachers, and educational policies counseling internationally. The interested reader may obtain a brief description of the results of this effort at Barr and Stephenson (2011), while the key points of the operational definition – as described in a collation of CT resources for teachers (ISTE and CSTA 2011) – present CT as a problem-solving process that includes (but is not limited to) the following characteristics:

- Formulating problems in a way that enables people to use a computer and other tools to help solve them
- Organizing and analyzing data logically
- Representing data through abstractions, such as models and simulations
- Automating solutions through algorithmic thinking (a series of ordered steps)
- Identifying, analyzing, and implementing possible solutions with the goal of achieving the most efficient and effective combination of steps and resources
- Generalizing and transferring this problem-solving process to a wide variety of problems

The "CS Principles" course designed by the College Board and NSF constitutes another interesting approach since it focuses on CT practices and is based on the seven "big ideas" of computer science (The College Board 2010):

1. Computing is a creative activity.
2. Abstraction reduces information and details to facilitate focus on relevant concepts.
3. Data and information facilitate the creation of knowledge.
4. Algorithms are used to develop and express solutions to computational problems.
5. Programming enables problem-solving, human expression, and creation of knowledge.
6. The internet pervades modern computing.
7. Computing introduces a global impact.

From the United Kingdom and the Royal Society, with support from the Royal Academy of Engineering, finally, comes the following definition:

> Computational thinking is the process of recognizing aspects of computation in the world that surrounds us and applying tools and techniques from Computer Science to understand and reason about both natural and artificial systems and processes. (Royal Society 2012)

Royal Academy's definition brings CT closer to Denning's (2009) CS definition and its application for the study of the natural and artificial world.

The evolution in the definition of CT shows that the concept is still under discussion. Despite its evolution, CT remains rather blurred in the broad education community. More modern definitions tend to relate CT with the application of computation as a problem-solving and an epistemological means in several disciplines. A useful outcome from the above review of the definitions would be the union of the proposed sets of CT dimensions. Hence, the set of dimensions, which the various definitions seem to highlight, include:

- Creative problem-solving
- Algorithmic approach to problem-solving
- Problem solution transfer
- Logical reasoning
- Abstraction
- Generalization
- Representation and organization of data
- Systemic thinking
- Evaluation
- Social impact of computation

A reasonable question that arises, therefore, is whether CT is an umbrella concept describing the loose assembly of pedagogically significant dimensions of CS, a practice of CS, or a distinctive entity concerning a specific kind of thinking in the context of computation. In the next section, we'll try to shed some light on the CT concept commenting on the criticism against it. This will also help justify the analysis scheme that will be used in the research part of the paper.

10.2.2 Criticism of the CT Concept as a Vehicle to Reclaim CS Role in Education

There is rather limited, however serious, criticism expressed against the CT concept, to date. In his article, Easterbrook (2014) mentions a few critiques that have been made with respect to the vagueness of the term. He also claims that, because computational thinkers are oriented toward problems that can be tackled with computers, problems such as ethical dilemmas, value judgments, and societal change could be ignored or downgraded to simpler, computationally approachable versions. Especially in the case of sustainability practice, he proposes systems thinking as the necessary bridge for the utilization of CT, as it provides a domain ontology for reasoning and thinking critically about sustainability. Furthermore, Denning (2009) expressed deep concerns about the risk of degrading CS to one of its key practices (CT) and of replacing the usual misconception that "CS=Computer Programming" with the new inadequate equation, that is *"CS=CT"*. As Denning notes, CS is not just a tool in the hands of scientists but a way to do science. Computers are a tool, not the object of study, and computation is an essential method of doing science since it concerns the understanding and control of information processes (artificial and natural). Advances in CS, therefore, allow scientists to envision new problem-solving strategies and experiment with new solutions, both in the real and virtual worlds. The work of organizations like Partnership for Advanced Computing in Europe (http://www.prace-ri.eu/), which awards scientists who advance science using computing, is illustrative for the significance of high-performance computing in scientific and technology advancement. As Denning (2009) highlights, this rationale stems as early as 1975 when Physics Nobel Laureate Ken Wilson promoted the idea that simulation and computation are ways to do science, previously not available. In collaboration with other scientists from several fields, Ken Wilson advocated the creation of supercomputing centers worldwide, aiming to confront grand scientific challenges by computation. One could claim thus that, at the core of Denning's view lies the belief that, while both, computation and CT are essential to the advancement of science, computation, as a concept, is more fundamental than CT. Denning's point of view that CS provides the way to "the study of information processes, natural and artificial" endorses the value of CS in general education since, in a world where CS is ubiquitous, those who possess computing skills are expected to be better problem-solvers and more adaptable to the modern socioeconomic environment and, consequently, more competent at dealing with great challenges. Furthermore, since computation is considered a new inevitable way to do science, supplemental to theory and experiment, the development of CT emerges as a strategic advantage for future scientific progress in any society. Through this perspective, Denning promotes "The Great Principles Framework" as a complete approach to the integration of CS Education in K-12 (Denning and Martell 2015). According to Denning (2009), "The Great Principles Framework" is a way to express CS as a field of science based on deep and enduring fundamental principles. The framework has two parts: core principles and core practices. The core principles are grouped into seven categories:

- Computation (meaning and limits of computation)
- Communication (reliable data transmission)
- Coordination (cooperation among networked entities)
- Recollection (storage and retrieval of information)
- Automation (meaning and limits of automation)
- Evaluation (performance prediction and capacity planning)
- Design (building reliable software systems)

The core practices are areas of skill and ability in which computing people can display various levels of performance such as beginner, competent, and expert.

There are four core practices: *(a) Programming, (b) Engineering of systems, (c) Modeling, and (d) Applying*. Denning considers that CT can be seen either as a style of thought that runs through the practices or as a fifth practice. It is the ability to interpret the world as algorithmically controlled conversions of inputs to outputs.

10.3 Research Framework

As can be deducted from the discussion on CT meaning, the essence of CT lies in the ability of someone to approach problem-solving the way computer scientists do. The above view does not attribute special innate talents to computer scientists but rather refers to acquired skills and conceptual models, reflecting specific educational background, as well as experience gained by solving complex problems with computers. While anyone – with proper training – can use software applications in the context of his/her work or for entertainment purposes, the metaphors and ways of thinking of CS must be taught explicitly (Guzdial 2008). For a systemic and sustainable integration of CT in formal education, resources that would first persuade educational policy-makers and, later on, allow teachers to integrate CT in the realm of their knowledge both in principle and classroom are required (Barr and Stephenson 2011). Many initiatives have produced such resources and have been progressing to the integration of CS in education so far, yet CT and its relation to CS remain rather unclear and controversial.

10.3.1 Research Rationale

To contribute to the disambiguation of CT and map the gap between CS and CT educational potential, the present work explores the conceptual interpretation of CT in widely known K-12 curricula. More specifically, the paper explores the expected understanding of CT by the teachers as this is depicted in the pedagogical translation (Deng 2009) of the curricula into learning activities. To obtain this, directed qualitative content analysis (Seker and Guney 2012) on sets of learning activities of selected curricula-initiatives will be conducted, using a complex coding scheme, based on current theoretical conceptions of CT, the great principles of computing framework (Denning 2009), and selected educational dimensions. The aims of the analysis include (a) to promote CT conceptual disambiguation, (b) to deductively confirm that CT concerns the application of CS to other school subjects and gain a better understanding of the nature of this relation, and (c) to search for evidence supporting Denning's (2009) view according to which, while CT is one of the key practices of CS, it is not adequate to cover all the principles and practices of the discipline.

In the present work, the authors adopt Deng's (2009) view that a school subject is introduced at schools as a distinct representation of a content embodied in curriculum documents or materials (e.g., curriculum frameworks, syllabuses, textbooks, digital repositories of learning objects). Curriculum developers – often implicitly – apply a (subject dependent) *theory of content*, that is, a way of deliberately selecting, arranging, transforming, and framing the content so that it serves the educational purposes of a school subject. The theory of content is also applied in the process of selecting specific teaching, learning, and assessment methods for the corresponding school subject. Consequently, for an efficient teaching and the development of the full educational potential of a school subject, apart from familiarity with the content per se (content knowledge), teachers are required to have knowledge and understanding of the corresponding inherent theory of content as well as of the related curricular, learning and instructional issues (pedagogical content knowledge) (Shulman 1986). The formation of CT as a school subject, therefore, imposes the use of a theory of content in the relevant curriculum development process. The exploration of both CT theory of content and its understanding by teachers arises as a key research issue. Curriculum theory discerns three levels of curriculum-making: the *institutional (or abstract/ideal)*, the *programmatic (or analytic/technical)*, and the *classroom (or enacted)* (Deng 2009).

The *institutional curriculum* expresses the desired, anticipated, long-term outcomes of the school subject in social, cultural, and national levels (Doyle 1992a, b). In the case of CT, the institutional curriculum can be inferred from the general philosophy sections of the CT national curricula or the model curricula and the curriculum proposals of the various associations and initiatives. Any institutional CT curriculum is, more or less, in alignment with Wing's arguments on the value of the development of CT in general education.

The *programmatic curriculum* describes specific content (topics, concepts, problems, case studies, etc.) which have been selected, organized, and framed in a way that both meet the institutional curriculum expectations and are also consistent with the modern pedagogical approaches (e.g., inquiry, collaborative, interdisciplinary problem-based learning), as well as with the education research findings (didactics of the specific subject matter). In order to cope with this complex challenge, programmatic curriculum designers have to formulate a content theory. The programmatic curriculum constitutes a set of technical and analytical documents for use in schools. Thus, programmatic curricula about CT have the form of curriculum frameworks accompanied by guidelines for instruction and assessment. A selection of them is described in Sect. 10.3.2.1.

Finally, the *classroom curriculum* is formed by the transformation of the programmatic curriculum into instructional events and learning activities, at school. Classroom curriculum is, therefore, mainly defined by teachers who are expected to have a good comprehension of the content of the programmatic curriculum in order to be able to interpret it and translate it into instructional and learning activities, taking into consideration (a) the directions of the institutional curriculum, (b) their students' existing knowledge and experiences, and (c) their school context. Thus, CT classroom curriculum concerns the development of teaching and learning designs (e.g., scenarios, scripts, lesson plans) and materials, aiming at engaging students in the construction of their own knowledge and competences related to CT, as well as at the implementation of these designs in the classroom. The pedagogical translation concerns the development of educational experiences by selecting key issues (e.g., concepts, problems), their appropriate pedagogical representations, instructional and assessment methods, and resources (e.g., class exercises, creative examples, careful explanations). Obviously, the classroom level curriculum is the key to a successful implementation of any institutional and programmatic CT curriculum, while the educational potential of CT is practically determined by the classroom curriculum.

According to the above analysis, in order to better understand the nature of CT as a school subject, it is crucial to explore the curriculum content with respect to the three levels of curriculum-making, described above. Moreover, by analyzing the learning activities included in a classroom level curriculum, one may better conceive the teachers' understanding of CT institutional and programmatic curricula as well as the corresponding theory of content. Thus, in this paper, by analyzing a number of learning activities that accompany programmatic curricula in several curriculum proposals, the authors seek to explore the contemporary anticipated understanding of the CT curriculum content theory by the teachers and how it relates to the programmatic and the institutional curriculum conceptions of CT. Toward this direction, content analysis to CT learning activity sets, using as coding categories the dimensions of CT that are described in the programmatic and institutional curricula, will be applied. Furthermore, content categories in relation to pedagogy dimensions and CT concept disambiguation – as it will be analyzed in the following section – have been defined. More specifically, the analysis can potentially provide answers to the following questions:

RQ1. *Are all the theoretical dimensions of CT represented in the classroom curriculum?*

RQ2. *Which other school subjects are utilized for the development of CT in schools?*

RQ3. *Which teaching/learning methods and resources are proposed for the development of CT?*

RQ4. *Are there any dimensions (practices and key concepts) of the CS great principles framework proposed by Denning that are not covered by the CT classroom curricula (in other words, is the eq. CT = CS valid)?*

It is worth pointing out that, since the learning activities that will be analyzed within the framework of this work are not necessarily products of work of ordinary teachers, the relevant content analysis is expected to reveal a rather optimistic aspect of the teachers' understanding of CT content.

10.3.2 Research Methodology

Qualitative content analysis (Krippendorff 1980; Neundorf 2002) was employed in this research to explore the CT classroom curriculum nature as this is projected by the learning activities designs proposed in various curricula and initiatives. More specifically, a directed (deductive) content analysis (Elo and Kyngas 2007) with a predefined coding scheme, which was developed through the relevant theoretical analysis (the review of the CT conceptions), was used. The directed content analysis uses an existing theoretical framework to determine the initial coding scheme, while the results of the analysis validate or extend this theoretical framework (Hsieh and Shannon 2005 cited in Seker and Guney 2012). Content analysis requires the selection of the unit of analysis to start. In this study, the unit of analysis is the learning activity design (or lesson plan, or learning scenario, or learning script) that explicitly addresses CT for the K-12 grades.

10.3.2.1 Building the Collection of the Learning Activities Designs

To build the collection of learning activity designs, we first collected a set of curricula and initiatives concerning CT in K-12 education using (*1*) information derived from the literature review and (*2*) known electronic search engines, to look for initiatives offering teaching/learning resources and curricula concerning the integration of CT in the educational process. The results included CS curricula and initiatives which made clear reference to the term and relevant concepts, as well as activities including core dimensions of CT, without referring directly to the term. The sources were filtered based on their quality, intentional focus on CT, and their role as models for many other less known initiatives. The final set of curricula and initiatives as well as the numbers of the corresponding learning designs/scenarios that were selected for analysis are summarized in Table 10.1.

The sources presented in Table 10.1 contain many learning scenarios from which 58 – the ones that more strongly responded to the criteria defined above – were selected for further study and content analysis. More analytically:

- *Teaching London Computing* (https://teachinglondoncomputing.org/). TLC program is funded by the Mayor of London and the Department of Education to provide free classroom resources, workshops, and continuing professional

Table 10.1 CT learning design sources included in the analysis

Source	Selected learning designs
Teaching London computing	21
Barefoot program	14
CS unplugged	14
Computational thinking toolkit by ISTE and CSTA	9

development courses to support computing teachers in London. Studying the links between computing and a variety of other subjects, as well as how CT techniques apply across the curriculum, a partnership between the Queen Mary University of London and King's College London has developed a series of activities and educational material. The goal, among others, was to engage computer science and a series of other topics, like English, math, biology, physics, history, philosophy, language, music, design and technology, art, and dance, in an interdisciplinary way. The resources provided include CS and interdisciplinary learning activities for developing CT or even a series of fun activities and booklets based around puzzles or magic tricks that teach computing topics and CT. The resource hub of TLC provides a large number of scenarios in total, 21 of which were selected for analysis based on the activities' relevance with CT.

- *Barefoot Program* (http://barefootcas.org.uk/). The Barefoot Program, supported by the Department for Education, Computing at School (CAS) community, and British Telecommunications (BT), provides many high-quality resources aligned to the national curriculum, to support primary teachers of all UK nations to teach CS. For the Barefoot Program, although CT has wide applications across other disciplines and clearly helps in problem-solving procedures, it's obviously apparent and probably most effectively learned, through the rigorous, creative processes of writing code. From the 39 available resources, we selected 14 scenarios for the purposes of our study. The activities were selected after filtering the available resources to result in those related to CT topics.
- *CS Unplugged* (http://csunplugged.org/). CS Unplugged is a project carried out by CS Education Research Group at the University of Canterbury to provide a collection of learning activities (games, puzzles, etc.) that teach CS without the required use of a computer. The material is shared under a Creative Commons license and has been used around the world for over 20 years. The activities are aimed at the 5–12-year-old age group and mainly concern data representation, algorithms, procedures, intractability, cryptography, and the human face of computing, most of them embedding many concepts and approaches of CT. The main goal is that young students will experience the kind of questions and challenges computer scientists experience, without having to learn to program first. After the adoption of computing and CT in many classrooms, the collection is nowadays widely used. From about 30 available scenarios, 14 were selected for our study because they explicitly mention CT or some of the CT concepts.
- *Computational Thinking Toolkit* was created by ISTE and CSTA, with the support of NSF, in order to prepare and conduct a CT approach for K-12 students and teachers (Sykora 2014). The toolkit describes CT as a cross curricular initiative and a problem-solving skill. To guide teachers by example and support them in their attempts to teach CT, the toolkit introduces full learning experiences as well as CT scenarios, cultivating certain characteristics, dispositions ans attitudes as essential dimensions of CT. All the 9 CT learning scenarios of the toolkit concerning K-12 education were included in our analysis.

The detailed list of the selected learning scenarios for analysis is accessible at https://goo.gl/cG6LE1.

10.3.2.2 Defining the Coding Scheme

In a directed content analysis, the coding process requires a predetermined categorization matrix (coding scheme). In this study the coding scheme was developed with respect to (a) the union of CT dimensions that are proposed by the various initiatives (Table 10.2), (b) the key concepts and practices of CS as Denning proposed them, (c) pedagogy features (learning, teaching methods, social mode, unplugged or on the computer), and (d) interdisciplinary features (related subject matters other than CS). In this section, the dimensions of the coding scheme as well as the categories of each dimension are described. To serve the analysis purposes, we used a number of dimensions, only a subset of which, actually the ones mentioned in the research findings section, are presented below.

a. CT Dimensions

CT is analyzed to several dimensions by the several learning scenarios providers. The authors tried to find correspondences and define the union of these dimension

Table 10.2 CT dimensions across the initiatives and unified

TLC	Barefoot	ISTE-CSTA	Unified dimensions
Algorithmic thinking	Algorithms	Algorithms and procedures	Algorithmic thinking – AL
Abstraction	Abstraction	Abstraction	Abstraction – AB
Evaluation	Evaluation	Problem decomposition	Generalization – GE
Decomposition	Decomposition	Data collection	Logical reasoning – LR
Pattern matching	Patterns	Data analysis	Pattern matching – PM
Modeling	Logic	Data representation	Problem decomposition – PD
Sequencing		Automation	Problem translation – PT
Logical reasoning		Simulation	Evaluation – EV
Generalization			Representation – RE
Translating problems			Data collection – DC
Understanding people			Data representation – DR
Testing			Data analysis – DA
			Modeling – MO
			Simulation – SIM
			Automation – AUT
			Sequencing – SE
			Testing – TE
			Understanding people – UP

schemata, in order to use them as coding categories. The exploration of the distribution of scenarios throughout these dimensions and the relationships among them will provide crucial information for the understanding of CT at the classroom curriculum level. Table 10.2 lists the dimensions per initiative and the unified set used for the scenarios' analysis. The unified list is the union of the partial lists. CT dimensions are not explicitly mentioned by CS Unplugged project, so there is not a corresponding column in Table 10.2. A careful observation of Table 10.2 reveals that systems thinking is not considered as a CT dimension in the analyzed initiatives despite the fact that it is mentioned as such in other sources (Easterbrook 2014). This is a programmatic curriculum level omission since systems thinking is a key contribution of CS to problem-solving and the study of complex dynamic systems.

b. CT Approaches

Barefoot names a set of CT approaches and ISTE-CSTA a set of dispositions that correspondingly accompany and support the CT development (Table 10.3). CT approaches were used as a coding category in our analysis in order to explore their presence in the classroom curricula. The ISTE-CSTA dispositions are in partial correspondence to CT approaches so we consider the later as a means to help students develop these dispositions.

c. Curriculum Subject

As CT is supposed to concern the application of CS in other disciplines, the relevant scenarios are reasonably expected to be interdisciplinary. The authors used the scenarios' curriculum subject as another coding dimension to confirm the above CT conception. Furthermore, the distribution of the scenarios to the various subjects could reveal subjects that are not represented at all or others that are more common and popular. Finally, the inspection of the potential problems and concepts of each subject used in the scenarios will result in information concerning the extent of the designers' understanding of CT. The category elements of this coding dimension include school subjects (e.g., mathematics, science, history).

d. Grade

The grades proposed as target groups by each scenario were used as a coding category with the elements PK, G1, G2, . . ., G12. The distribution of scenarios to the grades and more specifically the inspection of the available number of scenarios for

Table 10.3 The CT approaches category elements

CT approach	ISTE-CSTA disposition
Tinkering	
Creating	Tolerance for ambiguity, ability to deal with open-ended problems
Debugging	Confidence in dealing with complexity
Persevering	Persistence in working with difficult problems
Collaborating	The ability to communicate and work with others to achieve a common goal or solution

each CT dimension by grade will provide useful information about which dimensions the learning designers find more appropriate and/or easier to cultivate for each grade.

e. CT tools

CT tools category includes the titles of software and/or hardware that are used/ proposed in the scenarios. This category will permit us to explore the kind of learning technologies and CS tools that are used for the cultivation of CT in the scenarios. The categories in this dimension arose during the analysis and contain values such as Scratch, Unplugged, data logging, robotics kit, automation kit, etc.

f. Learning Approach

The learning and instructional approaches employed in the scenarios can provide information about the pedagogical beliefs of the designers. The categories which arose in this dimension are inquiry learning, problem-based learning, project-based learning, game-based learning, role playing, etc.

g. Great Principles of Computing, Concepts, and Practices

This category consists of the great principles of computing concepts and practices, as those are described by Denning and Martell (2015). This coding permits us to explore the validity of the eq. CT = CS. The categories of the specific dimension match the principles and practices of the framework mentioned in Sect. 10.2.2. Computational thinking is considered a key practice and is not used as a coding category, since it will undoubtedly be contained in a CT learning scenario.

10.4 Findings

Each of the learning designs included in the collection under analysis has been categorized in terms of the coding scheme by two of the authors, who are experienced K-12 computer science teachers with undergraduate and postgraduate studies in CS. After the coding, the content analysis data exploration was implemented. The findings related to each coding category are presented in the following sections.

10.4.1 Findings Regarding the Age of the Proposed Target Groups

The study of the analyzed learning designs' distribution according to the grades is of great interest for two reasons. The first reason concerns the availability of scenarios for all dimensions and ages, while the second one relates to the indication of the developmental appropriateness of each CT dimension, for each age group (e.g.,

Table 10.4 Frequency distribution of ages which the proposed scenarios address

	TLC	Barefoot	ISTE-CSTA	CS Unplugged	SUM
PK	1	7	2	0	10
G1	1	6	2	0	9
G2	1	9	2	4	16
G3	1	5	0	6	12
G4	1	7	1	12	21
G5	1	6	1	13	21
G6	1	6	1	13	21
G7	6	0	2	14	22
G8	16	0	2	14	32
G9	16	0	3	14	33
G10	17	0	3	14	34
G11	20	0	3	14	37
G12	20	0	3	14	37

which age should the cultivation of abstract thinking start at?). To answer these questions, the contingency table between ages/grades and sources of scenarios (Table 10.4), as well as the contingency table between CT dimensions and ages/grades (Table 10.5), were studied. Following is an analysis for each of the tables.

To calculate the frequencies in Table 10.4, we have included each scenario in all the grades it corresponds to, according to its designers. For example, a scenario referring to age 7+ appears as an option in the G7–G12 rank. Consequently, we can interpret the content of a cell as the number of the theoretically alternative scenarios that each source provides for a specific grade. The last column (SUM) refers to the overall scenarios that are available per grade and typically is not a part of the contingency table but facilitates the analysis. Considering Table 10.4, it becomes obvious that:

a. Each learning design source addresses a different basic age group. This is what the editors of the scenarios claim, nevertheless. This is also statistically confirmed since the Fischer's exact test in Table 10.1 ($p < 0{,}001$) shows a correlation between the variables "groups" and "scenarios sources." More specifically, the selected activities from the CS Unplugged set concern classes G4+, Barefoot offers activities involving lower grades (PK-G6), while TLC provides choices for all classes, but their number becomes remarkably larger from G7 and beyond. The situation is different in the ISTE-CSTA proposals where there is an even distribution of the proposed activities across all classes, albeit the number of options is limited.
b. Although CT seems to involve all ages, a careful observation of the SUM column reveals a significant reduction of options in younger groups/lower grades. One reason for this may be that children need to advance in the understanding of sciences independently before they are ready to integrate CS as an epistemological tool in their study. It is also likely that developing scenarios concerning the various CT dimensions for younger ages constitutes a far more demanding task

and may therefore require a systematic study of the relevant didactics field. The ISTE-CSTA example suggests that, although it may require a more laborious and cautious designing process, bringing CT teaching to younger ages may not be an impossible venture. This claim is consistent with the concerns arisen by Armoni (2012) and efforts highlighting the existence of relevant opportunities (Fessakis et al. 2013).

Another question that would be interesting to explore is the existence of a correlation between the CT dimensions and the students' age of introduction to the sample scenarios. To approach this question, we study the contingency table between the dimensions of CT and grades, related to the scenarios analyzed. Each entry in Table 10.5 represents the frequency of the CT dimension of the

Table 10.5 Correlation between the dimensions of CT and ages, related to the scenarios analyzed

	PK	G1	G2	G3	G4	G5	G6	G7	G8	G9	G10	G11	G12
Algorithmic thinking – AL	6	6	12	9	13	11	15	15	22	22	21	24	24
Abstraction – AB	5	4	6	3	9	8	10	8	13	12	13	15	15
Generalization – GE	3	2	4	4	10	11	13	11	17	15	15	15	15
Logical reasoning – LR	5	5	8	6	12	13	16	13	16	14	13	14	14
Pattern matching – PM	2	2	3	2	5	6	7	5	8	8	8	8	8
Problem decomposition – PD	8	7	12	8	14	15	17	12	15	16	16	16	16
Problem translation – PT	0	0	1	1	5	5	6	6	6	6	7	8	8
Evaluation – EV	2	1	2	2	4	4	4	4	8	8	8	8	8
Representation – RE	3	2	6	6	11	13	14	13	15	15	15	15	15
Data collection – DC	1	1	1	0	0	0	0	0	0	1	1	1	1
Data representation – DR	2	2	5	5	12	12	15	15	18	18	19	20	20
Data analysis – DA	1	1	1	1	1	2	3	3	3	3	3	4	4
Modeling – MO	3	2	5	4	5	4	6	4	7	7	7	9	9
Sequencing – SE	0	0	0	0	0	0	0	0	2	3	3	3	3
Simulation – SIM	1	1	2	1	3	2	3	3	5	5	6	7	7
Automation – AUT	2	2	2	1	2	1	4	3	3	2	2	2	2
Testing – TE	2	2	3	1	0	1	1	1	3	3	3	3	3
Understanding people – UP	0	0	0	0	0	0	0	3	3	3	4	4	4

corresponding row for the grade/age of the corresponding column. So, for example, in the collection analyzed, there are six scenarios in which algorithmic thinking is cultivated (AL) in the first grade (G1), while the number of available scenarios for the cultivation of abstraction (AB) in grade 12 (G12) is 15.

An observation of Table 10.5 shows that not all dimensions of CT are evenly distributed across ages. In general, it seems that in most grades, algorithmic thinking prevails. Some dimensions are underrepresented, while others appear in excess, in scenarios for specific ages. For example, engaging with patterns in early grades is an activity of great importance in the didactics of mathematics (for the development of pre-algebraic concepts) as well as in general for the cultivation of the generalization and abstraction capacities and could also be carried out with the use of computers for the simultaneous development of CS concepts. Moreover, data collection and analysis, combined with CS, provide opportunities for the development of concepts from mathematics, statistics, social sciences, data science, and other fields. Surprisingly, there are a limited number of data collection scenarios, despite their apparent correlation with data analysis. Similarly, simulation scenarios could be exploited in early grades. The relatively small number of scenarios regarding these CT dimensions indicates that there is room for development in the collections and it contradicts programmatic curricula intentions. The lack of sequencing (SE) scenarios may indicate that this dimension should be merged with another, e.g., with algorithmic thinking (AL), or that it may be a synonym of some other CS practice, such as programming. Finally, the lack of scenarios involving the dimension understanding people (UP) before G6 indicates that further reflection on how artificial intelligence concepts could be introduced in early stages is needed, while this also constitutes a direction for educational research development. For the rest of the table (excluding zero cells), what generally applies is that while the age factor ascends, the number of available scenarios for each dimension increases, a fact that was expected after the uneven distribution in Table 10.4. The statistical investigation of the correlation of the rows and columns of Table 10.5 had a negative result. It therefore seems that, in the analyzed scenarios, CT dimensions have no general relationship to age. Fischer's exact test per cell in Table 10.5 detected as significant the deviations of the displayed frequencies from the theoretically expected ones, in cells with the value 0 (Table 10.5). Other, partial correlations, positive or negative, of some CT dimensions to specific age groups are not assessed as generalizable from the sample.

Summarizing the results of the analysis regarding the age dimension, it is noted that the imbalanced distribution of the CT learning designs across the different age groups is also reflected in the various dimensions of CT. Some dimensions, despite their obvious usefulness and the available themes, are not implemented by any scenario in some ages. There is, therefore, both the need and the room for developing scenarios that exploit dimensions such as data analysis, pattern matching, modeling, simulation, automation, and people understanding. The difficulties may have to do with how well-versed the designer is in the didactics of the subjects, for the specific ages, with his/her awareness of the content of other subjects, his/her interest, etc. An interdisciplinary approach to the issue – that is, the collaboration of educational design specialists from various subjects, CS included – would probably help.

10.4.2 Findings Regarding the Comprehension of CT Concept by the Scenario Designers

Table 10.6 shows the distribution of scenarios per CT dimension. The application of statistical tests (X^2 and Kolmogorov-Smirnov) produced the result that the distribution is normal with $\mu = 15.72$ and sigma $= 10.68$, Skewness (Pearson) $= 0.389$, and Kurtosis (Pearson) $= -1.2$.

This implies that most of the CT dimensions are exploited by scenarios, the number of which is smaller than the mean μ. Table 10.6 reveals that the dimensions of CT that more frequently appear in scenarios ($f > \mu$) are algorithmic thinking, problem decomposition, abstraction, data representation, logical reasoning, generalization, representation, and modeling. It seems that it may be easier to write scenarios for certain CT dimensions, either because the specific dimensions are more comprehensible to the designers or because their didactics is a more widely studied field. It also appears that there is plenty of room for scenario design in dimensions such as simulation (SIM), data analysis (DA), automation (AUT), and understanding people (UP). The CT dimensions that appear to be more popular in the sample scenarios indicate a conservative perception of the concept by the designers, as, through them, CT is primarily viewed as being identical to algorithmic thinking (AL) and relevant dimensions. This case will also be investigated through the search of correlations between the various CT dimensions. Interesting questions that arise at this point include the following: *Are there any correlations among CT dimensions? Are there any dimensions that frequently occur together while some other couples seem incompatible* (i.e., *the appearance of a dimension implies the*

Table 10.6 Distribution of scenarios per CT dimension

CT dimension	Total scenarios
Algorithmic thinking – AL	38
Problem decomposition – PD	29
Abstraction – AB	27
Data representation – DR	26
Logical reasoning – LR	25
Generalization – GE	24
Representation – RE	22
Modeling – MO	19
Simulation – SIM	13
Pattern matching – PM	12
Evaluation – EV	10
Problem translation – PT	8
Data analysis – DA	8
Automation – AUT	7
Testing – TE	6
Understanding people – UP	4
Sequencing – SE	3
Data collection – DC	2

lack of another)? Which are the groups of dimensions that appear together? These questions could shed some more light on the interpretation and understanding of CT by instructional designers, issue.

Table 10.7 summarizes the correlations among CT dimensions as those appear in the scenarios analyzed. The correlations among the dimensions were initially tested by the X^2 test, but Table 10.7 shows the statistically significant associations with their direction according to the Spearman rho coefficient. A positive correlation implies a frequent simultaneous occurrence of the respective dimensions in a scenario. Conversely, a negative correlation means that the presence of a dimension is an important factor to predict the absence of the other. Correlations between dimensions indicate conceptual affinity. In Table 10.7 the reader can see that algorithmic thinking (AL) appears independently of all others, which practically means that it cannot help us predict the occurrence of another dimension. Abstraction (AB) is negatively related to pattern matching (PM) and positively to generalization (GE), evaluation (EV), data analysis (DA), and modeling (MO). It thus appears that AB is most often associated with generalization and modeling activities, as well as with data analysis scenarios. The negative correlation with pattern matching is probably due to the lack of knowledge of the relationship which connects patterns firstly to generalization and eventually to abstraction. While it is reasonable that, in the context of CT, abstraction is approached through modeling and analysis, since this is the way that CS contributes to the cultivation of abstraction in other sciences, one should not miss the point that there are alternative paths in CS

Table 10.7 Table of the correlations among the CT dimensions

	Positive correlation (significance level a = 0.05)	Negative correlation (significance level a = 0.05)
AL	–	–
AB	GE, EV, DA, MO	PM
GE	AB, EV	–
LR	PD	–
PM	–	AB, MO
PD	LR	SIM, UP
PT	RE, DR	–
EV	AB, GE	–
RE	PT, DR	–
DC	DA, SE, AUT	–
DR	PT, RE, SIM	–
DA	AB, DC, SIM, AUT	–
MO	AB, SIM	PM
SE	DC	–
SIM	DR, DA, MO	–
AUT	DC, DA	–
TE	–	–
UP	–	PD

that lead to the development of abstraction. Pattern matching, process, and data structure abstraction (e.g., recursive computations and graphics) are a few of them. There is a correlation between generalization (GE) and both abstraction (AB) and evaluation (EV), while one would also expect to see a correlation with pattern matching (PM), etc. Logical reasoning (LR) is positively correlated with problem decomposition (PD), while one would also expect it to be associated with AL and DA, at least. Pattern matching is negatively associated only with abstraction (AB) and modeling (MO). In general, the correlations identified could be characterized reasonable but incomplete. For example, the correlation among automation (AUT), data collection (DC), and data analysis (DA) is reasonable and expected, but the lack of correlation between pattern matching (PM) and generalization (GE) or simulation (SIM) indicates that the collection scenarios are not fully covering the range of computational thinking scope. Thus, the teachers' and/or educational designers' understanding of CT is susceptible to significant improvement. Furthermore, some dimensions appear independent and somewhat distant and isolated from others (AL, TE, UP) and therefore need to be further studied to determine ways they can be exploited in combination. By the MCA analysis on the CT dimension table and selecting as axes the first two analysis factors that explain the 50.5% inertia of the observations cloud, we can graphically summarize the dimensions and scenarios in both dimensions, as shown in Chart 10.1.

In the first quadrant of Chart 10.1, it can be seen that the placement of AB-1 almost diagonally is equidistant from group G1 = (AB-1, DR-1, MO-1, SIM-1, SE-1) of the fourth quadrant as well as from G2 = (AB-1, LR-1, GE-1, EV-1, RE-1, PD-1, PT-1,) of the first quadrant. Group G3 = (PM-1, TE-1, AUT-0, SIM-0, DA-0, MO-0, AB-0) is also distinguished in the second and third quadrant. The isolated

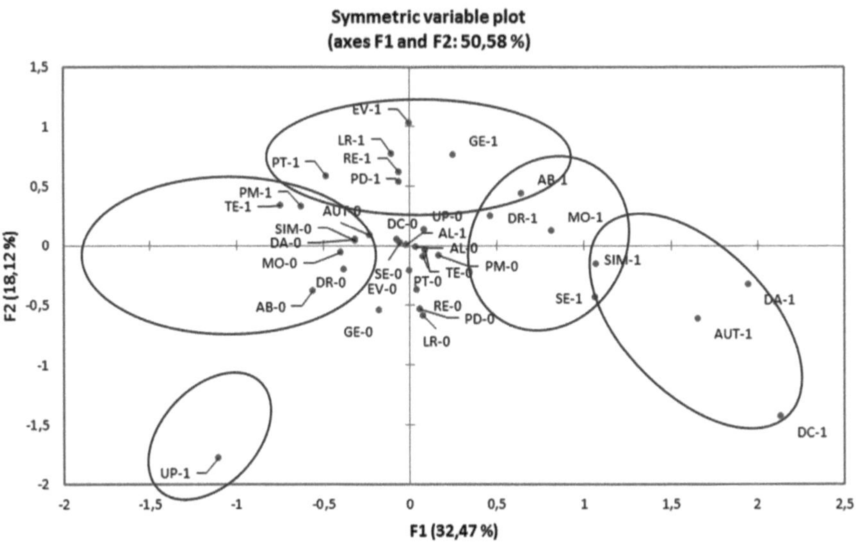

Chart 10.1 Correlations between CT dimensions and scenarios in two dimensions

category UP-1 can be seen in the third quarter. Present in the center of the axis system are the most independent and most frequently occurring dimensions, with the leading all AL, almost at the origin of the system. The groupings of the dimensions are consistent both with the correlations between the CT dimensions (Table 10.7) and the comments on the underrepresentation of particular combinations of dimensions, e.g., PM and GE and AB, DA and AL, etc., in the scenarios. Chart 10.2 illustrates the distribution of the scenarios in the categories.

The MCA analysis highlights scenarios groups with specific combinations of dimensions as well as the lack of scenarios for other combinations, as described above. It therefore appears that the current understanding of computational thinking by the scenario designers omits an important range of scenarios.

As far as the CT approaches are concerned, the distribution in Table 10.8 shows that the most popular approach is tinkering while debugging is quite distant from the rest. Such a distribution seems reasonable since construction is more difficult than tinkering, while persevering and commitment are prerequisites in CS.

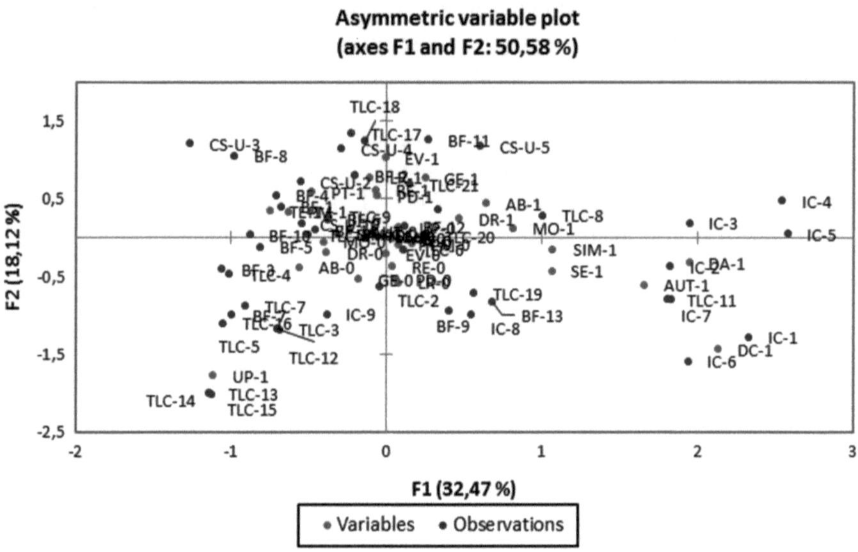

Chart 10.2 Illustration of the correlation between CT dimensions and scenarios, in two dimensions at MCA

Table 10.8 Distribution of frequencies of the scenarios' CT approaches

CT approaches	Scenarios
Tinkering	26
Persevering	19
Collaborating	15
Creating	11
Debugging	9

10.4.3 Findings Regarding the Conception of CT as Application of CS for Producing Knowledge in Other Disciplines

The first question to be tackled under this category is *(i) Which disciplines, others than CS, and at what frequency, were involved in the analyzed CT scenarios?* Table 10.9 shows that the majority of scenarios combine CT with mathematics, a significantly smaller number with arts and language, while dramatically fewer are the scenarios relating to other school subjects such as science (where CT could be exploited in a variety of scenarios for building simulations, for the collection and analysis of data deriving from experiments, for the visualization of concepts and data, etc.) and social science (where there are opportunities for exploiting CT in the collection and analysis of data, systems modeling, etc.).

Also missing are scenarios inspired by the digital humanities field. This imbalance demonstrates that CT cannot be automatically developed by instructional designers, but it needs more effort and collaboration of interdisciplinary teams in order for it to unfold in all its aspects. The interdisciplinary approach seems to be the key to the transformation of the CT programmatic curricula into the respective classroom ones. Current capacities of instructional designers in the application of CS in other disciplines might not suffice, as also do the CS skills themselves. The solution probably lies in the collaboration with instructional designers of other disciplines.

Table 10.9 Distribution of scenarios in the various disciplines

School subject	Scenarios
Mathematics	19
Arts	9
Language – English	8
Algebra	6
Life science – biology	5
Technology – design	5
Interdisciplinary	5
Geometry	4
Science	4
Social studies	4
Geography	3
History	3
Literature	2
Nutrition	1
Philosophy	1
Physical education	1
Dance	1

The second question that is addressed in this category is *(ii) Is there any correlation between CT dimensions and the disciplines involved in the scenarios?*

The attempt to detect correlations between the disciplines involved in the scenarios and the CT dimensions highlights some interesting facts about which dimensions are consistent with and exploited by certain school subjects in all the scenarios analyzed. To be more specific, the correlation analysis with the coefficient Spearman rho revealed as significant (a = 0.05) the correlations of Table 10.10.

Most of the results were expected and support the view that CT concerns the use of CS as an epistemological tool in other disciplines together with theory and experiment. Yet, there are correlations that are missing or seem strange. For example, correlations among science and the modeling and simulation dimensions, life science-biology and simulation as well as between technology-design and automation dimensions are absent, while the negative correlation of the pairs, mathematics and simulation and technology-design and algorithmic thinking, comes as a surprise.

Table 10.10 also shows that the common way to connect the various disciplines with the CT dimensions passes through the data collection and analysis paths, while other, more profound correlations may find it difficult to reach the level of scenarios.

Table 10.10 Correlations between school subjects and CT dimensions

	Correlation to CT dimension	
School subject	**Positive (+)**	**Negative (−)**
Mathematics	Logical reasoning – LR	Simulation – SIM
Geometry	Problem decomposition – PD Problem translation – PT	
Algebra	Pattern matching – PM	
Science	Data collection – DC Data analysis – DA	
Life science – biology	Data collection – DC Modeling – MO Automation – AUT	
Technology – design	Understanding people – UP	Algorithmic thinking – AL
Language – English	Problem decomposition – PD Representation – RE	
Literature	Data collection – DC	
Social studies	Data collection – DC Data representation – DR Data analysis – DA Automation – AUT	
History	Data representation – DR Sequencing – SE	
Interdisciplinary	Data collection – DC Data analysis – DA Simulation – SIM Automation – AUT	

10.4.4 Findings on the Potential Difference Between CT and Denning's Fundamental Concepts and Practices Framework

The coding in this dimension permits the testing of Dennings' thesis that CT is a key practice of CS and that CT curricula cannot cover all the CS principle categories. Table 10.11 shows the results of the analysis of the selected CT learning scenarios according to the principles framework.

As the data reveal, Denning is probably right since CT scenarios do not cover fundamental principle categories such as coordination (e.g., concurrent fork and join programming, synchronization among information-processing agents, etc.). The number of scenarios concerning communication (reliable transfer of data among systems) is rather small, as well as the number of scenarios concerning the recollection and evaluation (time and memory complexity) issues that constitute key CS problems. The frequencies of CS practices in Table 10.11 are very small, but this fact is rather expected for CT scenarios in K-12 since the real CS practices are relevant to CS majors.

10.4.5 Findings on Other Educational Parameters

Table 10.12 shows the tools that are utilized in the CT learning scenarios and the corresponding frequencies in a number of scenarios.

Strikingly, most scenarios do not require any kind of technology for their implementation (Unplugged) even if they do not belong to the CS Unplugged collection. As far as the rest of the scenarios are concerned, a number of them have been particularly created for implementation with the Scratch educational programming environment, some others for use with any programming language

Table 10.11 Covering of the principles and practices of computing framework

Principle categories	Scenarios
Computation	36
Communication	1
Coordination	0
Recollection	2
Automation	9
Evaluation	5
Practices	**Scenarios**
Design	0
Programming	1
Engineering of systems	0
Modeling	1
Applying	0

Table 10.12 Distribution of the frequencies of the digital tools in the CT scenarios

CT tools	Scenarios
Unplugged	42
Scratch	6
Any language	4
Simulation	3
Robotics	2
WWW browser	2
Spreadsheet	1

Table 10.13 Frequency distribution of the learning approaches in the CT scenarios

Learning approach	Scenarios
Problem-based learning	25
Project-based learning	17
Game-based learning	15
Inquiry learning	11
Role playing	5

and only two for use with educational robotics. Contrary to the authors' expectations, the table reveals a complete absence of scenarios that exploit LOGO language and educational data logging and automation devices or alternative robotics and automation platforms, such as Arduino or Raspberry PI. This specific finding indicates that CT is being treated as the theoretical aspect of CS, while it also shows a tendency to distinguish CS from computers. CT, however, requires the use of CS in other disciplines, in an authentic way. Consequently, in the context of observations and experiments, for example, one would expect to come across science scenarios that use sensors and computers to collect and process data from the natural environment. Or, even scenarios on how sensors work and how the digitization of signals is performed, scenarios on the use of meteorological stations, ultrasonic devices, three-dimensional printers, etc. The projection of the scenarios on the basis of the tools they exploit suggests directions toward which the development of CT scenarios can be further extended.

As far as the learning approaches employed in the scenarios is concerned (Table 10.13), they are consistent with the modern views of teaching. They could, however, be richer or more specialized and include more approaches such as learning by design, learning by making, and collaborative learning.

10.5 Summary, Discussion, and Conclusion

The current need for disambiguation of the CT concept has triggered and subsequently fueled an intensive discussion about the role of CS in general education. The efforts to clarify the meaning of CT resulted in the advancement of the understanding of CS's significance as a core school subject. This concentration of interest, in combination with the inclusion of CS in the field of STEM education, makes the presence and the role of both CS and CT in general education increasingly important and urgent. The work presented in this chapter is an effort to contribute to the clarification of the CT concept and its relation to CS Education. More specifically, the chapter's structure launches with a review of the historical evolution of the CT concept and the related criticism formulated, and it is then followed by the exploration of the conceptual interpretation of CT in the learning activity designs of, widely known, K-12 CT curricula and initiatives. The deductive content analysis implemented for this purpose was based on a complex coding scheme which was developed after reviewing the current theoretical conceptions of CT.

The content analysis of the CT learning scenarios reveals interesting findings, concerning the pedagogical translation of CT into classroom curriculum (Deng 2009), some of which are summarized here.

In the programmatic level analysis, an omission concerning systems thinking as a CT dimension was detected, despite the fact that it constitutes an accepted field of CS application for interdisciplinary problem-solving and the study of complexity (Fessakis and Kirodimou 2016). As far as the *age dimension* is concerned, there is an imbalanced distribution of the available CT learning designs across the different age groups. In general, the number of the available CT learning scenarios increases with age. This relation is also reflected in the context of the CT dimensions. The lack of scenarios that are suitable for younger ages points the need for didactics research concerning CS and CT in first grades, in parallel with other school subjects' educational research, constituting an interdisciplinary approach to the issue of using CS as an epistemological tool.

Regarding the *comprehension of the CT* concept by the learning scenarios designers, the analysis concluded that some CT dimensions appear to be much more popular than others. This indicates a rather conservative perception of the CT concept by the designers, according to which, CT is primarily viewed as being identical to algorithmic thinking and some relevant dimensions. More scenarios are needed in various dimensions such as data analysis, pattern matching, modeling, simulation, automation, and people understanding. Additionally, the correlations among the CT dimensions identified (simultaneous presence of the dimensions in the scenarios) appear reasonable but incomplete. It therefore appears that the current understanding of CT by the scenario designers leaves an important range of scenarios out. Thus, the teachers' and/or educational designers' understanding of CT is susceptible to significant improvement.

A theoretical review of the CT concept reveals that current CT conceptions concern the use of CS as an epistemological tool for problem-solving in other scientific fields, alongside with theory and experiment. Thus, CT as a subject matter is expected to be developed using *interdisciplinary* learning scenarios combining CS with several other fields. The analysis of the sample scenarios shows that most of them combine CT with mathematics, a significantly smaller number involve arts and language, while dramatically fewer are the scenarios relating to other school subjects such as science and social science. This limitation supports the argument that in order for CT to unfold in all its aspects, CT instructional/learning designers need to collaborate in interdisciplinary teams. In addition, further research, focused on the exploration of CS impact in other disciplines, is necessary, in order to highlight the CT factor. The analysis of the connections of CT dimensions to the various school subjects (Table 10.10) reasonably reveals that a common way to connect the various disciplines with CT dimensions passes through the data collection and analysis paths. Table 10.10 also reveals some other profound links that need to be further developed since they can form the base of interesting and fruitful CT learning scenarios. Surprisingly, programmatic curricula do not include significant CT dimensions, such as systems thinking and digital citizenship, areas of significant CS applications that should affect general education.

Finally, the research findings support Denning's view that CT is rather a key practice of CS, since the sample scenarios content analysis revealed that fundamental CS principle categories such as coordination are not covered at all, while categories such as communication, recollection, and evaluation are sporadically covered.

Based on the findings of the content analysis, answers to the research questions could be formulated as follows:

RQ1. *Are all the theoretical dimensions of CT represented in the classroom curriculum?*

Excluding the systems thinking dimension that is omitted from the programmatic level of the analyzed curricula, the general answer is "Yes"; not all dimensions appear in equal frequency though. CT scenario designers mainly focus on algorithmic thinking. Many important CT dimensions such as pattern matching and automation are not covered to a satisfactory extend and certainly not in a balanced and spiral, across the grades, way.

RQ2. *Which other school subjects are utilized for the development of CT in schools?*

Several school subjects are involved, albeit mostly through data analysis. Some are utilized in an incomplete way, while mathematics is the most common subject. Concerning the rest, progress needs to be done and a more contemporary study of the links of CS with other subjects through the computational sciences is necessary. The digital humanities field lags far.

RQ3. *Which teaching/learning methods and resources are proposed for the development of CT?*

The employed teaching/learning methods are fairly progressive, yet the tools/resources study reveals a significant shortfall in educational key technologies, e.g., educational robotic kits, automation, specific programming languages, and mobile programming.

RQ4. *Are there any dimensions (practices and key concepts) of the CS great principles framework proposed by Denning that are not covered by the CT classroom curricula (in other words, is the eq. CT = CS valid)?*

The process revealed dimensions of the CS framework that are either not covered at all or are only partially covered by the scenarios analyzed. We could, thus, claim that the content analysis supports Denning's view that CS<>CT and that the role of CS in education is more profound than that of CT.

Taking into account the main findings of the content analysis, it becomes obvious that classroom translation of CT concept by the designers of the sample learning designs aligns to the institutional and programmatic conception. According to this conception, CT concerns the use of CS as part of the methodology of other disciplines, but it lacks many areas of application and is rather limited and unsystematic in others. It appears that, during the process of their transformation into learning scenarios, both the conceptual content of CT and its dimensions are limited enough. Some dimensions are underrepresented, and potential connections to several school subjects remain untapped. Thus emerges the need for the conduction of research toward the direction of exploring the relation of CS to the production of knowledge in modern disciplines. This research will inform teacher's training. It is not easy for instructional designers to have a deep knowledge of the epistemological uses of CS in different disciplines such as science, literature, etc. Most examples relate to CS applications in mathematics, an area which current scenario designers seem to be more familiar with. An interdisciplinary approach to the issue could therefore solve the problem. Interdisciplinary groups of learning designers and systematic mapping of the scenarios to authentic CT applications in various scientific fields is required, to unfold the scope of the concept. Furthermore, didactics research on the key dimensions of CT for younger ages is a prerequisite for the development of spiral structured curricula for CT and CS. Thus, the presented research supports the concurrent work of Yadav et al. (2017) according to which the preservice teacher's education programs need to be redesigned, to develop teachers CT competences and prepare them to incorporate CT in K-12 classrooms. The present research provides information about the difficulties and future needs of CS and CT teachers, as well as indications about the required research directions. Besides from arguing for the significance of teacher training and the value of an interdisciplinary approach to the production of CT learning scenarios, the study also records the CT dimensions that need to be further studied, as well as the relations of CS to other schools subjects that have not been educationally exploited yet.

Moreover, the claim that there are fundamental concepts of CS that are not covered by the CT approach is strengthened. Hence, in order to optimize the results, further research on educational computing and the didactics of CS needs to be conducted, along with the spread of CT in general education.

The limitations and constraints of the present research include the subjective categorization of learning designs and the limited set of CT scenarios sources. Furthermore, since the learning activities analyzed are not necessarily products of ordinary teachers' work, the results may present a more optimistic version of the reality of the teachers' general population. A direct examination of teachers, with respect to their understanding of CT content and CT curriculum theory of content, (Loughran et al. 2012) is proposed as a future work. Empirically, very little is known about the realization of CT in schools and classrooms (Yadav et al. 2011). Empirical studies are therefore needed, to investigate the classroom realization of CT. To conclude, CT is vital for science and technology progress and a valuable competency for the modern citizen in general. It concerns the study of using CS for solving problems across all disciplines, including math, science, and the humanities. In addition, CT tends to see the world in terms of a series of generic problems that have computational solutions (solution patterns) (Easterbrook 2014). Although CT concept helps utilize CS in general education, it only offers a limited solution to the problem. The exclusive integration of CT in the separate school subjects' curricula could lead to the teaching of some applications of computing by non-properly trained CS-educated teachers (Leyzberg and Moretti 2017) and may not serve the policy goal of attracting new students to CS departments and STEM carriers (Hutchison 2012; Lang et al. 2013; DesJardins 2015). In addition, CT as a part of CS key practices does not take advantage of the full potential of CS in education. The interdisciplinary nature of CT may inspire some educational policy-makers to use it as a possible solution of the CS teachers' shortage problem (Kosturko 2016; Goode 2007) since CT could be taught by properly trained teachers of several other school subjects. The introduction of CT in education as part of other school subjects' curricula will not optimally confront the CS integration in general education in the long term. The theoretical analysis and the findings of this research contribute to the long and demanding work needed to meaningfully and sustainably integrate CS in general education for the interest of society.

References

Abelson, H., & Disessa, A. A. (1980). *Turtle geometry: The computer as a medium for exploring mathematics*. Cambridge, MA: MIT Press.

Armoni, M. (2012). Teaching CS in kindergarten: How early can the pipeline begin? *ACM Inroads, 3*(4), 18–19. https://doi.org/10.1145/2381083.2381091.

Barr, V., & Stephenson, C. (2011). Bringing computational thinking to K-12: What is involved and what is the role of the computer science education community? *ACM Inroads, 2*(1), 48–54. https://doi.org/10.1145/1929887.1929905.

Brinda, T., Puhlmann, H., & Schulte, C. (2009). Bridging ICT and CS: Educational standards for computer science in lower secondary education. *ACM SIGCSE Bulletin-ITiCSE '09, 41*(3), 288–292. https://doi.org/10.1145/1562877.1562965.

Cuny, J. (2011). Transforming computer science education in high schools. *Computer, 44*(6), 107–109. https://doi.org/10.1109/MC.2011.191.

Deng, Z. (2009). The formation of a school subject and the nature of curriculum content: An analysis of liberal studies in Hong Kong. *Journal of Curriculum Studies, 41*(5), 585–604. https://doi.org/10.1080/00220270902767311.

Denning, P. (2009). The profession of IT: Beyond computational thinking. *Communications of the ACM, 52*(6), 28–30. https://doi.org/10.1145/1516046.1516054.

Denning, P. J., & Martell, C. H. (2015). *Great principles of computing*. Cambridge, MA: MIT Press.

DesJardins, M. (2015). Explainer: What it will take to make computer science education available in all schools. *The Conversation*. Retrieved April 8, 2017, from https://goo.gl/45RA4l

diSessa, A. A. (2000). *Changing minds: Computers, learning and literacy*. Cambridge: MIT Press.

Doyle, W. (1992a). Curriculum and pedagogy. In P. W. Jackson (Ed.), *Handbook of research on curriculum* (pp. 486–516). New York: Macmillan.

Doyle, W. (1992b). Constructing curriculum in the classroom. In F. K. Oser, A. Dick, & J. L. Patry (Eds.), *Effective and responsible teaching: The new syntheses* (pp. 66–79). San Francisco: Jossey-Bass.

Easterbrook, S. (2014). From computational thinking to systems thinking: A conceptual toolkit for sustainability computing. In M. Höjer, P. Lago, & J. Wangel (Eds.), *Proceedings of the 2014 conference ICT for sustainability* (pp. 235–244). Amsterdam: Atlantis Press. https://doi.org/10.2991/ict4s-14.2014.28.

Elo, S., & Kyngas, H. (2007). The qualitative content analysis process. *Journal of Advanced Nursing, 62*(1), 107–115.

Fesakis, G., & Serafeim, K. (2009). Influence of the familiarization with “scratch” on future teachers’ opinions and attitudes about programming and ICT in education. *ACM SIGCSE Bulletin, 41*(3), 258–262. https://doi.org/10.1145/1595496.1562957.

Fessakis, G., & Kirodimou, E. (2016). Improving the teachers’ understanding of complex systems through dynamic systems modelling and problem solving. *International Journal for Mathematics in Education, 7*, 63–96.

Fessakis, G., Gouli, E., & Mavroudi, E. (2013). Problem solving by 5–6 years old kindergarten children in a computer programming environment: A case study. *Computers & Education, 63*, 87–97. https://doi.org/10.1016/j.compedu.2012.11.016.

Goode, J. (2007). If you build teachers, will students come? The role of teachers in broadening computer science learning for urban youth. *Journal of Educational Computing Research, 36*(1), 65–88. https://doi.org/10.2190/2102-5G77-QL77-5506.

Grover, S., & Pea, R. (2013). Computational thinking in K-12: A review of the state of the field. *Educational Researcher, 42*(1), 38–43. https://doi.org/10.3102/0013189X12463051.

Guzdial, M. (2008). Paving the way for computational thinking. *Communications of the ACM, 51* (8), 25–27. https://doi.org/10.1145/1378704.1378713.

Henderson, P. B., Cortina, T. J., Hazzan, O., & Wing, J. M. (2007). *Computational thinking*. In Proceedings of the 38th ACM SIGCSE technical symposium on Computer Science Education (SIGCSE ‘07) (pp. 195–196). New York: ACM Press.

Hsieh, H., & Shannon, S. E. (2005). Three approaches to qualitative content analysis. *Qualitative Health Research, 15*(9), 1277–1288.

Hutchison, L. F. (2012). Addressing the STEM teacher shortage in American schools: Ways to recruit and retain effective STEM teachers. *Action in Teacher Education, 34*(5–6), 541–550. https://doi.org/10.1080/01626620.2012.729483.

Isbell, C., Stein, L., et al. (2009). (Re)defining computing curricula by (re)defining computing. *ACM SIGCSE Bulletin, 41*(4), 195–207. https://doi.org/10.1145/1709424.1709462.

ISTE & CSTA (2011). *Computational thinking: Teacher resources* (2nd ed.). Retrieved January 6, 2017, from http://www.iste.org/docs/ct-documents/ct-teacher-resources_2ed-pdf.pdf?sfvrsn=2

Kosturko, L. (2016). Computer Science education support surging: But who’s going to teach it? *SAS Curriculum Pathways*. Retrieved April 8, 2017, from https://goo.gl/qacjUI

Krippendorff, K. (1980). *Content analysis: An introduction to its methodology*. London: Sage.

Lang, K., Galanos, R., Goode, J., Seehorn, D., Trees, F., Phillips, P., & Stephenson, C. (2013). *Bugs in the system: Computer science teacher certification in the US*. The Computer Science Teachers Association and The Association for Computing Machinery.

Leyzberg, D., & Moretti, Ch. (2017). *Teaching CS to CS teachers: Addressing the need for advanced content in K-12 professional development*. In *Proceedings of the 2017 ACM SIGCSE technical symposium on Computer Science Education* (SIGCSE ‘17) (pp. 369–374). New York: ACM. https://doi.org/10.1145/3017680.3017798.

Loughran, J. J., Berry, A., & Mulhall, P. (2012). *Understanding and developing science teachers’ pedagogical content knowledge*. Rotterdam: Sense Publishers.

National Research Council (U.S.). & Committee for the Workshops on Computational Thinking. (2010). *Report of a workshop on the scope and nature of CT*. Washington, DC: National Academies Press. Retrieved January 3, 2017, from http://public.eblib.com/choice/publicfullrecord.aspx?p=3378614

Neundorf, K. (2002). *The content analysis guidebook*. Thousand Oaks: Sage.

Papert, S. (1991). *Mental storms: Children, computers and powerful ideas* (E. Stamatiou, Trans.). Athens: Odysseas.

Papert, S. (1996*). An exploration in the space of mathematics educations. International Journal of Computers for Mathematical Learning, 1*(1), 95–123. Retrieved January 7, 2017, from http://www.papert.org/articles/AnExplorationintheSpaceofMathematicsEducations.html

Rosenbloom, P. (2004). A new framework for computer science and engineering. *Computer, 37* (11), 31–36. https://doi.org/10.1109/MC.2004.186.

Royal Society. (2012). *Shut down or restart: The way forward for computing in UK schools.* Retrieved January 11, 2017, from http://royalsociety.org/uploadedFiles/Royal_Society_Content/education/policy/computing-in-schools/2012-01-12-Computing-in-Schools.pdf

Seker, H., & Guney, B. G. (2012). History of science in the physics curriculum: A directed content analysis of historical sources. *Science & Education, 21*(5), 683–703. https://doi.org/10.1007/s11191-011-9416-6.

Shulman, L. S. (1986). Those who understand: Knowledge growth in teaching. *Educational Researcher, 15*(2), 4–31.

Sykora, C. (2014). *Computational thinking for all.* Retrieved December 29, 2016, from https://www.iste.org/explore/articleDetail?articleid=152

The College Board. (2010). *AP Computer Science principles. Claims and evidence statements.* Retrieved from http://www.csprinciples.org/home/about-the-project

Wing, J. (2006). Computational thinking. *Communications of the ACM, 49*(3), 33–36.

Wing, J. (2011, June). Research notebook: Computational thinking –What and why? *The Link Magazine of Carnegie Mellon University's School of Computer Science*. Retrieved January18, 2017, from http://www.cs.cmu.edu/link/research-notebook-computational-thinking-what-and-why

Yadav, A., Zhou, N., Mayfield, C., Hambrusch, S., & Korb, J. T. (2011). *Introducing computational thinking in education courses*. In Proceedings of the 42nd ACM technical symposium on Computer science education (pp. 465–470). ACM Press. doi:https://doi.org/10.1145/1953163.1953297

Yadav, A., Mayfield, C., Zhou, N., Hambrusch, S., & Korb, J. T. (2014). Computational thinking in elementary and secondary teacher education. *ACM Transactions on Computing Education, 14* (1), 1–16. https://doi.org/10.1145/2576872.

Yadav, A., Stephenson, C., & Hong, H. (2017). Computational thinking for teacher education. *Communications of the ACM, 60*(4), 55–62. https://doi.org/10.1145/2994591.

Chapter 11
Introducing Coding and Computational Thinking in the Schools: The TACCLE 3 – Coding Project Experience

Francisco José García-Peñalvo, Daniela Reimann, and Christiane Maday

11.1 Introduction

There is a general trend worldwide to make computer science a basic skill (García-Peñalvo et al. 2017; Llorens Largo et al. 2017). This is related to future generations of workers that should know, at least, the basic laws of a computer-based society and, without demerit to humanities or social sciences, trying to reduce the current gap with STEM (science, technology, engineering, and mathematics) (CEDEFOP 2015) careers.

Current society is software-driven (Manovich 2013). A very common situation in countries with a high rate of unemployment is they have unfilled positions for engineers and technicians for the industry and digital services. This means that a growth in the demand of positions related to technology and scientific knowledge, particularly engineering, but not only, is not reflected in the increase of students in such university degrees.

In the European Union, more than 800,000 professionals skilled in computing/informatics by 2020 are expected; many educators, parents, economists, and politicians are starting to think that students need some computing and coding skills (Balanskat and Engelhardt 2015).

In EEUU there are different studies that recommend the creation of a well-defined set of K-12 computer science standards based on algorithmic/computational thinking concepts (Tucker et al. 2006; Wilson et al. 2010).

F. J. García-Peñalvo (✉)
Research Institute for Educational Sciences, University of Salamanca, Salamanca, Spain
e-mail: fgarcia@usal.es

D. Reimann · C. Maday
Karlsruhe Institute of Technology, Institute of Vocational and General Education, Karlsruhe, Germany

M. S. Khine (ed.), *Computational Thinking in the STEM Disciplines*,
https://doi.org/10.1007/978-3-319-93566-9_11

On the other hand, new devices (Alonso de Castro 2014; Sánchez-Prieto et al. 2013, 2014), from smartphones and tablets to electronic learning toys and robots, find new audiences with increasingly young children. This causes new challenges for teachers (Sánchez-Prieto et al. 2016a, b, 2017), for example, how to define developmentally appropriate activities and content for children of different ages (Bers et al. 2014).

These new devices caused that the real life in the physical space is represented in the virtual space in all its facets, e.g., the place where I am, the activities I am undertaking, with whom I communicate and interact, and what I buy. Data traces in the virtual space, which capture more and more what we do, are stored, networked, and sent to third parties (Boyd and Ellison 2008; Guettat et al. 2010). At the same time, the subjects in the digitized world always receive more accurate proposals and offers of assistance systems from the virtual space (Chajri and Fakir 2014; Colomo-Palacios et al. 2017). The virtual affects the physical reality to an increasing extent, but the virtual space is not a "neutral world," but it is driven by corporations and their business interests.

Whereas information technology (IT) literacy is the capability to use today's technology in one's own field, the notion of IT fluency adds the capability to independently learn and use new technology as it evolves (National Research Council Committee on Information Technology Literacy 1999) throughout one's professional lifetime. Moreover, IT fluency also includes the active use of algorithmic thinking (including programming) to solve problems, whereas IT literacy is more limited in scope.

The most frequent approach to teaching digital literacy has been to gradually encourage the learning of programming, and the term code literacy (Prensky 2008) has been coined to refer to the process of teaching children programming tasks, from the simplest and most entertaining to the most complex; this way the student's progress is centered on the difficulty of the tasks and in their motivating characteristic. This means a link between the learning with the response to a stimulus instead to the child's learning and cognitive capabilities, following the traditional behaviorist theories (Zapata-Ros 2015).

However, there exist an alternative in the constructionism approach, yet considered by Papert (1980) in his researches based on the Logo programming language, that conveys the idea that the child actively builds knowledge through experience and the related "learn-by-doing" approach to education. Papert wanted to create "a mathematics children can love rather than inventing tricks to teach them a mathematics they hate," because Papert's leitmotifs were thinking about thinking and the freedom to achieve one's potential (Stager 2016).

The term computational thinking was made popular by Jeannette M. Wing (2006), with her definition "computational thinking involves solving problems, designing systems, and understanding human behavior, by drawing on the concepts fundamental to computer science." Aho (2012) simplified this concept defining it as the thought processes involved in formulating problems, so "their solutions can be represented as computational steps and algorithms." García-Peñalvo (2016f) defined computational thinking as the application of high level of abstraction and an algorithmic approach to solve any kind of problems.

European Erasmus+ TACCLE 3 – Coding project (García-Peñalvo 2016a, b,c, d; TACCLE 3 Consortium 2017) focusses on supporting school teachers and developing their confidence to deliver the new computing curriculum including coding and computational thinking approaches.

In this chapter, TACCLE 3 – Coding is introduced, and in this framework, the experience of using wearables with target groups in higher education (pedagogy, engineering pedagogy) as well as in elementary teacher training is going to be presented.

11.2 TACCLE 3 – Coding Project

TACCLE 3 – Coding is a European Union Erasmus+ KA2 Programme project that supports primary school staff and others who are teaching computing to 4–14-year-olds. It started at September 2015 and will end at October 2017.

The project consortium is coordinated by GO! Het Gemeenschapsonderwijs (Belgium) and composed of the following partners: the Pontydysgu Limited (United Kingdom), Scholengroep 1 Antwerpen (Belgium), Karlsruher Institut Für Technologie (Germany), Hariduse Infotehnoloogia Sihtasutus (Estonia), Tallinn University (Estonia), University of Salamanca (Spain), Aalto-Korkeakoulusaatio (Finland), and Itä-Suomen yliopisto (Finland).

All the information and the project outcomes and deliverables are available at the project website http://www.taccle3.eu, and they are licensed under a Creative Commons Attribution-NonCommercial-ShareAlike 4.0 International License.

Also, training courses will be available for both in-service and future teachers.

Many European countries are introducing computing and coding as core curriculum topics (Balanskat and Engelhardt 2015). Some have already done so; many others are intending to. Inevitably the detail of the curricula will be different in each country, but there is a substantial overlap – almost all of the curricula available so far include programming, control technology, and computational/logical thinking, so TACCLE 3 has started with these (García-Peñalvo et al. 2016).

Figure 11.1 shows the main page of the projects website. From this, users may access to different kinds of resources organized by the following categories:

- Using logic
- Algorithms
- Creating + debugging programs
- Controlling things

In Fig. 11.2, the tabs on the top menu correspond to the curriculum areas and underpin the schemes of work that in turn form the basis for the lessons you will be delivering in the classroom. Under each heading, you will find a variety of ideas, lessons, and materials directly related to classroom activities.

One the most interesting TACCLE 3 resources is the activity/lesson. It is published in the form of blog post. Each post explains the basic concept followed

Fig. 11.1 Main page of the TACCLE 3 website: Source (TACCLE 3 Consortium 2017)

by the aim of the lesson which in turn contributes to one or more of the attainment targets in the computing curriculum.

The outline of the activity follows this scheme:

Title

1. *Overview*

   Brief description
   Age
   Level
   Twenty-first-century skills
   Tips to adapt the lesson (e.g., to older/younger students, students with special needs, etc.)

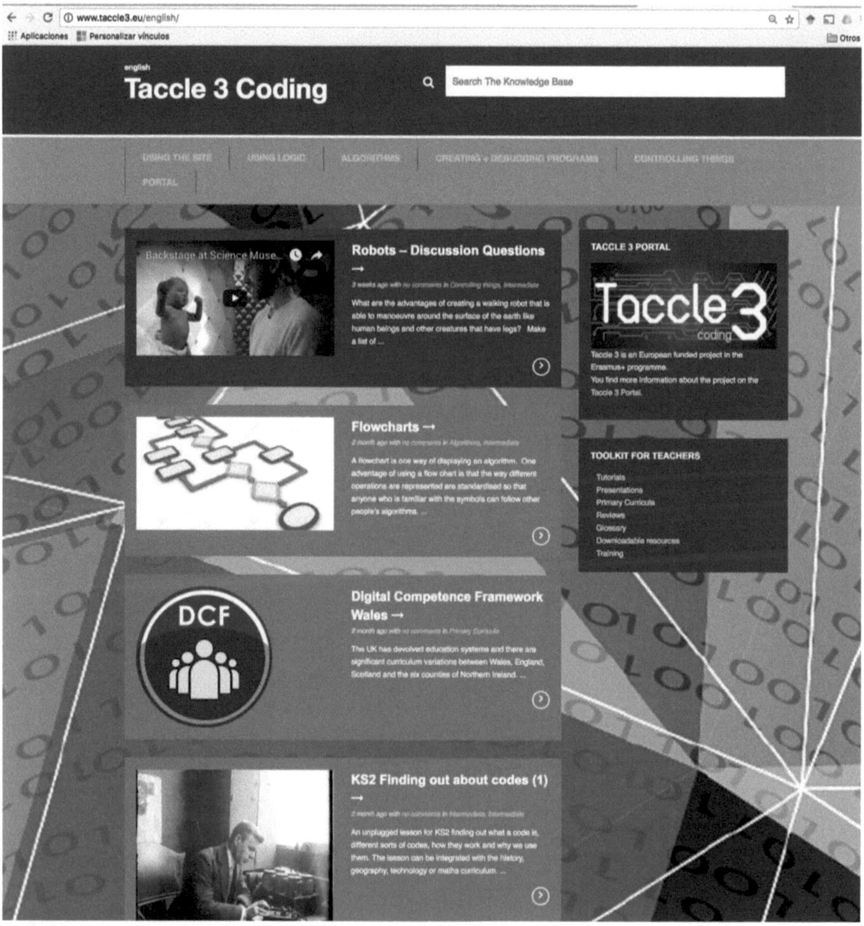

Fig. 11.2 Categories of the available resources

Material

2. *Aim of the activity*
3. *Needed tools and resources*
4. *Practical activity description*

Figure 11.3 shows an example of a TACCLE 3 activity oriented to introduce the decomposition process, breaking down a problem into smaller manageable parts. Decomposition helps in solving complex problems and managing large projects.

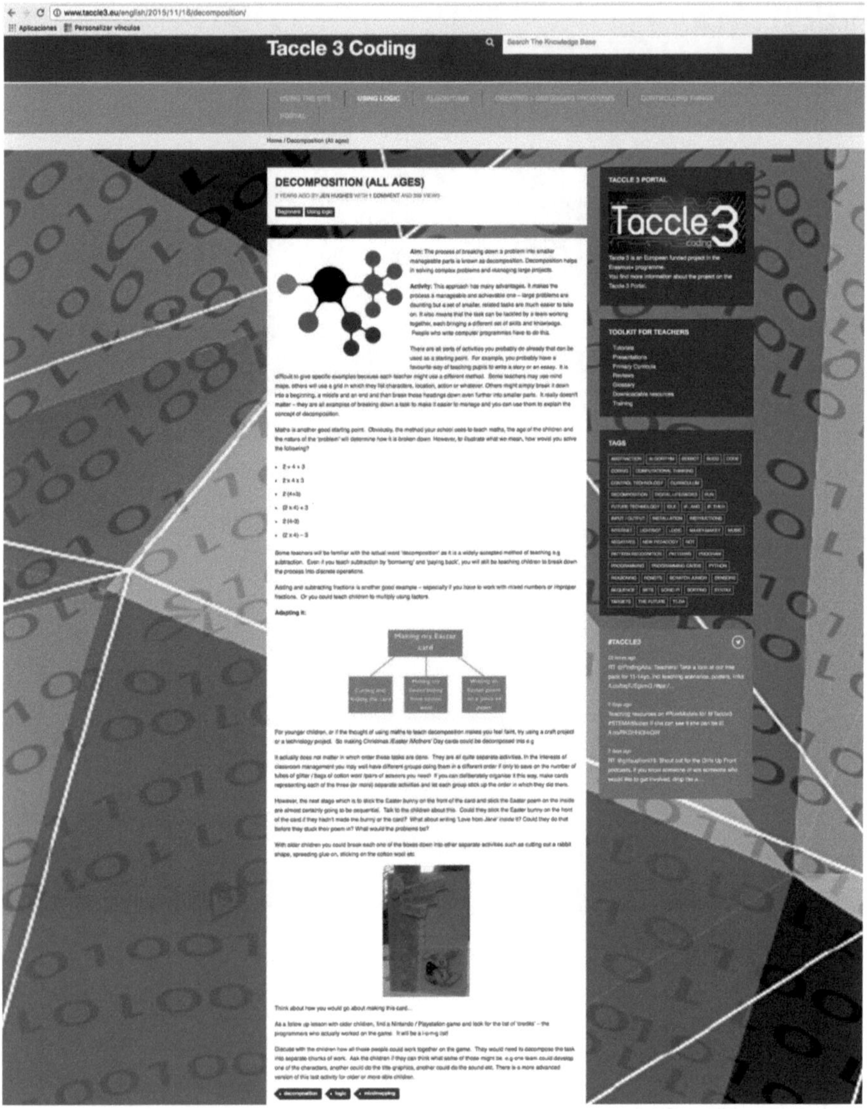

Fig. 11.3 Decomposition activity in TACCLE 3 – Coding project. (Source http://www.taccle3.eu/english/2015/11/16/decomposition/)

11.3 Developing Smart Textile Objects

To make the abstract learning contents of coding more graspable and usable for teachers at primary school level, the concepts are linked to imagination and phantasies of young children, who can invent and realize their own project ideas to be developed by the learners themselves. The latter is done in project-based learning scenarios (Estruch and Silva 2006; Markham et al. 2003), using embedded sensor- and actuator-based systems, using the Arduino LilyPad technology introduced by Buechley (2014), extended by a visual interface to facilitate programming using icons in a free drag-and-drop environment (Amici) (Kafai 2014).

The technology chosen opens up to link the ideas and imagination to computational thinking (Wing 2006, 2008, 2011) and acting through more art- and craft-based, creative processes. Examples of smart textile projects (and the sketching of electronic circuits) are used. Smart textiles, which are also referred to as "wearables," are a generation of clothes and accessories with embedded microcomputers and offer various possibilities for learning about computational modeling. The system, worn on the body, can respond with behavior programmed by the children themselves. They manipulate and change technology. Using, e.g., conductive yarn (as connector), sensors, motors, and LED lights as well as sewable circuit boards (Arduino LilyPad introduced by Buechley), smart textiles create a link between sensual-haptic materials (Fernández et al. 2016; Scopes and Smith 2010), precise computer control, and creative concepts. New interfaces – sewed, woven, or stitched – can be experienced between body, clothing, and the environment. It can be stitched together with conductive thread to create interactive garments and accessories. In conjunction with the open-source Arduino technology, they open up opportunities for cross-disciplinary teaching of the subjects of art, design, computer science, and music, for example, to address learning in the context of storytelling wearables (Tan 2005), wearable music (Rosales 2012), or sounding artifacts (Trappe 2012).

The Arduino LilyPad technology consists of hard components as well as a programming interface which can be connected to an icon-based interface to be used by younger children at primary school level.

The LilyPad can "sense information about the environment using inputs like light and temperature sensors and can act on the environment with outputs like LED lights, vibrator motors, and speakers" (http://lilypadarduino.org). Kafai (2014) highlighted the LilyPad Arduino kit being a shapable set of technologies, bringing together crafting, design, and technology, supporting individual learning processes.

11.4 Curriculum Modules for Primary School Teacher Training

The learning activities developed include a teacher training, as well as a tutorial for beginners to programming, which introduces the teacher both to the handling of the LilyPad Arduino hardware and to the application of the Amici user interface and can be used as instructions for teaching processes related to interactive clothing. Also, the development of creative themes is addressed, to support imagination and self-initiated learning. The teacher training is based on the modules identified to develop a project.

The teachers get familiar with the hardware, such as the electronic components, main board, connectors (including unusual wires made of ink or yarn), and sensors and actuators. The teachers use the same modules for project-based learning with physical computing as the school kids in hands-on workshops.

Since, however, the handling of the software and hardware used in the project is documented only insufficiently in Germany, it was decided to write down in a structured way the experiences gained. Although the resultant tutorial does not claim to discuss all software and hardware issues, relevant problems need to be explained in detail. The tutorial was developed on the basis of the EduWear manual compiled by the "Digital media in Education (dimeb)" research group of the University of Bremen (http://goo.gl/a8c2L7).

The following lesson plans for classroom sessions are linked to each other and based on one another. They form the teaching units on developing sensor- and actuator-based systems/developing a project with Arduino LilyPad and Amici software.

Module 1 Getting Familiar with Hardware
This module is part of a series of lesson plans to introduce children (from grade 5 up) to smart textile objects, based and the programming of sensors and actuators set up in an electronic circuit. After the series of lessons 1–6, the learners will be able to develop, connect, and program a sensor- and actuator-based interactive system and contextualize it in a project. Also, there are lessons to introduce the development of electronic circuits through painting connectors (wires) using conductible ink. In those lessons, the learners design and paint electronic systems, which can be integrated in an interactive book project.

a. Aims: familiarizing with the terms and related hardware and understanding the components as a networked system
b. Terms to be introduced: sensor, actuator, connector, main board LilyPad, input, output, and meaning/function in a circuit/interactive system.
c. Methods: relation to sensory perception/the human senses and/or learners to represent the components physically
d. Develop photo work sheets for identification of hardware components, including learning material including exercises

Module 2 Developing an Electronic Circuit
In module 2 of the LilyPad Arduino-based Smart textile introduction series, learners learn to develop a circuit, cable it and make it run by themselves. This way pupils learn: 1. to develop a circuit, cable it and make it run by themselves, 2. how to cable the components using crocodile clips. There are exercises based on work sheets to arrange the components and cables, so that an LED glows continuously or an LED shines on and switches off, come along with the module.

Module 3 Developing an Interactive System: Programming Arduino LilyPad
In this module, the learners learn to program Arduino LilyPad main board by using the icon-based drag-and-drop programming environment Amici. In the session pupils are introduced to Amici software through work sheets with exercises related to LED on/off or for a particular time, in the context of an interactive system. The aim is to make transparent computational thinking and modeling behavior by

developing a program. It intends the pupils to understand the computer as a shapable, controllable medium.

Module 4 Programming Arduino LilyPad: Getting Familiar with Amici
The main aim of this module is to make transparent computational thinking/modeling, and algorithms control computer, in order to understand the computer as a shapable, controllable medium to the learners by doing, testing, and debugging. In this module, the issue of testing and debugging is introduced to the pupils by making them develop, test, and debug a program by themselves. The interrelation of such processes which belong together is addressed. Aims are to develop a program, test, and debug it. The issue of bugs and debugging is addressed (original etymological meaning of the bug, esp. for younger kids!). There are exercises to arrange the components and cables, so that a LED glows or so that a LED shines on and switches off.

Module 5 Developing a Project with Arduino LilyPad and Amici
In this module, the learners are encouraged to develop an idea for an interactive project, based on sensors and actuators they know from the previous lessons. By developing an idea for an interactive project, have them identify the tasks to fulfill and the realization by themselves. Co-construction of knowledge is supported and learned through the working and design processes. This learning activity deals with using logic and algorithms.

Module 6 Painting Electronic Circuits
This module deals with particular connectors. Painting electronic circuits can be used as a vehicle to technology education in early age groups. Conductible ink in a pen is used for electronic components in the context of drawing images. As Buechley has stressed, "electronics aren't just for experts and engineers. Kids and amateurs should be able to play, too." Buechley (2014) designed paper-based electronics for "sketching" and folding. Teachers like to get and test learning materials which are ready to use in the classroom but also designed flexible enough to be amended individually according to their own purposes, needs, target groups, and ideas. In the following example, learning material is presented. Using conductible ink, the issue of "algorithms" as an endless set of activities which, after its realization, lead to a solution is introduced for primary school level. Therefore, the paper cards (Figs. 11.4 and 11.5) were developed. In Fig. 11.4 the cooking of a pan cake is used as an example for an algorithm.

Fig. 11.4 Drawn algorithm in the form of a game. Learner to put together the images in the right order

Fig. 11.5 If every step is put together correctly, the LED glows

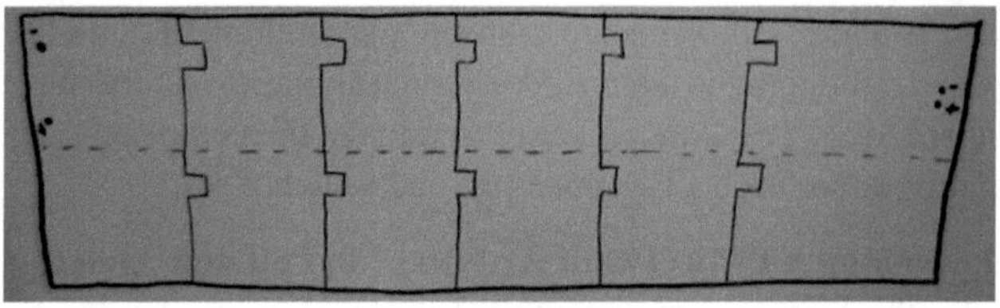

Fig. 11.6 A blanc paper algorithm puzzle is sketched for individual use

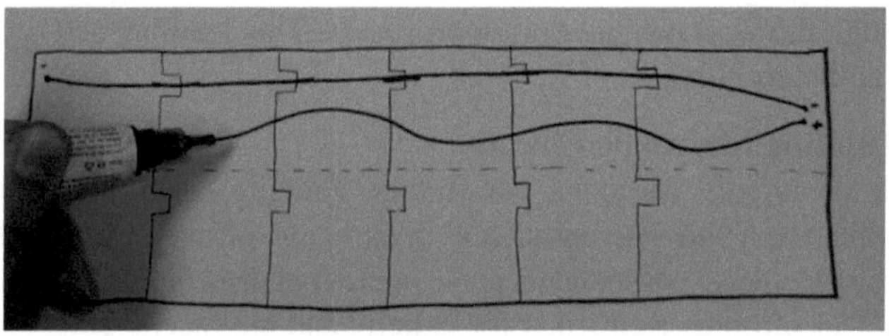

Fig. 11.7 The connection is done using electronic ink

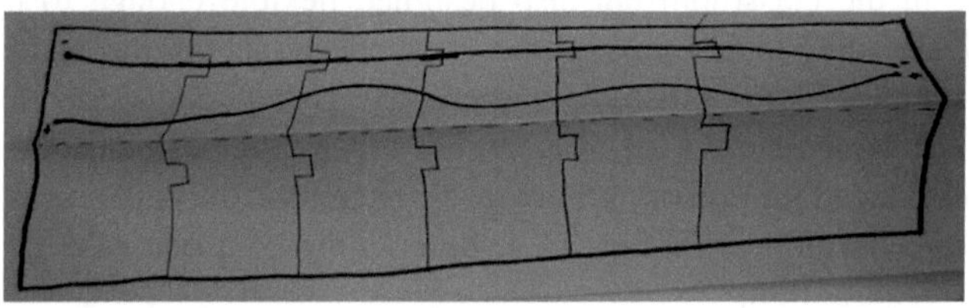

Fig. 11.8 Fold along the dotted line

For initiating the process, a blanc algorithm puzzle is handed out to the pupils (Fig. 11.6).

In step 4, a connection between the ends has to be drawn (Fig. 11.7).

Afterward, it needs folding along the dotted line (Fig. 11.8).

In step 6, pieces are cut apart. Obviously, there is only one correct order of the parts. Here you can see that there will be no electric connection (Fig. 11.9).

In step 7, the parts are folded and numbered. At the front, an algorithm can also be written or drawn (e.g., a recipe) (Fig. 11.10).

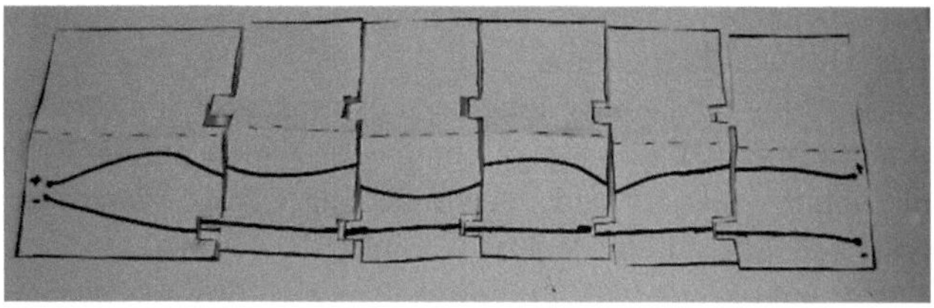

Fig. 11.9 Cutting pieces apart

Fig. 11.10 Fold parts and number them

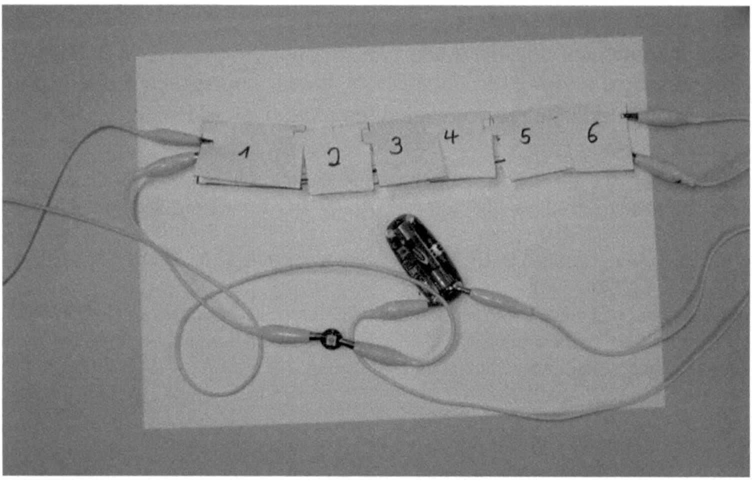

Fig. 11.11 Cable an actuator and a battery

In the 8th step, an actuator and a battery have to be wired to the end and to the starting point. Then the actuator will react if the algorithm is laid in the correct order (Fig. 11.11).

11.5 Conclusions

Introducing computational thinking and a solid base of coding is the educational agenda of many countries worldwide. The challenge is to do it in the right way so that the objective is not confused and really influences the acquisition of key twenty-first-century competences; trying to avoid these contents will become in another subject an over-saturated curriculum.

Methods of teaching which have long been overtaken such as the reduction on a tool-oriented and resource-based use of computers, which may now be overtaken, despite all the interdisciplinary and interconnecting efforts, remain as a reality in today's schools, colleges, and outer school contexts.

The presented approaches which were well received by the pedagogical target groups are available for teaching computational modeling at school and university as well as in outside school settings. They can be absorbed and used to ensure a sustainable and systematic integration of computer science contents and embed them into the curricula, crossing the borders of disciplines and school subjects, such as computer science/IT, textile, art, and design education.

In this sense TACCLE 3 project looks for sharing experiences and resources to achieve the pursued goal involving the right actors.

Teachers that are interested in participating in TACCLE 3 – Coding may do it in several ways:

- Visiting the website to access to the resources.
- Writing news related to coding in the schools.
- Making learning activities/lessons.
- Making resource reviews (products, tools, books, courses, etc.) oriented to other teachers. There exists a recommended template (García-Peñalvo 2016e).

Acknowledgments We thank Leah Buechley for sharing the inspiring TED talk and Heidi Schelhowe for making available the software Amici and for testing and modifying eduwear tutorials.

This work has been supported by EU Erasmus+ Programme, KA2 project "TACCLE 3 – Coding" (2015-1-BE02-KA201-012307).

This project has been funded with support from the European Commission. This communication reflects the views only of the author, and the Commission cannot be held responsible for any use which may be made of the information contained therein.

References

Aho, A. V. (2012). Computation and computational thinking. *Computer Journal, 55*(7), 832–835. https://doi.org/10.1093/comjnl/bxs074.

Alonso de Castro, M. G. (2014). Educational projects based on mobile learning. *Education in the Knowledge Society, 15*(1), 10–19.

Balanskat, A., & Engelhardt, K. (2015). *Computing our future. Computer programming and coding priorities, school curricula and initiatives across Europe*. Retrieved from Brussels, Belgium: http://fcl.eun.org/documents/10180/14689/Computing+our+future_final.pdf/746e36b1-e1a6-4bf1-8105-ea27c0d2bbe0

Bers, M. U., Flannery, L., Kazakoff, E. R., & Sullivan, A. (2014). Computational thinking and tinkering: Exploration of an early childhood robotics curriculum. *Computers and Education, 72*, 145–157. https://doi.org/10.1016/j.compedu.2013.10.020.

Boyd, D. M., & Ellison, N. N. (2008). Social network sites: Definition, History, and scholarship. *Journal of Computer-Mediated Communication, 13*(1), 210–230. https://doi.org/10.1111/j.1083-6101.2007.00393.x.

Buechley, L. (2014). *Crafting the Lilypad Arduino*. Retrieved from http://makezine.com/2014/07/18/leah-buechley-crafting-the-lilypad-arduino/

CEDEFOP. (2015). *EU Skills Panorama (2014) STEM skills Analytical Highlight*. Retrieved from http://skillspanorama.cedefop.europa.eu/sites/default/files/EUSP_AH_STEM_0.pdf

Chajri, M., & Fakir, M. (2014). Application of data mining in e-commerce. *Journal of Information Technology Research, 7*(4), 79–91. https://doi.org/10.4018/jitr.2014100106.

Colomo-Palacios, R., García-Peñalvo, F. J., Stantchev, V., & Misra, S. (2017). Towards a social and context-aware mobile recommendation system for tourism. *Pervasive and Mobile Computing, 38*, 505–515. https://doi.org/10.1016/j.pmcj.2016.03.001.

Estruch, V., & Silva, J. (2006). Aprendizaje basado en proyectos en la carrera de Ingeniería Informática. *Actas de las XII Jornadas de la Enseñanza Universitaria de la Informática (JENUI 2006), Deusto, Bilbao, 12–14 de julio de 2006* (pp. 339–346).

Fernández, C., Esteban, G., Conde-González, M. Á., & García-Peñalvo, F. J. (2016). Improving motivation in a haptic teaching/learning framework. *International Journal of Engineering Education (IJEE), 32*(1B), 553–562.

García-Peñalvo, F. J. (2016a). A brief introduction to TACCLE 3 – Coding European Project. In F. J. García-Peñalvo & J. A. Mendes (Eds.), *2016 International Symposium on Computers in Education (SIIE 16)*. IEEE, USA.

García-Peñalvo, F. J. (2016b). *Presentación del Proyecto TACCLE3 Coding*. Paper presented at the Workshop EI<18. Educación en Informática sub 18, Salamanca, España. http://repositorio.grial.eu/handle/grial/653

García-Peñalvo, F. J. (2016c). *Presentation of the TACCLE3 Coding European Project*. Retrieved from http://repositorio.grial.eu/handle/grial/654

García-Peñalvo, F. J. (2016d). Proyecto TACCLE3 – Coding. In F. J. García-Peñalvo & J. A. Mendes (Eds.), XVIII Simposio Internacional de Informática Educativa, *SIIE 2016* (pp. 187–189). Salamanca, España: Ediciones Universidad de Salamanca.

García-Peñalvo, F. J. (2016e). *Template for TACCLE 3 resources reviewing*. Retrieved from https://doi.org/10.6084/m9.figshare.3545033.v1

García-Peñalvo, F. J. (2016f). What computational thinking is. *Journal of Information Technology Research, 9*(3), v–viii.

García-Peñalvo, F. J., Reimann, D., Tuul, M., Rees, A., & Jormanainen, I. (2016). *An overview of the most relevant literature on coding and computational thinking with emphasis on the relevant issues for teachers*. Belgium: TACCLE 3 Consortium. doi:https://doi.org/10.5281/zenodo.165123.

García-Peñalvo, F. J., Llorens Largo, F., Molero Prieto, X., & Vendrell Vidal, E. (2017). Educación en Informática sub 18 (EI<18). *ReVisión, 10*(2), 13–18.

Guettat, B., Chorfi, H., & Jemni, M. (2010). Automatic update of e-learning environments based on heterogeneous traces. In *Proceedings of the 2nd International Conference on Education Technology and Computer (ICETC 2010)* (pp. V4-512-V514-516). EEUU: IEEE.

Kafai, B. (2014). *Connected code. Why children need to learn programming*. Cambridge, MA: MIT press.

Llorens Largo, F., García-Peñalvo, F. J., Molero Prieto, X., & Vendrell Vidal, E. (2017). La enseñanza de la informática, la programación y el pensamiento computacional en los estudios preuniversitarios. *Education in the Knowledge Society, 18*(2), 7–17. https://doi.org/10.14201/eks2017182717.

Manovich, L. (2013). *Software takes command*. New York: Bloomsbury.

Markham, T., Larmer, J., Education, B. I. F., & Ravitz, J. (2003). *Project based learning handbook: A guide to standards-focused project based learning for middle and high school teachers*. Buck Institute for Education.

National Research Council Committee on Information Technology Literacy. (1999). *Being fluent with information technology*. Washington, DC: National Academy Press.
Papert, S. (1980). *Mindstorms: Children, computers, and powerful ideas*. New York: Basic Books.
Prensky, M. (2008). *Programming is the new literacy*. Retrieved from http://www.edutopia.org/literacy-computer-programming
Rosales, A. (2012). Wearable music. In *Creating sound effects and music by playing*. Retrieved from Berlin.
Sánchez-Prieto, J. C., Olmos-Migueláñez, S., & García-Peñalvo, F. J. (2013). Mobile Learning: Tendencies and Lines of Research. In F. J. García-Peñalvo (Ed.), *Proceedings of the first international conference on Technological Ecosystems for Enhancing Multiculturality (TEEM'13) (Salamanca, Spain, November 14–15, 2013)* (pp. 473–480). New York: ACM.
Sánchez-Prieto, J. C., Olmos-Migueláñez, S., & García-Peñalvo, F. J. (2014). Understanding mobile learning: Devices, pedagogical implications and research lines. *Education in the Knowledge Society, 15*(1), 20–42.
Sánchez-Prieto, J. C., Olmos-Migueláñez, S., & García-Peñalvo, F. J. (2016a). *A TAM based tool for the assessment of the acceptance of mobile technologies among teachers*. Retrieved from Salamanca, Spain: http://hdl.handle.net/10366/127435
Sánchez-Prieto, J. C., Olmos-Migueláñez, S., & García-Peñalvo, F. J. (2016b, August 22–26). *Technologically reluctant teachers. A TAM based study on compatibility and resistance to change among pre-service teachers*. Paper presented at the ECER 2016, Dublin, Ireland.
Sánchez-Prieto, J. C., Olmos-Migueláñez, S., & García-Peñalvo, F. J. (2017). MLearning and pre-service teachers: An assessment of the behavioral intention using an expanded TAM model. *Computers in Human Behavior, 72*, 644–654. https://doi.org/10.1016/j.chb.2016.09.061.
Scopes, P., & Smith, S. P. (2010). Integrating haptic interaction into an existing virtual environment toolkit. In J. Collomosse & I. Grimstead (Eds.), *Theory and practice of computer graphics*. UK: The Eurographics Association.
Stager, G. S. (2016). Seymour Papert (1928-2016). Father of educational computing. *Nature, 537*, 308–308.
TACCLE 3 Consortium. (2017). *TACCLE 3: Coding Erasmus + Project website*. Retrieved from http://www.taccle3.eu/
Tan, X. L. (2005). *Storytelling wearables*. Retrieved from http://we-make-money-not-art.com/xiao_li_tans_st/
Trappe, C. (2012). Creative access to technology: building sounding artifacts with children *Proceedings of the 11th International conference on interaction design and children, IDC'12 (Bremen, Germany — June 12–15, 2012)* (pp. 188–191). New York: ACM.
Tucker, A., Deek, F., Jones, J., McCowan, D., Stephenson, C., & Verno, A. (2006). *A model curriculum for K-12 computer science: Final report of the ACM K–12 task force curriculum committee* (2nd ed.). New York: ACM.
Wilson, C., Sudol, L. A., Stephenson, C., & Stehlik, M. (2010). *Running on empty: The failure to teach K-12 computer science in the digital age*. New York: Association for Computing Machinery (ACM).
Wing, J. M. (2006). Computational thinking. *Communications of the ACM, 49*(3), 33–35. https://doi.org/10.1145/1118178.1118215.
Wing, J. M. (2008). Computational thinking and thinking about computing. *Philosophical Transactions of the Royal Society a-Mathematical Physical and Engineering Sciences, 366*(1881), 3717–3725. https://doi.org/10.1098/rsta.2008.0118.
Wing, J. M. (2011). Computational thinking. In G. Costagliola, A. Ko, A. Cypher, J. Nichols, C. Scaffidi, C. Kelleher, & B. Myers (Eds.), *2011 I.E. symposium on visual languages and human-centric computing* (pp. 3–3).
Zapata-Ros, M. (2015). Pensamiento computacional: Una nueva alfabetización digital. In *RED, Revista de Educación a distancia,* 46.

Chapter 12
Case Studies of Elementary Children's Engagement in Computational Thinking Through Scratch Programming

Sze Yee Lye and Joyce Hwee Ling Koh

12.1 Introduction

In the recent years, there has been burgeoning interest in teaching programming to K-12 students (Grover and Pea 2013; Kafai and Burke 2013), with countries like Estonia and the United Kingdom making programming a compulsory subject in schools. In school-based programming activities, K-12 students typically create interactive animations with child-friendly programming languages such as Alice (Cooper 2010) and Scratch (Resnick et al. 2009).

Programming achieves more than just having students to write lines of codes. It is essentially a problem-solving activity (Jonassen 2011; Kafai and Burke 2013; Palumbo 1990). It has been proposed that programming exposes students to computational thinking where they apply programming concepts such as abstraction and modularization for problem-solving and engage in programming practices such as testing and debugging (Brennan and Resnick 2012; Wing 2006). Wing (2006) argues that computational thinking "represents a universally applicable attitude and skill set everyone, not just computer scientists, would be eager to learn and use" (p. 33). Computational thinking also parallels the twenty-first-century competencies and skills such as problem-solving, creativity, and innovation (Ananiadou and Claro 2009; Binkley et al. 2012).

Despite the heightened interest in K-12 programming, little attention has been given to understand how K-12 students engage in computational thinking as they program. Such kinds of studies are pertinent in K-12 settings, especially in elementary school contexts where young children may need more support when engaging in programming activities (Grover and Pea 2013). Armed with these findings,

S. Y. Lye (✉)
ICT Department, Teck Whye Primary School, Singapore, Singapore

J. H. L. Koh
Higher Education Development Centre, University of Otago, Dunedin, New Zealand

M. S. Khine (ed.), *Computational Thinking in the STEM Disciplines*,
https://doi.org/10.1007/978-3-319-93566-9_12

educators can better support students with varying programming abilities to engage in computational thinking during programming lessons. To address this research gap, this paper seeks to examine elementary school students' programming behaviours through multiple case studies. Specifically, this study provides an in-depth case analysis of three Grade 4 students' programming strategies through on-screen recordings, interviews, and examination of their programming projects. The implications of these findings for the development of children's computational thinking in K-12 contexts are discussed.

12.2 Literature Review

12.2.1 Computational Thinking

Despite the rising popularity of computational thinking, the definition of computational thinking is still highly contested (Barr and Stephenson 2011; Brennan and Resnick 2012; Grover and Pea 2013). Some of these definitions may not involve the use of programming tools. For example, the National Research Council (NRC), in its outline for the "Framework for K-12 Science Education" (NRC 2012), describes mathematics and computational thinking as the representation of variables and their relationship with computer tools. On the other hand, the International Society for Technology in Education (ISTE) considers computational thinking as algorithmic thinking with automation tools and data representation with the use of simulation.

As the focus of this study is to examine computational thinking through programming for K-12 students, we are using the computational thinking framework proposed for Scratch by Brennan and Resnick (2012). Scratch is a popular programming language used in K-12 settings (e.g., Baytak and Land 2011; Kafai et al. 2010; Tangney et al. 2010). With respect to Scratch, Brennan and Resnick (2012) suggest three dimensions of computational thinking: computational concepts, computational practices, and computational perspectives. Table 12.1 summarizes the key ideas on

Table 12.1 Computational thinking dimensions

Dimension	Description	Examples
Computational concepts	Concepts that programmers use	Variables Loops
Computational practices	Problem-solving practices that occur in the process of programming	Being incremental and iterative Testing and debugging Reusing and remixing Abstracting and modularizing
Computational perspectives	Students' understandings of themselves, their relationships to others, and the technological world around them	Expressing and questioning about the technology world

these three dimensions. These dimensions are appropriate for understanding how K-12 students approach programming as Scratch is similar to other contemporary visual programming languages for the younger children (e.g. Alice). These languages use simple-to-understand programming blocks, and the program output is typically presented in the form of animated objects. Therefore, this framework is likely to be suitable for considering computational thinking for programming contexts in K-12 education.

12.2.2 Computational Thinking Concepts in K-12

Most K-12 programming studies examine the issues related to the computational thinking dimension of computational concepts. In these classes, students complete programming tasks after being introduced to the relevant computational concepts such as sequence, loops, parallelism, events, conditionals, and operators (e.g. Burke 2012; Denner et al. 2012). These programming classes are based on the learning theory of constructionism which "attaches special importance to the role of constructions in the world as a support for those in the head, thereby becoming less of a purely mentalist doctrine" (Papert 1994, p. 143).

Essentially, the students in these studies learn programming while actively constructing their programming projects. Numerous studies assess students' acquisition of different computational concepts (Brennan and Resnick 2012) by examining students' created artefacts (Burke 2012; Denner et al. 2012), their tests, or quiz results (e.g. Lewis 2011; Lin and Liu 2012; Martin et al. 2013; Su et al. 2014) or students perceptions through interviews (Meerbaum-Salant et al. 2013) and surveys (Feng and Chen 2014; Sáez López et al. 2016). These studies unanimously report positive results where students could grasp the associated computational concepts.

Some other studies report that questioning by teachers (Feng and Chen 2014; Su et al. 2014), help from peers (Denner et al. 2012; Lewis 2011; Martin et al. 2013), or assistance from parents (Lin and Liu 2012) to be some forms of support that K-12 students need during programming to help acquire computational concepts. Yet, no studies examined if students differed in terms of their computational thinking during programming and if this influenced how they approach programming.

12.2.3 Computational Thinking Practices in K-12

Only four studies focus on the computational practice of testing and debugging for K-12 students (Baytak and Land 2011; Berland et al. 2013; Fessakis et al. 2013; Wyeth 2008) with data collected through the log of programming activities (Berland et al. 2013), field observation (Bers et al. 2014; Fessakis et al. 2013), and video recordings (Baytak and Land 2011; Wyeth 2008).

For the computational thinking practice of testing and debugging, it was found that the students were more likely not to plan but "rather tried commands one by one" (Fessakis et al. 2013, p. 94). When immediate feedback on their programs was given, students were encouraged to explore and experiment with their programs more frequently (Berland et al. 2013; Fessakis et al. 2013; Wyeth 2008). Analysis of elementary school children's programming behaviour with Logo found that they were more like to "identify programming actions at the level of individual programming statements" and "fail to 'chunk' program code in relation to the goal it is intended to accomplish" (Lehrer et al. 1999, p. 297). This resembled the behaviour of adult novice programmers who tend to focus on small chunks within programs during testing and debugging (Bednarik 2012; Vessey 1985; Wiedenbeck et al. 1993).

During testing and debugging, young children were also found to engage in help-seeking from knowledgeable others (Feng and Chen 2014). Baytak and Land's (2011) observation of 5th graders saw them seeking help from both teachers and the peers after noticing the discrepancies between their actual and intended programming output. The children studied did not see the teacher as the only expert and would also seek help from their peers. This could be because of the largely constructionist approaches adopted for school-based programming classes. Seeking help concurred with the concept of community in constructionism (Papert 1980) in which the community members (e.g. students and teachers) can act as "collaborators, coaches, audience, and co-constructors of knowledge" (Kafai and Resnick 1996, p. 6).

12.2.4 Research Gap in K-12 Programming Studies

Existing studies provide some brief insights about how elementary school students are exhibiting computational thinking. With majority of the studies focusing on computational concepts, there is generally a lack of studies investigating the cognitive aspects of programming for the younger children (Grover and Pea 2013) which embody the important computational practices that support computational thinking. This is a glaring gap in current K-12 programming research.

Interestingly, with the emergence of help-seeking as a programming behaviour of young children, it exacerbates the need for educators to understand their programming behaviours so that instructional support for programming classes can be better designed to foster computational thinking. Our review shows that programming classes are typically conducted in a one-size-fits-all manner. Understanding the possible variations in students' programming behaviours is therefore important for determining how instruction may be differentiated as needed. Understanding young children's programming behaviours could also provide insights about how students are problem-solving during programming. Such studies would be of immense interest to K-12 educators and researchers as developing problem-solving skills during programming is a motivating factor for the increased interest in K-12 programming (Barr and Stephenson 2011; Resnick et al. 2009).

12.3 Research Purpose

This study, thus, seeks to advance our current limited knowledge on how young students are approaching computational thinking during programming. Specifically, the paper aims to answer the following research questions:

When given a programming task,

1. What programming behaviours did students exhibit?
2. Are there any differences among students' programming behaviours?

12.4 Methodology

To answer the research questions, a multiple case study research design was adopted. This research design "is an in-depth description and analysis of the bounded system" (Merriam 2009, p. 40). Multiple case studies enable in-depth analyses and comparison of the characteristics and nuances of programming behaviours among young children by using each child as a unique case. It provides a way for the initial vocabulary of young children's programming behaviour to be established for further analysis in future.

12.4.1 Study Context

12.4.1.1 The Programming Class

The study was conducted in a typical 6-year Singapore elementary school. This school conducted a 5-week Scratch programming class for all Grade 4 students as after-school enrichment activity. Students attended classes once a week, and each session lasted 2 hours. The purpose of this activity was to expose the students to programming. Scratch was chosen as a programming language as the school felt that it would be easier for the children to pick up this easy-to-use programming language. Scratch is an intuitive visual programming language produced by MIT for children to create interactive stories and games (see Fig. 12.1). In Scratch, students would snap blocks together in jigsaw syntax to form scripts. Scripts would be used to control sprites or objects.

Students were taught various programming concepts through different programming tasks between lessons one to four. During the first two lessons, students were introduced to fundamental programming concepts such as sequences and events

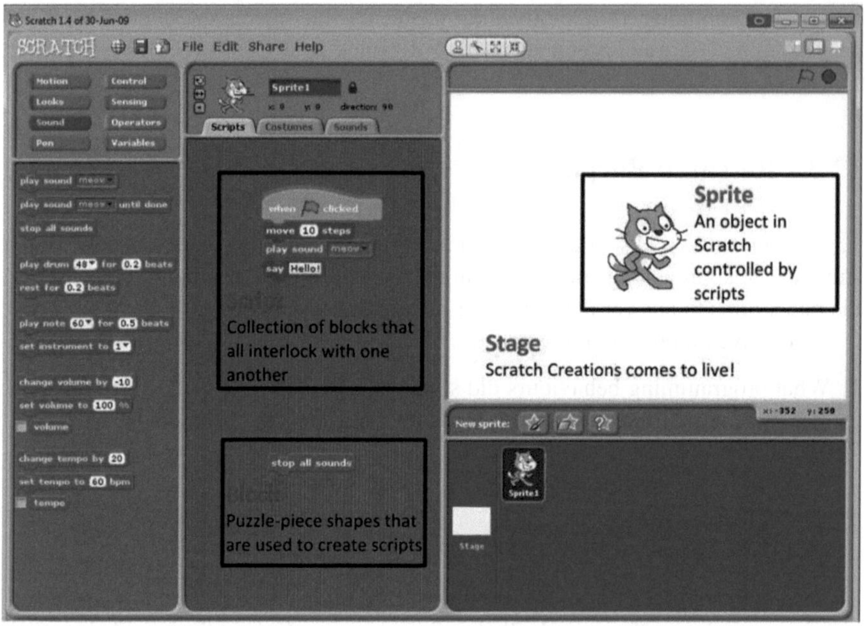

Fig. 12.1 Scratch layout

(e.g. *When Green flag clicked* block) through the creation of animated stories. During the next two lessons, the students were asked to create games with more complex programming concepts such as conditionals (e.g. *if* block) and operators (e.g. variable-related blocks). During the last lesson, the students could choose to create any program they wanted. The programming task analysed in this paper took place during the fourth lesson.

12.4.1.2 The Programming Task

In this task, the students were to create a scoring mechanism for a bouncing ball game that they created during the third lesson where players had to move an object (e.g. the plank) to catch a bouncing ball (see Fig. 12.2). In this task, students had to create the scoring mechanism where:

1. The score would be reset to 0 each time the game was restarted.
2. The score would increase incrementally based on the rules set by the student.

This task was selected for analysis as it was a complex programming activity. In terms of the computational thinking framework outlined in Table 12.1, it involves the computational concepts of operators such as variables. In this task, students need to use the variable-related blocks such as *set score* and *change score* to set the rules for controlling the score variable (see Fig. 12.2). An example of a possible solution is

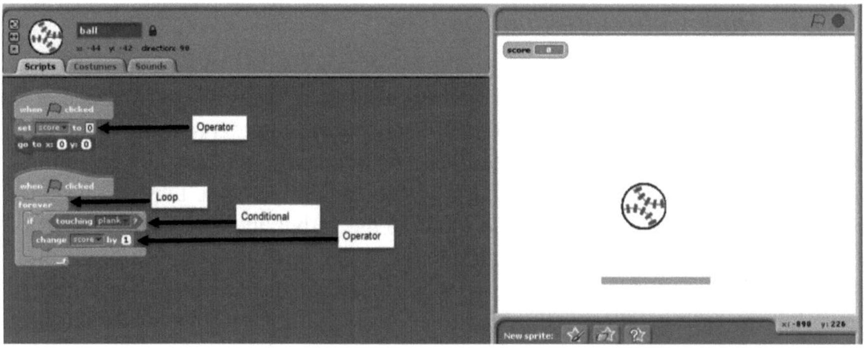

Fig. 12.2 Possible solution to programming task

shown in Fig. 12.2 where *set score to 0* is triggered when the green flag on the top right-hand corner is clicked by the player, whereas *change score by 1* awards the player one point each time the ball touches the plank. Students also had to use nested conditional looping (i.e. the *if touching plank* block is nested within the *forever* block; see Fig. 12.2). If there was no nested conditional looping (i.e. the removal of *forever* block), the *if* block would only execute once only during the start of the game, and the score would not be increased accordingly when the ball touched the plank. Such kinds of nested looping were found to be challenging for the students (Lee 2010; Meerbaum-Salant et al. 2013).

Prior to this task, the students had already learned the computational thinking concepts of sequences, loops, and conditionals and had created programs that implemented these concepts. The concept of operators (e.g. variable-related blocks) was new in this task.

12.4.2 Selection of Cases

Each case was purposefully selected based on the maximum variation strategy which was common in qualitative studies (Creswell 2013; Merriam 2009). These cases represented students with different performance outcomes in the programming task. The program of Student A did not meet the requirements at all, while that of Student C met all the requirements. Student B's program met requirements partially as the scoreboard in his game could not be reset to zero when a new game was played. This case selection strategy provided varied perspectives of programming based on task performance outcomes. It would add to the current limited knowledge on how elementary school students approach programming and whether there may be any differences among students with different performance outcomes in programming tasks.

12.4.3 Data Collection and Analysis

This data collection method is based on methods used in the studies of Berland et al. (2013) and Baytak and Land (2011) where students' programming logs were documented and, at times, videorecorded. In our study, students' programming logs were documented as a video with a screen capture software that was installed on their computers. The software also captured the verbal utterances of the students. A 20-minute on-screen recording with utterances was captured for each of the three students which spanned the duration they spent on the scoring task. Through capturing the on-screen recording and utterances, we hope to obtain insights into the dimensions of computational practices applied by the students. These video recordings formed the primary data for the study. To triangulate these data collected, 10-minute individual interviews were conducted with each student after the five lessons. During the interview, students were asked about their programming experiences. The program that each student created was also collected and analysed to understand the outcomes of students' computational practices and their application of computational concepts.

Content analysis was used to analyse the on-screen recordings. Each recording was broken into moves as units of analysis. Each move represented a specific programming behaviour. These units were coded using directed content analysis where there were predetermined categories from existing literature reviews and categories that emerged from the data (Hsieh and Shannon 2005). In this study, we first drew reference from the computational practices specified by Brennan and Resnick (2012) for computational thinking which comprised of experimenting and iterating, testing and debugging, reusing and remixing, and abstracting and modularizing. When examining students' programming behaviours, we found it that it was difficult to distinguish practices such as experimenting and iterating from testing and debugging. From our data, we classified these practices as Coding and Examining (see Table 12.2) as we observed students to be either inserting and

Table 12.2 Activities during programming

Programming behaviour	Observed N	% of total	Standardized residual
Examining – Students examining the program output or blocks	85.00	37.61	6.61
Coding – Students coding by inserting or manipulating blocks or sprites	67.00	29.65	3.63
Verbalizing – Students making utterances about the contents of blocks	29.00	12.83	−2.69
Helping – Students helping others to program or seeking help from peers or resources	24.00	10.62	−3.52
Non-programming tasks – Activities that are not related to programming such as playing games and interacting with peers on non-programming issues	21.00	9.29	−4.02
Total	226	100.00	

manipulating blocks or sprites or examining their blocks and programming output. The category of Coding was also used by Wyeth (2008) and Berland et al. (2013) to describe students' programming moves.

Throughout this process, we observed students to be making verbalizations and seeking or giving help as they engaged in Coding and Examining. Therefore, the categories of Verbalizing and Helping were added. In addition, students also engaged in Non-programming tasks as they played games after completing their assigned programming tasks and interacted with peers on on-programming issues.

While Brennan and Resnick proposed that students' computational practices also comprised reusing and remixing as well as abstracting and modularizing, we felt that these could be better assessed through examining the students' programming process. As the purpose of content analysis was to code students' programming behaviour move by move, we did not include these categories within the coded programming behaviours but analysed if students exhibited these practices as we study their programming process in the Findings section. To establish inter-rater reliability, 10% of the units were coded by a second rater, and a Cohen's kappa of 0.87 was derived, indicating adequate reliability. The student interviews were analysed for dominant themes using content analysis.

To answer the research questions, the frequencies of the coded categories were counted up and analysed statistically. The first research question was examined with the chi-square goodness-of-fit test to identify the dominant programming activities. The Pearson's chi-square test for association was used to analyse the differences in programming behaviour among students. These results were triangulated against the themes that emerged from the interviews as well as the students' programming products.

12.5 Findings

12.5.1 Students' Programming Behaviours

As evident in the recordings, the students were engaged in five kinds of programming behaviours (see Table 12.2).

The chi-square goodness-of-fit test found significant differences among the frequencies of the categories coded (χ^2 (4, N = 226) = 74.26, $p < 0.05$). As per Agresti (2007), cells with a standardized residual greater than +2 indicated that its observed frequency was above-expected, whereas a residual smaller than −2 indicated that the observed frequency was below-expected. Thus, Table 12.2 shows that students' engagement in Examining and Coding during programming were above-expected, accounting for more than 50% of the coded units, whereas the rest of the behaviours were below-expected. Examining occurred most frequently among the units coded with students executing their programs to observe if there was any mismatch between the intended and actual output. Students could also be Examining by mousing over blocks to determine if these should be applied or modified. Besides

Fig. 12.3 Verbalizing on existing blocks

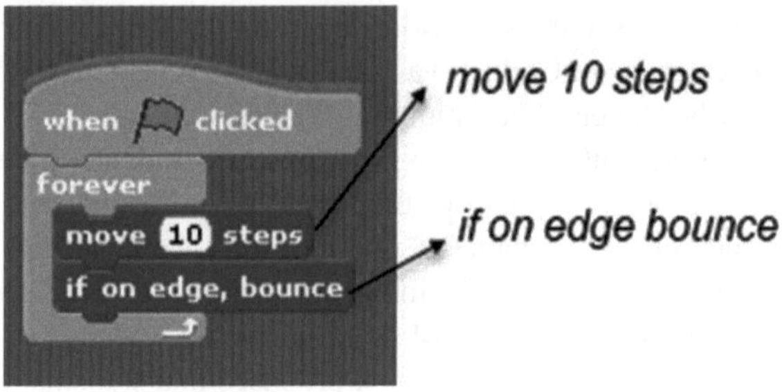

Examining, students also paid considerable attention to Coding, which accounted for nearly 30% of the coded units. These behaviours involved students inserting or manipulating blocks or sprites.

Among the programming behaviours that had below-expected frequencies, Verbalizing and Helping accounted for close to 23% of the coded units. Verbalizing referred to students' verbal utterances that were made in correspondence to their examining and coding activities. They could be thinking aloud about possible programming steps (e.g. "Make variable") or verbalizing particular blocks in their program. In Fig. 12.3, for example, a student verbalized "Move 10 steps" and "If on edge bounce" as he was examining the blocks he had used in the program.

About 11% of the units coded involved students helping or seeking help from others. When faced with problems during programming, some students would seek help from the knowledgeable others (e.g. peers or teachers) or refer to content resources such as the teacher's worked examples. When help was sought by peers, students were found to be providing step-by-step instructions, answering peers' questions, pointing out possible mistakes, or helping their peers to do the programming.

About 9% of the coded units involved students engaging in Non-programming tasks. This included interacting with others on non-programming matters, playing Scratch games created by themselves or their peers. These activities typically occurred when students have completed their programming-related activities ahead of time or when students have encountered repeated failed attempts to solve their programming bug and decided to do something other than their programming activity.

12.5.2 *Differences in Programming Behaviours*

A chi-square test of independence found significant differences among students' programming behaviours (χ^2 (8, N = 226) = 37.69, $p < 0.05$). From Table 12.3, it can be seen that the frequency of occurrences of Examining and Coding were within expectation for all three students, indicating that they were largely on task. Student A had above-expected frequencies for Helping and Non-programming tasks, while that

Table 12.3 Comparison of activities

	Examining	Coding	Verbalizing	Helping	Non-programming tasks	Total
Student A – Failed to meet requirements						
Actual	19.00	12.00	1.00	11.00	13.00	56.00
Expected	21.06	16.60	7.19	5.95	5.20	56.0
%	33.93	21.43	1.79	19.64	23.21	100.00
Standardized residual	−0.66	−1.55	−2.85	2.53	4.14	
Student B – Partially met requirements						
Actual	30.00	27.00	9.00	2.00	6.00	74.00
Expected	27.83	21.94	9.50	7.86	6.88	74.00
%	40.54	36.49	12.16	2.70	8.11	100.00
Standardized residual	0.63	1.57	−0.21	−2.70	−0.43	
Student C – Met all requirements						
Actual	36.00	28.00	19.00	11.00	2.00	96.00
Expected	36.11	28.46	12.32	10.19	8.92	96.00
%	37.50	29.17	19.79	11.46	2.08	100.00
Standardized residual	−0.03	−0.14	2.69	0.35	−3.21	
Total count	85.00	67.00	29.00	24.00	21.00	226.00
%	37.61	29.65	12.83	10.62	9.29	

for Verbalizing was below-expected. This contrasted sharply with Student C whose observed frequency for Verbalizing was above -expected, while the observed frequency for Non-Programming tasks was below-expected. Student B was the only student with below-expected frequency for Helping. The specific programming behaviours of each student are described in the following sections.

12.5.2.1 Student A

Student A is an 11-year-old boy who claimed to have learned Scratch programming from his after-school care centre. However, he did not manage to fulfil the programming task as he did not create any script with the operator-related blocks. The program he created did not have the required scoring mechanism at all.

Even though Student A's engagement in Coding was within expectation, analysis of his programming process showed that he was not coding according to the task specifications but redesigned the programming task to make it more manageable (see Fig. 12.4). Working from a bouncing ball program that he created during the last lesson, he inserted another sound block (*play sound HumanBeatBox until done*) below an existing sound block *(play sound HipHop until done)* rather than create the scoring mechanism. He also created a script with two blocks so that the user can stop

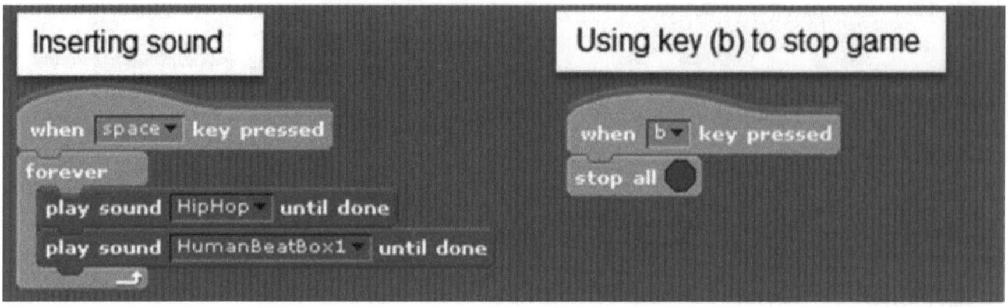

Fig. 12.4 Student A's scripts for the redesigned manageable task

the game by pressing the b key. In terms of computational thinking concepts specified by Brennan and Resnick (2012), the student has applied the concepts of sequence and events. He was unable to use operator-related blocks (e.g. *set score*) or create blocks with nested conditional looping. He engaged in the computational programming practices of experimenting and iterating to add media effects at this point. The simplicity of the task led to little need for engagement in the practices of testing and debugging.

From Table 12.3, his observed frequency for Helping was above-expected. He actively sought help from peers or used content resources such as worked examples in his attempts at the programming task. He explicitly requested his peers to "just come and do it [the scoring task] now". His peer obliged and made all the required changes for the scoring task.

The student then sought to fix the bouncing ball problem in his program. The ball was supposed to move downwards when hitting the upper plank 1 and upwards when hitting the lower plank 2. The program output in Fig. 12.5a shows the ball bouncing upwards when hitting plank 1. An examination of the students' original program in Fig. 12.5b shows an error in specifying the *point* block. For a ball to point downwards *if touching plank 1*, the nested point block should be specified as *point in direction 180*. Conversely, for the ball to move upwards *if* touching plank 2, the nested point block should be specified as *point in direction 0*. The student attempted to engage in some aspects of the computational thinking practices of reusing and remixing when he used the teacher's bouncing ball template (see Fig. 12.5c) to improve his program. This template was originally provided as some of the students were not able to make the ball bounce properly in the previous lesson. The teacher provided the template to help the students focus on programming the scoring mechanism during this lesson. With the help of this resource, the student engaged in the computational practices of experimenting and iterating as well as testing and debugging to fix the problem. Figure 12.5d shows him modifying his program accordingly by changing the *turn* block inputs from 60° to 45° as per the template. This change did not solve the problem as the problem was with *the point in direction* block. For example, the ball was still pointed upwards when it hit plank 1. He did not succeed at the first try and had to refer the template again. Figure 12.5e shows his second attempt where he changed the inputs in the *point* block for both plank 1 and

(a) Program output

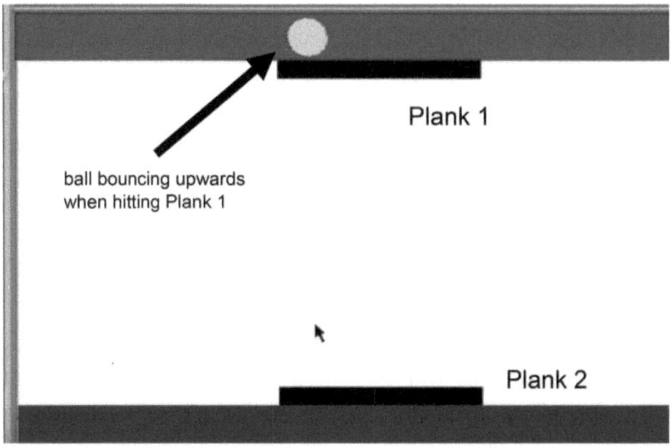

(b) Student's Original Program

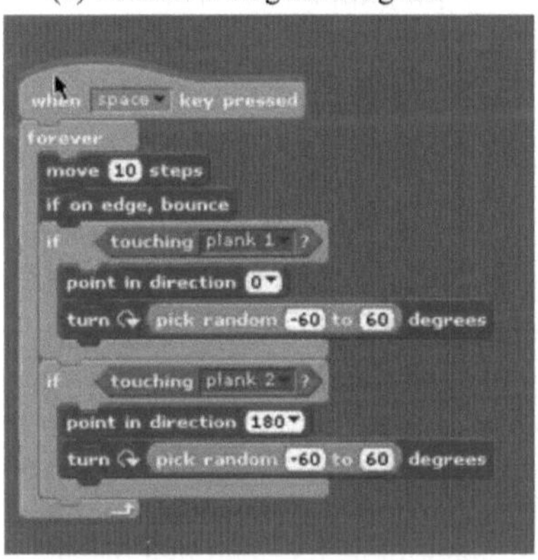

(c) Bouncing ball template

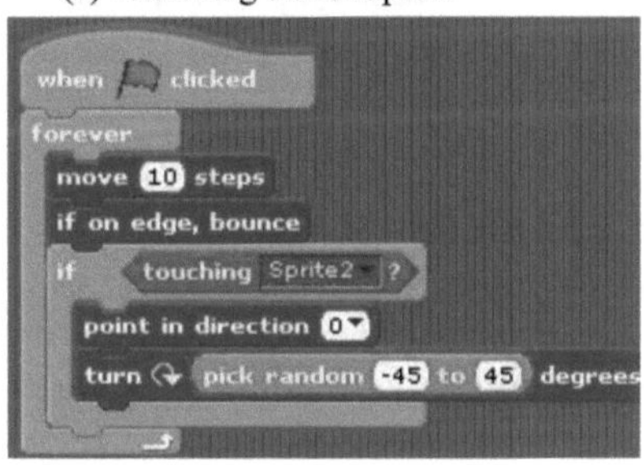

(d) Modification 1 – Changing inputs to turn block

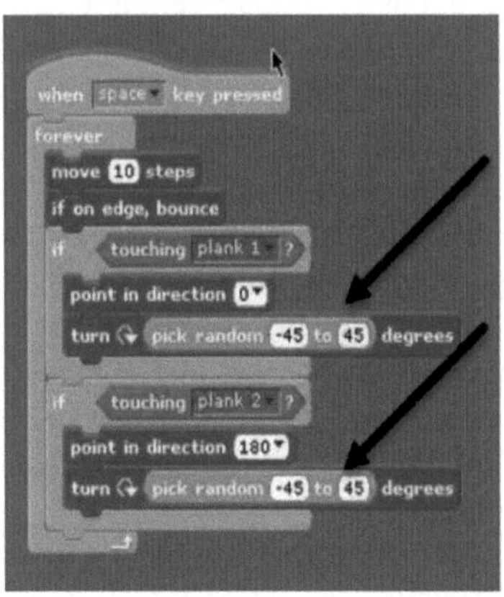

(e) Modification 2 – Changing inputs to point

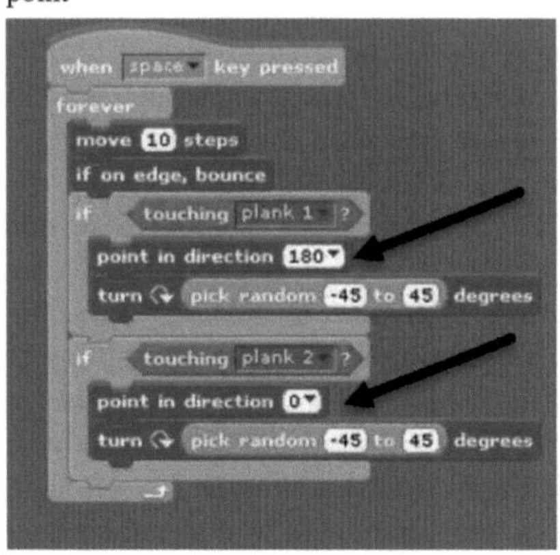

Fig. 12.5 Modification based on bouncing ball template

plank 2. This enabled the ball to bounce properly as the orientation of the ball was set correctly at downwards and upwards for plank 1 and plank 2, respectively.

About 23% of the units coded for Student A was for Non-programming tasks, the highest among the three students. Analysis of the videos showed that when he was not working on the programming task, he was either interacting with his friends on non-programming matters or playing his self-created Scratch games either individually or with his peers. When interviewed, Student A expressed a trial-and-error approach to programming where:

> I see the output. I click on the blocks and see what they can do.... I try to fix them again and try to make the program correct.

It appeared that he made surface-level changes to blocks by changing its inputs to match the teachers' template but did not attempt to shift or manipulate blocks dynamically. He refocused the programming tasks and chose to do what appeared manageable to him, relying heavily on peer support for the tasks that were beyond him. His higher-than-expected focus on Non-programming tasks could also be a means of avoiding tasks that may seem difficult to him.

12.5.2.2 Student B

Student B is an 11-year-old boy who did not have prior programming experience. His program partially fulfilled the task requirements because it failed to reset the scoring mechanism of the game to zero when the game started.

From Table 12.3, it can be seen that Student B was the only student who had below-expected frequencies for Helping as he largely depended on himself to solve programming problems. A major problem he encountered was that his scoreboard remained at zero and could not change. Screen recordings revealed that despite the seven unsuccessful attempts, Student B persevered by Coding, Examining his program output, and modifying the program continuously without seeking external help. At times, he would verbalize his thoughts as indicated by the units coded under Verbalizing. In terms of computational thinking practices, Student B can be characterized as one who engaged intensively in experimenting and iterating as well as testing and debugging.

Figure 12.6 shows Student B's coding process. In his original program (see Fig. 12.6a), Student B's score could not change because he applied a block that *set score to 0* rather than *change score*. In Modification 1 (see Fig. 12.6b), he applied nested conditional looping by shifting the *if* block within *forever* and *if touching Sprite 2*. He also attempted to replace *set scores to 0* to *change scores by 1*. He examined the output and found that the score still did not change as he had different conditions nested within the *if touching Sprite 2* block.

He then attempted to understand his program by verbalizing the blocks he just modified (*If touching Sprite 1, change score*). In Modification 2 (see Fig. 12.6c), he shifted the *turn* right by 60° block to the forever loop to isolate the *change score*

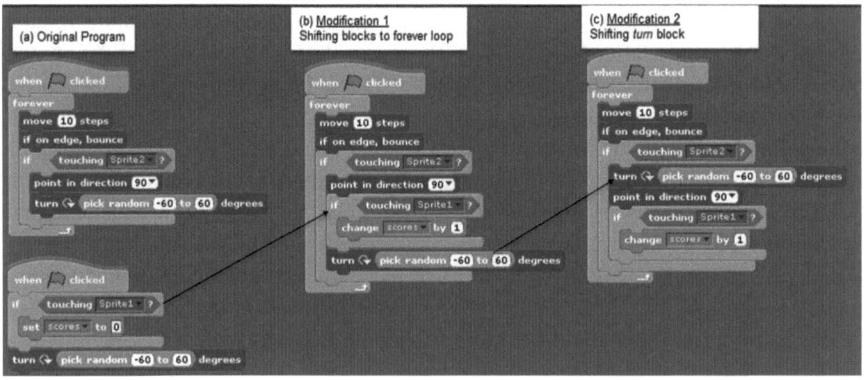

Fig. 12.6 Making changes

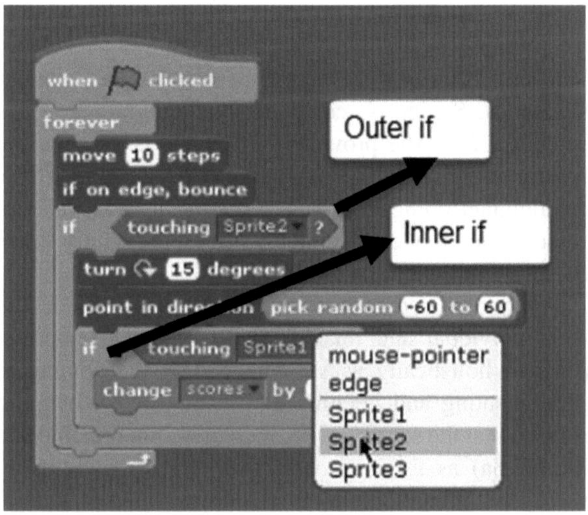

Fig. 12.7 Student B's final program

block. Seeing that the output is not still changing, he again verbalized the blocks related to scoring (*If touching Sprite 1, change score*).

Student B eventually solved the problem when he realized that he had specified the wrong conditions as the two blocks containing the *if touching* statements referred to different sprites (see Fig. 12.7). The self-realization of the buggy block was evident through his utterance ("This thing is Sprite 2"). He then edited the program accordingly by making the sprite consistent in both the outer *if* and inner *if* blocks (see Fig. 12.7). However, Student B failed to realize that the inner *if* block was redundant and the same solution could be attained by just adding the *change score* block after the *point in direction* block. Even though Student B has fixed the changing score problem, he did not manage to complete the other task requirement of resetting the score to zero during a game replay.

In contrast to Student A, Student B made some attempts to examine related blocks. For example, after Modification 2 (see Fig. 12.6c), he verbalized the related blocks that might cause the score to change (*If touching Sprite 1, change score*). From this utterance, we can infer that he was reading one block after another and considering how the two blocks in close proximity could affect each other. This was further triangulated in his interview in which he described that he read his program in "step-by-step" mode as an approach to programming. Nevertheless, we observe that Student B typically focused his attention on the blocks he had just modified, thereby limiting his examination to a small chunk of program. His utterances, which tend to focus on single blocks, e.g. *Set score*, showed that he could be reading chunks of blocks in isolation. This could be a reason why he failed to realize that there was a redundant inner *if* block.

12.5.2.3 Student C

Student C is an 11-year-old girl who had some experience with programming. Like Student B, Student C completed her programming task with minimal help from others. In fact, she provided help to peers even as she was completing her own programming task. The standardized residuals of Table 12.3 show that Student C predominantly engaged in Verbalizing as she was Coding and Examining. This could be her way of articulating her programming strategy in order to stay focused on the given task.

Analysis of screen recordings showed that Student C was thinking aloud about both individual and related blocks, which exemplified attempts to analyse her program holistically as she engaged in the computational thinking practices of experimenting and iterating and testing and debugging. For example, Student C realized that the score was not increasing but remained at zero during game play (see Fig. 12.8a) as the *set score to 0* is nested within the *forever* block. She thus verbalized, "Set Score to 0 outside the forever thingy," and then shifted the *set score to 0* block outside the *forever* block (see Fig. 12.8b). In contrast to Student B, these moves showed that Student C had a more holistic view of her program as she did not pay attention to blocks within close proximity of each other. In this example, she was cognizant that the *forever* block could affect the *set* block even though these were not close to another.

Student C was also observed to be making strategic programming moves rather than inserting blocks by trial and error. In one iteration of her program, she discovered a bug where her *change score* block was nested within a *forever* block (see Fig. 12.9a). This caused the score to change infinitely. She realized how the *forever* block was related to the ever-changing score in the program. Thus, she removed the forever loop (see Fig. 12.9b).

Student C also ensured that there were no redundant blocks in her program as she removed sprites in her previous moves. For example, in Fig. 12.10, there were

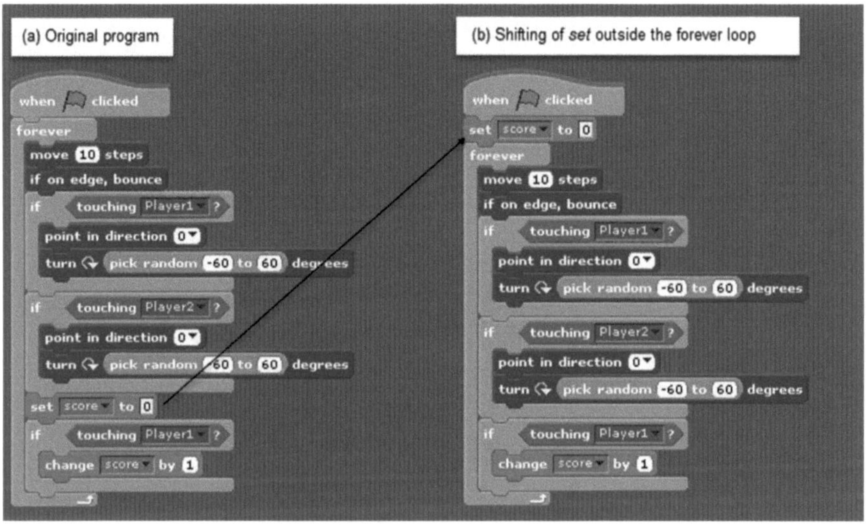

Fig. 12.8 Verbalizing proposed solution

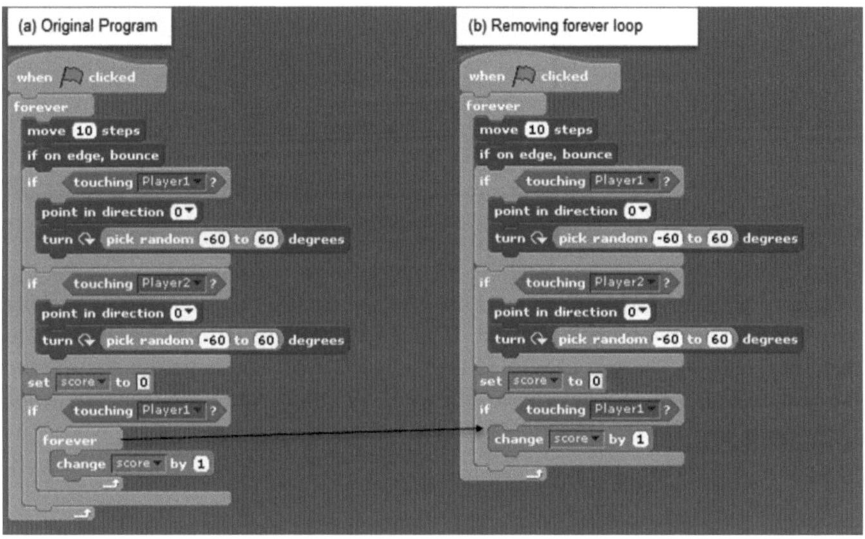

Fig. 12.9 Making a strategic move

missing conditions in the last two *if* blocks. While the presence of these blocks did not affect the running of the program, these would never be executed. Through her utterance "I know how to make this shorter", we deduced that she was aware that these blocks were redundant. She then removed these blocks from the program accordingly.

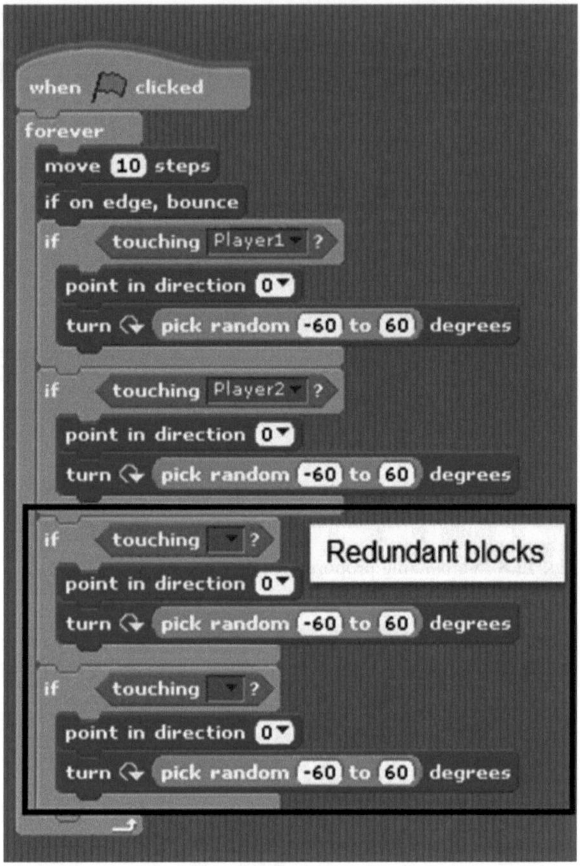

Fig. 12.10 Removing redundant blocks

Interviews with Student C showed that she approached programming by seeking to understand her program as she said, "I have to look at the script". As compared to the other students, she was able to go beyond surface-level changes by considering the relationships among blocks across the whole program. Among the three students, she was the only student who fulfilled all the task requirements.

12.6 Discussion

This study analysed possible differences in the programming behaviours of elementary school students through three purposive naturalistic case studies of students with different performance outcomes in a common programming task. Three kinds of programming behaviours were observed.

12.6.1 *Programming Behaviours*

12.6.1.1 Student A: Trial-and-Error Programming

It was found that Student A approached programming through changing block inputs by trial and error. In terms of computational thinking practices, it appeared that Student A was not deficient in carrying out experimenting and iterating as well as testing and debugging. However, this study shows that the mere act of carrying out these computational thinking practices did not imply that quality programming was happening. One reason for Student A's programming behaviour could be explained with literature review findings about children where they were likely to be programming in trial and error due to the immediate feedback afforded by the programming languages (Berland et al. 2013; Fessakis et al. 2013; Wyeth 2008). Nevertheless, Student A also exhibited some forms of problem avoidance as he reframed the programming task into what he appeared to be more confident of doing, depended on external help, and engaged in Non-programming tasks to a larger extent than the other students. Student A also did minimal manipulation of blocks and did not attempt to create nested loops, possibly indicating his poor grasp of domain knowledge. Domain knowledge refers to the "generic theories and principles upon which the system is designed" (Jonassen 2011, p. 80), and it involves both declarative and procedural knowledge of the system of interest (Alexander 1992). Declarative knowledge is about "knowing that", while procedural knowledge is about "knowing how" (Alexander 1992). Student A appeared to lack declarative knowledge related to computational thinking concepts as he did not know which blocks to use. His attempts at copying the teacher's template demonstrated his weak procedural knowledge as he did not appear to have had deep understanding of the programming intent of the template.

12.6.1.2 Student B: Piecemeal Programming

As compared to Student A, Student B appeared to have stronger understanding of computational thinking concepts as he was able to use sequences, loops, operators, and conditionals. Student B can be characterized as a self-dependent but piecemeal programmer as he largely examined programming blocks in close proximity to each other and did not examine the interrelationships among blocks. While such kinds of programming behaviour did not surface in K-12 programming studies, this is nevertheless common among novice programmers at tertiary level (e.g. Vessey 1985). By adopting such a programming approach, Student B failed to have a thorough understanding of his program. Therefore, in terms of computational thinking practices, it can be seen that Student B's execution of experimenting and iterating as well as testing and debugging was analytically superior to that of Student A. There were some attempts to understand bugs and make reasoned changes to his program even though it was piecemeal in nature.

12.6.1.3 Student C: Holistic Programming

In comparison, Student C can be characterized as one who attempted to program through analysing and understanding her program holistically. It appeared that frequent verbalization supported Student C as she attempted to make reasoned rather than haphazard changes to the program. Interestingly, Student C also attempted to make her program efficient by cleaning up redundant blocks. Among the three students, Student C showed good understanding of computational thinking concepts. Her attempts at the computational thinking practices of experimenting and iterating as well as testing and debugging was undergirded by a holistic understanding of where she was heading with her program. Such an approach is superior to the trial-and-error and piecemeal approaches adopted by Students A and B, respectively.

These descriptive findings provided us with insights of how these students were problem-solving during programming. As the fostering of problem-solving skills is one of the driving forces behind the heightened interest in K-12 programming and computational thinking (Kafai and Burke 2013; Resnick et al. 2009), the following instructional implications need to be considered:

12.6.2 Instructional Implications

12.6.2.1 Develop Domain Knowledge

The programming strategy adopted by Student A indicated that some students could need more time to acquire the domain knowledge for programming. Essentially, these students could be unclear of the functions of the programming block. Without such basic domain knowledge, they might be just experimenting with blocks and not engaging in any meaningful problem-solving. Besides explicit teaching, students can be pointed to instructional resources such as videos, cue cards, or the Help menu in Scratch.

Other kinds of instructional resources could include partial worked examples from which student can be taught to reuse and remix for their given programming task. This appeared to be something that Student A referred to. In fact, reusing and remixing is widely adopted by programming professionals (Haefliger et al. 2008).

In this study, Student A actively sought help from his peers, but they responded by doing the task for him. Peer support, as scaffold, could be encouraged with peers guiding but not taking over programming tasks. Such kinds of peer support are in line with the community in constructionism (Papert 1980) where community members (e.g. students) can act as “collaborators, coaches, audience, and co-constructors of knowledge” (Kafai and Resnick 1996, p. 6). Evidently, peer support is an important element in K-12 programming (Denner et al. 2012; Lewis 2011; Martin et al. 2013). Nevertheless, young children may not instinctively know how to ask good questions. Teachers may have to engage in modelling (Palinscar and Brown 1984; Schünemann et al. 2013) and provide students with question prompts to do so (Choi et al. 2005; Ge and Land 2003).

The integrative use of teacher, peers, and instructional resources can function as a distributed scaffolding system (Puntambekar and Kolodner 2005) for building students' domain knowledge. This can better cater to students who have different zones of proximal development (Brown et al. 1993; Li and Lim 2008; Puntambekar and Kolodner 2005) with respect to programming.

12.6.2.2 Help Students to Develop a Holistic View of the Program

The study results showed that Student B lacked a holistic view of the program as he focused on the blocks that concerned the changes he was making at a point in time. As such, he failed to consider how the different parts of the program were interrelated and was not able to build "a causal model of the program structure and the error in it" (Vessey 1985, p. 490). This is unlike better programmers who are more likely to "seek the relations of objects, which leads to a connected view of [the] program" (Wiedenbeck et al. 1993, p. 807).

For students like Student B, they could be taught to comprehend their programs by thinking aloud and verbalizing their programming steps like Student C. Such kinds of verbalization could help them better understand their program steps. This thinking-aloud strategy is highly recommended in other subject areas such as reading comprehension (Kucan and Beck 1997) where it was suggested that "breaking into the reading process by asking readers to think aloud, or verbalize, as they read focuses their attention and requires that they spend more time thinking about what they are reading" (p. 292).

Another way to develop a holistic view of the program is through program comprehension exercise. Students could be given programs and asked to predict their outputs and reasons for their predictions. The chosen programs could be based on the students' own buggy programs which could also be used to demonstrate how blocks in the various parts of the programs are related. In this study, the teacher did provide opportunities for the students to predict output during whole-class discussion or when students approached teacher for help. As the students only had three programming lessons prior to undertaking the task for this study, more program comprehension exercises may be needed to help the students to develop a holistic view of the program.

12.6.3 Help Students to Develop the Practice of Abstraction and Modularization

Although Student C completed the assigned task, her solution could be further enhanced through abstraction and modularization which is "building something large by putting together collections of smaller parts" (Brennan and Resnick 2012, p. 7). An example can be seen in Fig. 12.2 where different stacks of blocks were created for the resetting of score to 0 and the increment of scores when the ball touched the plank. Abstraction and modularization makes it easier for programmers to debug programs by compartmentalizing them into different stacks of codes or sprites. However, for Student C, the program was not compartmentalized, and the solution was implemented in just one lengthy stack of programming blocks (see Fig. 12.8). From this study, we speculate that abstraction and modularization may not come naturally for elementary students, even those who may be better programmers. There would, thus, be a need to make students aware of such good programming practices through role modelling. Developing abstraction and modularization is likely to sharpen students' problem-solving skills as it could help them to organize the programming problem to make troubleshooting easier.

12.7 Limitations and Future Research

In this paper, we studied how three students were programming, while they were a given a fairly complex programming task. Even though the task may be relatively complex for these young and inexperienced programmers, it nonetheless is not reflective of students' programming behaviours when confronted with more complex, open-ended programming tasks. The programming behaviours that emerged still need to be validated through further studies. Another limitation of the study was the selection of three purposive case studies based on students' programming outcomes. The intent of this study was to surface some initial insights, but further studies through larger samples of students are needed to validate the programming behaviours observed. The study is also limited in its examination of how students' prior programming experiences and motivations for programming may have influenced their programming behaviours. Even though both Student A and Student C appeared to have had some experience in Scratch programming, their programming outcomes and programming behaviours differed. Students' motivation and self-efficacy for programming could have influenced the efforts and outcomes. However, these factors were not examined deeply in this study as the focus was on students' programming behaviours. In future studies, the influence of instrumental aspects as well as students' perception of their learning environments could be taken into consideration for a more holistic understanding of students' programming behaviours. Another limitation of this study is that computational programming perspectives of the children were not examined as the focus of the study was to understand the children's programming behaviours. Future studies could examine this aspect through post-activity reflections for the children to articulate how their understandings of themselves, others, and the world have shifted through their engagement in Scratch programming.

12.8 Conclusion

Despite the heightened interest in how computational thinking may be applied in K-12 programming, only a few studies examined how young children program. This study attempted to bridge this gap through in-depth analysis of how three Grade 4 students were programming. The different programming behaviours of these three students indicate that young children may not approach programming in a similar manner. As such, K-12 programming teachers may need to scaffold these different programming behaviours with differentiated forms of instruction so that computational thinking can be acquired more easily by students with different efficacies for programming. The findings of this study provide initial insights for K-12 educators. More studies in this area are needed for interested educators and researchers to plan K-12 programming lessons effectively.

References

Agresti, A. (2007). *An introduction to categorical data analysis*. Hoboken, NJ: John Wiley.

Alexander, P. A. (1992). Domain knowledge: Evolving themes and emerging concerns. *Educational Psychologist, 27*(1), 33.

Ananiadou, K., & Claro, M. (2009). *21st century skills and competences for new millennium learners in OECD countries*. OECD Education working papers, 41. doi: https://doi.org/10.1787/218525261154.

Barr, V., & Stephenson, C. (2011). Bringing computational thinking to K-12: What is involved and what is the role of the computer science education community? *ACM Inroads, 2*(1), 48–54. https://doi.org/10.1145/1929887.1929905.

Baytak, A., & Land, S. M. (2011). An investigation of the artifacts and process of constructing computers games about environmental science in a fifth grade classroom. *Etr&D-Educational Technology Research and Development, 59*(6), 765–782. https://doi.org/10.1007/s11423-010-9184-z.

Bednarik, R. (2012). Expertise-dependent visual attention strategies develop over time during debugging with multiple code representations. *International Journal of Human-Computer Studies, 70*(2), 143–155. https://doi.org/10.1016/j.ijhcs.2011.09.003.

Berland, M., Martin, T., Benton, T., Smith, C. P., & Davis, D. (2013). Using learning analytics to understand the learning pathways of novice programmers. *Journal of the Learning Sciences, 22* (4), 564–599. https://doi.org/10.1080/10508406.2013.836655.

Bers, M. U., Flannery, L., Kazakoff, E., & Sullivan, A. (2014). Computational thinking and tinkering: Exploration of an early childhood robotics curriculum. *Computers & Education, 72*, 145–157. https://doi.org/10.1016/j.compedu.2013.10.020.

Binkley, M., Erstad, O., Herman, J., Raizen, S., Ripley, M., Miller-Ricci, M., & Rumble, M. (2012). Defining twenty-first century skills. In P. Griffin, B. McGaw, & E. Care (Eds.), *Assessment and teaching of 21st century skills* (pp. 17–66). Dordrecht, Netherlands: Springer.

Brennan, K., & Resnick, M. (2012). *New frameworks for studying and assessing the development of computational thinking*. Paper presented at the annual American Educational Research Association meeting, Vancouver, BC, Canada. http://web.media.mit.edu/~kbrennan/files/Brennan_Resnick_AERA2012_CT.pdf

Brown, A. L., Ash, D., Rutherford, M., Nakagawa, K., Gordon, A., & Campione, J. C. (1993). Distributed expertise in the classroom. In G. Saloman (Ed.), *Distributed cognitions:*

Psychological and educational considerations (pp. 188–228). Cambridge, UK: Cambridge University Press.

Burke, Q. (2012). The markings of a new pencil: Introducing programming-as-writing in the middle school classroom. *Journal of Media Literacy Education, 4*(2), 121–135.

Choi, I., Land, S., & Turgeon, A. (2005). Scaffolding peer-questioning strategies to facilitate metacognition during online small group discussion. *Instructional Science, 33*(5–6), 483–511. https://doi.org/10.1007/s11251-005-1277-4.

Cooper, S. (2010). The design of Alice. *ACM Transactions on Computing Education, 10*(4), 1–16. https://doi.org/10.1145/1868358.1868362.

Creswell, J. W. (2013). *Qualitative inquiry and research design: Choosing among five approaches* (3rd ed.). Thousands Oaks, CA: Sage.

Denner, J., Werner, L., & Ortiz, E. (2012). Computer games created by middle school girls: Can they be used to measure understanding of computer science concepts? *Computers & Education, 58*(1), 240–249. https://doi.org/10.1016/j.compedu.2011.08.006.

Feng, C.-Y., & Chen, M.-P. (2014). The effects of goal specificity and scaffolding on programming performance and self-regulation in game design. *British Journal of Educational Technology, 45* (2), 285–302. https://doi.org/10.1111/bjet.12022.

Fessakis, G., Gouli, E., & Mavroudi, E. (2013). Problem solving by 5–6 years old kindergarten children in a computer programming environment: A case study. *Computers & Education, 63*, 87–97. https://doi.org/10.1016/j.compedu.2012.11.016.

Ge, X., & Land, S. (2003). Scaffolding students' problem-solving processes in an ill-structured task using question prompts and peer interactions. *Educational Technology Research and Development, 51*(1), 21–38. https://doi.org/10.1007/BF02504515.

Grover, S., & Pea, R. (2013). Computational thinking in K–12: A review of the state of the field. *Educational Researcher, 42*(1), 38–43. https://doi.org/10.3102/0013189x12463051.

Haefliger, S., Von Krogh, G., & Spaeth, S. (2008). Code reuse in open source software. *Management Science, 54*(1), 180–193.

Hsieh, H.-F., & Shannon, S. E. (2005). Three approaches to qualitative content analysis. *Qualitative Health Research, 15*(9), 1277–1288.

Jonassen, D. (2011). *Learning to solve problems: A handbook for designing problem-solving learning environments*. New York: Routledge.

Kafai, Y., & Burke, Q. (2013). Computer programming goes back to school. *Phi Delta Kappan, 95* (1), 61–65.

Kafai, Y., & Resnick, M. (Eds.). (1996). *Constructionism in practice: Designing, thinking, and learning in a digital world*. Mahwah, NJ: Lawrence Erlbaum.

Kafai, Y., Fields, D. A., & Burke, Q. (2010). Entering the clubhouse: Case studies of young programmers joining the online scratch communities. *Journal of Organizational and End User Computing, 22*(2), 21–35. https://doi.org/10.4018/joeuc.2010101906.

Kucan, L., & Beck, I. L. (1997). Thinking aloud and reading comprehension research: Inquiry, instruction, and social interaction. *Review of Educational Research, 67*(3), 271–299. https://doi.org/10.3102/00346543067003271.

Lee, Y.-J. (2010). Developing computer programming concepts and skills via technology-enriched language-art projects: A case study. *Journal of Educational Multimedia and Hypermedia, 19*(3), 307–326.

Lehrer, R., Lee, M., & Jeong, A. (1999). Reflective teaching of logo. *Journal of the Learning Sciences, 8*(2), 245–289. https://doi.org/10.1207/s15327809jls0802_3.

Lewis, C. M. (2011). Is pair programming more effective than other forms of collaboration for young students? *Computer Science Education, 21*(2), 105–134. https://doi.org/10.1080/08993408.2011.579805.

Li, D. D., & Lim, C. P. (2008). Scaffolding online historical inquiry tasks: A case study of two secondary school classrooms. *Computers & Education, 50*(4), 1394–1410. https://doi.org/10.1016/j.compedu.2006.12.013.

Lin, J. M. C., & Liu, S. F. (2012). An investigation into parent-child collaboration in learning computer programming. *Educational Technology & Society, 15*(1), 162–173.

Martin, T., Berland, M., Benton, T., & Smith, C. P. (2013). Learning programming with IPRO: The effects of a mobile, social programming environment. *Journal of Interactive Learning Research, 24*(3), 301–328.

Meerbaum-Salant, O., Armoni, M., & Ben-Ari, M. (2013). Learning computer science concepts with scratch. *Computer Science Education, 23*(3), 239–264. https://doi.org/10.1080/08993408.2013.832022.

Merriam, S. B. (2009). *Qualitative research: A guide to design and implementation*. San Francisco, California: Jossey-Bass.

NRC. (2012). *A framework for K-12 science education: Practices, crosscutting concepts, and core ideas*. The National Academies Press.

Palinscar, A. S., & Brown, A. L. (1984). Reciprocal teaching of comprehension-fostering and comprehension-monitoring activities. *Cognition and Instruction, 1*(2), 117–175. https://doi.org/10.1207/s1532690xci0102_1.

Palumbo, D. B. (1990). Programming language/problem-solving research: A review of relevant issues. *Review of Educational Research, 60*(1), 65–89. https://doi.org/10.3102/00346543060001065.

Papert, S. (1980). *Mindstorms: Children, computers, and powerful ideas*. New York: Basic Books.

Papert, S. (1994). *The children's machine: Rethinking school in the age of the computer*. New York: Basic Books.

Puntambekar, S., & Kolodner, J. L. (2005). Toward implementing distributed scaffolding: Helping students learn science from design. *Journal of Research in Science Teaching, 42*(2), 185–217. https://doi.org/10.1002/tea.20048.

Resnick, M., Maloney, J., Monroy-Hernandez, A., Rusk, N., Eastmond, E., Brennan, K., et al. (2009). Scratch: Programming for all. *Communications of the ACM, 52*(11), 60–67. https://doi.org/10.1145/1592761.1592779.

Sáez López, J. M., González, M. R., & Cano, E. V. (2016). Visual programming languages integrated across the curriculum in elementary school: A two year case study using "scratch" in five schools. *Computers & Education*. https://doi.org/10.1016/j.compedu.2016.03.003.

Schünemann, N., Spörer, N., & Brunstein, J. C. (2013). Integrating self-regulation in whole-class reciprocal teaching: A moderator–mediator analysis of incremental effects on fifth graders' reading comprehension. *Contemporary Educational Psychology, 38*(4), 289–305. https://doi.org/10.1016/j.cedpsych.2013.06.002.

Su, A. Y. S., Yang, S. J. H., Hwang, W. Y., Huang, C. S. J., & Tern, M. Y. (2014). Investigating the role of computer-supported annotation in problem-solving-based teaching: An empirical study of a scratch programming pedagogy. *British Journal of Educational Technology, 45*(4), 647–665. https://doi.org/10.1111/bjet.12058.

Tangney, B., Oldham, E., Conneely, C., Barrett, S., & Lawlor, J. (2010). Pedagogy and processes for a computer programming outreach workshop—the bridge to college model. *IEEE Transactions on Education, 53*(1), 53–60.

Vessey, I. (1985). Expertise in debugging computer programs: A process analysis. *International Journal of Man-Machine Studies, 23*(5), 459–494. https://doi.org/10.1016/S0020-7373(85)80054-7.

Wiedenbeck, S., Fix, V., & Scholtz, J. (1993). Characteristics of the mental representations of novice and expert programmers: An empirical study. *International Journal of Man-Machine Studies, 39*(5), 793–812. https://doi.org/10.1006/imms.1993.1084.

Wing, J. M. (2006). Computational thinking. *Communications of the ACM, 49*(3), 33–35.

Wyeth, P. (2008). How young children learn to program with sensor, action, and logic blocks. *Journal of the Learning Sciences, 17*(4), 517–550. https://doi.org/10.1080/10508400802395069.

Chapter 13
Integrating Computational Thinking in School Curriculum

Mehmet Aydeniz

13.1 Introduction

There has been an increasing interest in making computer science a core subject in K-12 school curriculum and integrating computational thinking into STEM subjects. Several countries have developed national policies, standards, and guidelines for curriculum, teacher education, and assessment. In the United States, interest in computer science became a hot topic after NSF's CS 10K Project, whose aim was to prepare 10,000 computer science (CS) teachers by 2016 (Astrachan et al. 2011; The National Center for Women & Information Technology (NCWIT) 2012). NSF has sponsored different working groups to study the feasibility of infusing CS into K-12 education, conduct research about pedagogical aspects of computational thinking (CT), and develop assessment metrics for measuring CT skills across subject areas. These efforts resulted in a comprehensive definition of CT and ways to train teachers and develop the curriculum. The funding provided by the NSF motivated both curriculum development and teacher education efforts across the United States. The CS efforts in the United States can be put into four categories: (1) curriculum development programs (e.g., AP CS principles, Exploring Computer Science (ECS), code.org), (2) programs that target broadening participation in CS, (3) programs designed to certify teachers to teach computer science (Idaho Teach, UTeach, TEALS), and (4) integrating CT into STEM curriculum (e.g., NIELS, CTSiM, Scalable Game Design, Bootstrap).

Computational thinking entered into science education in the United States through introduction of the Next Generation Science Standards [NGSS] (National Research Council [NRC] 2013). NGSS places emphasis on science technology,

M. Aydeniz (✉)
College of Education, Health and Human Sciences, University of Tennessee, Knoxville, TN, USA
e-mail: maydeniz@utk.edu

M. S. Khine (ed.), *Computational Thinking in the STEM Disciplines*,
https://doi.org/10.1007/978-3-319-93566-9_13

engineering, and mathematics (STEM) practices. These practices include questioning, formulating evidence-based explanation, developing models, computational thinking, and engineering design. Among these practices, computational thinking has received significant attention from the computing industry and funding from federal agencies (Astrachan et al. 2011; Guzdial 2008; NRC 2011a, b; Wilson et al. 2010). Science and mathematics educators have applauded the investment in computational thinking and motivation to integrate computational thinking into science and mathematics curriculum. The motivation to integrate computational thinking into science and mathematics curriculum comes from the fact that computing concepts and practices have become an integral part of the work that professional scientists, mathematicians, and engineers do (Bailey and Borwein 2011; Denning 2017). If implemented effectively, computational thinking has the potential to significantly advance students' problem-solving and analytical thinking and data analysis and modeling skills (Barr and Stephenson 2011; Selby 2015). More, it is believed that the integration of computational thinking into school curriculum and students' engagement with application of computational practices and concepts in K-12 will better prepare them for the twenty-first-century economy and citizenship (Buckingham 2015; Smith 2016; Yadav et al. 2017; The World Bank 2016; Vee 2013).

Despite increasing interest in computational thinking and the potential academic benefits of integration for students, integration of computational thinking into school science and mathematics has proved to be challenging for several reasons. First, most schools cannot afford to add a computer science course to their existing curriculum as school STEM curriculum is already saturated (Cuny 2012, 2016). More, even when they do want to integrate computational thinking into school curriculum, they often have difficulty in finding a qualified teacher to teach the course (Astrachan et al. 2011; Barr and Stephenson 2011; UK Department for Education 2013; Yadav et al. 2017). The second problem related to the lack of qualified teachers to teach computational thinking in K-12 classrooms (Ericson et al. 2008). While teachers are familiar with and can adopt their instruction to several of the practices called for in the NGSS, they are not as familiar with engineering design and computational thinking nor do they have the knowledge and skills to teach computational thinking skills in their courses. Most teachers have not taken engineering or computer science courses as part of their college curriculum nor have they been exposed to pedagogies that will help them teach computational thinking in an effective manner. Problems with integration of computational thinking are not limited to teachers' familiarity with the concept. Third, discussions around definition of computational thinking, its place in school curriculum, and how to best achieve integration and how to design and deliver teacher training programs have not been settled yet (Barr and Stephenson 2011; Perkovic et al. 2010; Wilson et al. 2010). Despite these pressing questions, and ongoing discussions, STEM educators have designed and delivered curriculum and programs for teacher training and student learning in formal and informal contexts (Marling and Juedes 2016; Sengupta and Farris 2013; Sengupta and Wilensky 2009; Wilensky and Reisman 2006; the College Board, 2016).

The purpose of this chapter, therefore, is to conduct a meta-analysis of research studies conducted by computer science education community and in STEM education. This review is significant for several reasons. First, such review affords the opportunity to analyze how computational thinking has been defined and how concepts and practices have been adopted to science education. Second, it allows us to review professional development models that have been designed to help teachers develop pedagogical knowledge and skills for integration of CT in general and in science classrooms particularly along with their affordances and challenges. Third, it will give us the opportunity to see how CT concepts and practices have been measured both in integrated and stand-alone CT contexts. This review will also focus on the pedagogy of computational thinking in the context of science and intervention studies that focus on cognitive aspect of learning. Studies related to the affective aspect of computational thinking will not be addressed. The ultimate goal of this chapter is to provoke a discussion among STEM education community about the need to understand the urgency of the call and strategies for developing capacity to more rigorously integrate and thus more effectively implement computational thinking in K-12 classrooms.

The chapter starts with an introduction that highlights the developments that gave rise and motivation to integrate CT in K-12 classrooms, followed by a discussion of definition of computational thinking and its connections with science. Next, we discuss current efforts and models around CS teacher training and professional development in the United States. Then, we discuss research programs that have explored the connections between computational thinking and science learning. We end the chapter with a discussion of the implication of surging interest in and synergy around computer science and integration of computational thinking into STEM curriculum for STEM educators.

13.2 Defining Computational Thinking and Establishing Its Importance for STEM Education

Computational thinking has been defined in various ways and by several scholars and entities. Cuny et al. (2010) defines computational thinking as "the thought processes involved in formulating problems and their solutions so that the solutions are represented in a form that can be effectively carried out by an information-processing agent." The Royal Society defines computational thinking as "the process of recognizing aspects of computation in the world that surrounds us, and applying tools and techniques from computer science to understand and reason about both natural and artificial systems and processes" (Furber 2012, p. 29). The UK Department for Education has started an ambitious program around computational thinking. The UK Department for Education (2013) describes the purpose of the program among others as helping students: "To understand and apply the fundamental principles and concepts of computer science, including abstraction, logic, algorithms and data representation, to analyze problems in computational terms, and have repeated practical experience of writing computer programs in order to solve such problems, and to evaluate and apply information technology, including new or unfamiliar technologies, analytically to solve problems"(p. 1). The CS curriculum framework is structured in four key and progressive stages. Students at the fourth key stage are expected "To develop their capability, creativity and knowledge in computer science, digital media and information technology, to develop and apply their analytic, problem-solving, design, and computational thinking skills and to understand how changes in technology affect safety, including new ways to protect their online privacy and identity, and how to report a range of concerns" (p. 1). According to College Board (2017), computational thinking refers to the habits of minds by which learners combine the power of human curiosity, imagination, and creativity with the capabilities of intelligent machines to model natural phenomena, design systems, or solve complex problems using scientific data and heuristic reasoning. Wing (2010) defines computational thinking as the "thought process involved in formulating problems and their solutions so that the solutions are represented in a form that can be effectively carried out by an information-processing agent" (Wing 2010, p. 1). Accordingly, she suggested that computational thinking is a fundamental analytical thinking and problem-solving skill needed for everyone in a professional life that relies on computing tools for such practices as data representation, analysis, and modeling, not just for computer scientists. Consistent this view, the Computer Science Teachers Association (CSTA) states that "the study of computational thinking enables all students to better conceptualize, analyze, and solve complex problems by selecting and applying appropriate strategies and tools, both virtually and in the real world" (CSTA 2011, p. 9). While CT has been defined by several scholars and professional entities, no definition has made the impact on the field than Wing's (2010) definition of computational thinking (Teodte and Aydeniz 2015). Wing's definition has received some criticism as well.

Denning (2017), for instance, problematizes the definition of CT advocated by Wing (2010) and offers the term *computational design* instead. He states that "Computational designers spend much of their time inventing, programming, and validating computational models, which are abstract machines that solve problems or answer questions" (p. 17). He argues that "the thought processes of computational thinking should include those of skilled practitioners of the field where the computation will be used" (p. 15). Moreover, he attributes the origin of the term computational thinking to, the Nobel Laureate, theoretical physicist Kenneth Wilson, who, along with his colleagues, uses the term computational thinking to refer to the thought process used in "designing, testing, and using computational models" (p. 14). He further argues that both "computational science and computational thinking in science emerged from within the scientific fields—they were not imported from computer science" (p. 14). Further, he makes references to John von Neumann, "the polymath who helped design the first stored program computers, described algorithms for solving systems of differential equations on discrete grids" (p. 14). Denning (2017) maintains scientists and engineers have equally benefited from and contributed to the advancement of computational sciences. He places a significant emphasis on the power of simulations and the role of computational thinking in advancing science through simulations. Computational thinking and the power of computers have given rise to several advancements in science and mathematics ranging from "ability to simulate algorithms for solving systems of differential equations on discrete grids" to "ability to sequence and edit genes" using the power of computing (p. 13). Denning (2017) maintains that "computation has proved so productive for advancement of science and engineering that virtually every field of science and engineering has developed a computational branch "(p. 14). Denning is perhaps right in his assertion that science has made significant contributions to the field of computational thinking through its model-based practices.

Genome scientists, for instance, have studied how different traits are passed on from generations to generation using DNA information and to determine the evolutionary changes that are needed for a particular protein to change into another protein based on the underlying amino acid sequences using computational methods. Genetic scientists first used the Needleman-Wunsch algorithm, to compare sets of amino acid sequences with each other by using scoring matrices derived from the work of Dayhoff and more recently BLAST for performing fast, optimized searches of gene and protein sequence databases. Today, more sensitive searches utilizing gene/protein family-based Position-Specific Score Matrices (PSSMs) or Hidden Markov Models (HMMs) can be applied in order to perform comparative analyses and assign putative functions to unannotated proteins by establishing evolutionary relationships. These sensitive methods are usually used to identify domains which are functionally and evolutionary independent units of a protein. As a result of the tools and ways of thinking offered by computing, scientists have been able to discover 18 new candidate genes for autism through computational analysis of whole-genome sequences of more than 5000 people (Gholipour 2017).

Computational thinking has many aspects; some of the aspects are related to computing concepts, others are related to computational practices, yet others are related to soft aspects of computing. The main concepts of computational practices are abstractions, algorithms, debugging, automation, data and representations, and data analysis. We will not define and elaborate in each of these aspects here, but the table shown below can help our readers understand the definition of each of these aspects. Scientists, mathematicians, engineers, and statisticians use these concepts to design representational or predictive models, solve complex business or scientific problems, analyze large sets of data, and construct and communicate models of patterns within the data and increase the efficiency of computing systems.

Engaging in design using computational thinking requires certain dispositions and habits of mind. These dispositions include: "self-efficacy, confidence in dealing with complexity, persistence in working with difficult problems, tolerance for ambiguity, the ability to deal with open-ended problems, and the ability to communicate and work with others to achieve a common goal or solution." Einstein once said, "If I had an hour to solve a problem and my life depended on it, I would use the first 55 minutes determining the proper question to ask, for once I know the proper question, I could solve the problem in less than five minutes." This perspective on problem-solving requires open-mindedness and highlights the importance of soft skills in problem-solving. Therefore, in addition to teaching problem-solving and critical thinking skills, designing learning activities to help students develop these dispositions is critical to successful implementation of a computational thinking program. These are the types of skills that are needed in science and engineering as well. Therefore, these skills should not be left out when designing learning activities and assessing learning outcomes for students in the context of computer science or computational thinking. Taking a learning progression approach to designing such activities and assessment is critical to "the opportunity to learn" for students of all backgrounds, experiences, and levels.

13.3 Access and Equity in Computer Science Education

Over the years engineering and science have increased the processing power of computers, and powerful computers are now cheaper and thus accessible to anyone who wants to leave a mark in history of science, business, engineering, and technology. The increasing accessibility and increasing power of computers have, to some extent, democratized the scientific and engineering endeavors and opportunity to learn. As a result of accessibility to powerful computers and computer-based resources, "more people can be computational designers and tackle grand challenge problems" (Denning 2017, p. 17). Inspired by these facts, persistent economic opportunity gap for different sects of society and concern about economic competitiveness have encouraged the United States to initiate a bold initiative called Computer Science (CS) for All. The goal of this initiative is "to empower all American students from kindergarten through high school to learn computer science and be equipped with the computational thinking skills they need to be creators in the digital economy, not just consumers" (White House 2016, n.p). This initiative involves distributing $4 billion funding to states and $100 million directly to school districts in an effort "to expand K-12 CS by training teachers, expanding access to high-quality instructional materials, and building effective regional partnerships" (White House 2016, n.p).

Broadening participation in CS has been an issue in the United States. To broaden participation of marginalized student populations, NSF started a program called the Broadening Participation in Computing Alliance (BPC). The BPC Alliances were established between 2006 and 2009 to increase the number and diversity of college graduates in the computing and computationally intensive disciplines (NSF 2016). The Alliances are national and regional collaborations consisting of academic institutions, educators, professional societies, community organizations, and industrial partners that have vested interest in making CS accessible to all students with an explicit focus on underserved and underrepresented populations. The Alliances are charged with creating the "best practices, educational resources, advocacy networks, and forums needed to transform computing education" NSF 2013, n.p). While these efforts have generated a synergy and provided motivation to make CS accessible to all students, Margolis et al. (2012) argue, if we want to go beyond the issue of access and truly engage underrepresented student populations in CS in a meaningful and effective way, "we also must transform CS classroom culture and teaching in ways that engage and deepen how diverse students learn" (p. 72). In order to achieve this goal, they have developed, the "Exploring Computer Science" curriculum, which covers a range of CS topics through inquiry-based, hands-on, and culturally relevant pedagogy (p. 73) and now being implemented in a wide range of schools nationwide in the United States. This program is supplemented with teacher professional development program which will be discussed under professional development of CS teachers section of this chapter.

The issue of equity in CS has been a hot topic in other parts of the world as well. For instance, a recent report published by Kemp et al. (2016) highlights the issues of access and quality in CS education for different student populations in the United Kingdom. One of the findings of this report was that "Boys and mixed schools were more likely to offer computing than girls schools," and at The General Certificate of Secondary Education (GCSE), "19.6% of girls-only providers offered computing compared to 31.6% of boys-only and 29.1% of mixed providers. While 9.3% of girls-only providers offered computing at A-level 43.7% of boys-only and 24.5% of mixed providers offered computing at A-level" (p. 4).

Regarding equity in computer science, two core issues have been targeted by the CS community. First is the issue of access to computer science education. The discussion focuses on underrepresentation of females, students with special needs, and African-American and Hispanic students in CS. The main argument of this line of research is that there are no pathways and enough resources for these groups to enter CS programs. To address this issue, NSF has developed the broadening participation program to create a network and increase opportunities for these underrepresented groups to become part of CS community (Ashcraft and Blithe 2009; Burgstahler et al. 2012; Simard et al. 2010). The second line of research focuses on pedagogy of computing. Scholars who fall under this camp argue that the traditional CS curriculum ignores cultures and experiences of underrepresented student groups, and instruction fails to address the learning needs of these groups (Ben-Ari 2004; Brennan et al. 2011; Goode et al. 2014; Ryoo et al. 2016). Therefore, this group has intensified their efforts to develop a curriculum that values and integrates the cultural capital of the underrepresented groups and design and implement professional development for CS teachers to adopt inquiry-based, equitable instructional, and assessment strategies. One way to address the underrepresentation of women, students with special needs, and African-American and Hispanic students is the adoption of culturally relevant curriculum and use of culturally responsive instruction. The integration of CT into STEM courses, coupled with adoption of inquiry-oriented culturally responsive pedagogies, can definitely make CS more appealing to these underrepresented groups. This integration can potentially situate computing in a culturally relevant and personally meaningful context and thus increase student engagement and learning.

13.4 Integrating Computational Thinking into STEM Curricula

Arguably, computer science and the technologies it enables now are the driving force of world's growing economies, advanced defense systems, advanced communication technologies, scientific enterprise, and educational innovations (NSF 2013). Considering the pervasiveness of computing in current industrial and research fields, the ability to think computationally and use computing tools to make decisions, design systems, and solve problems has become essential to every discipline (Vee 2013). Thus, CT should penetrate into formal school curriculum and be part of informal STEM education programs for all ages and students of all backgrounds (Margolis and Goode 2016; Weintrop et al. 2016; Wilson et al. 2010). While the STEM education community and industry both agree that such integration should take place, there is no consensus on how such integration should take place, who should be responsible for making the integration possible, how to educate current and future teachers to integrate CT across STEM curriculum, and how to assess student learning in such an integrated learning context.

Several entities and individual scholars have been engaged in developing frameworks and pursuing critical research questions both for effective teaching of CT and for integration of CT into STEM curriculum. A commission led by the Association for Computing Machinery, Code.org, Computer Science Teachers Association, Cyber Innovation Center, and National Math and Science Initiative, in partnership with states and districts, contributed to the development of a framework called, "The K–12 Computer Science Framework" (American Computing Machinery 2016). This framework has provided a comprehensive list of concepts and practices that are associated with computer science. The *CS concepts* included in this framework include: "computing systems, networks and the internet, data and analysis, algorithms and programming, and impacts of computing" (p. 2). The CS practices include fostering an inclusive computing culture, collaborating around computing, creating computational artifacts, testing and refining computational artifacts, developing and using abstractions, recognizing and defining computational problems, and communicating about computing" (p. 2). The framework establishes a parallel between CS practices and science, engineering, and mathematical practices. This relationship has been discussed using modeling, problem-solving, abstractions, and computational thinking as the starting points.

Computational thinking is a necessary part of modern time scientists' professional practices. First, most scientist use computers to collect, classify, store, transform, and use empirical data needed for them to conduct their analysis (Emmott 2006). Second, most scientists use empirical data to develop, test, and refine models to represent, study, and predict the structure and behavior of natural, physical, and biological systems using the tools and power of computers. As such computer models serve as a core representational practice in the field of science (Denning 2017; Nersessian 1992; Lehrer and Schauble 2006; Basu et al. 2015). Third, scientists use the power of computing to represent or simulate the structure and behavior of biological and physical systems or design systems for data collection, testing their hypothesis, manipulation, and developing solutions to problems. As these points indicate, solving problems through computational thinking is a common practice among scientists.

There is a growing consensus which claims that integrating computational thinking into STEM curricula gives learners a more realistic view of how scientists and the professional in STEM fields function to achieve their research and professional goals and thus better prepares them for building the necessary knowledge and skills and habits of minds used in these fields (Augustine 2005; Dickes et al. 2016; Feurzeig et al. 2011; National Research Council (NRC) 2011a, b; Weintrop et al. 2016). The pioneer of computer science Papert (1991) claimed that learning programming in concert with other subjects can make learning more meaningful and easier (as cited in Basu et al. 2015). Similarly, science and mathematics provide "a meaningful context and a set of relevant problems" for using computational thinking, computing concepts, and practices (Hambrusch et al. 2009; Jona et al. 2014; Lin et al. 2009; Wilensky et al. 2014) as cited in Weintrop et al. 2016, p. 128). Weintrop et al. (2016) argue, "This reciprocal relationship—using computation to enrich mathematics and science learning and using mathematics and science contexts to enrich computational learning" (Weintrop et al. 2016, p. 128) is the driving force for the increasing motivation for integration-based computational thinking curricula. Computational thinking-based learning activities hold potential to engage student in such learning activities. Furthermore, there is significant amount of research that backs the claim that students learn best when they engage in design-based learning activities which focus on problem identification, formulation of possible solutions, testing of tentative solutions and use of representations for modeling, and reasoning about the behavior of the target system (Blikstein and Wilensky 2009; Kim et al. 2015a; b; Kolodner et al. 2003; Mehalik et al. 2008; Papert 1980, 1991; Penner 2000; Sengupta et al. 2013).

Some empirical studies have shown that middle and elementary school children can successfully use computational thinking tools to develop models of scientific phenomena. Moreover, these studies engaging students in such practices have shown to result in students developing both inquiry skills and a deep conceptual understanding of the targeted science ideas (Basu et al. 2015, 2017; Sengupta and Farris 2013: Sengupta et al. 2013; Taub et al. 2015; Wilensky 1995; Wilkerson-Jerde 2014). Therefore, there is a synergy among STEM education community both to improve the state of CS education in K-12 education and to integrate computational thinking into STEM curriculum.

While there has been a synergy to improve the state of CS in K-12 education and to integrate core computational thinking skills in STEM curricula, the primary focus of these efforts was first to "improve students' interest in CT through extracurricular activities, as opposed to aligning their learning activities with curricular topics in science or mathematics" (Basu et al. 2017, p. 2). More recently, scholars have made a conscious effort to integrate CT skills into STEM curricula (Guzdial 1995; Sherin 2001; Sengupta and Wilensky 2009; Sengupta et al. 2013;

Wilensky and Reisman 2006). Northwestern group has also successfully developed computational thinking-based lessons ranging from population biology, DNA sequencing, ohm laws, kinetic molecular theory, to chemical reactions (these lessons can be reached at http://ct-stem.northwestern.edu). These lessons have been successfully used in schools to engage student in computational thinking and science learning in a variety of science domains (Weintrop et al. 2016).

Despite these intentional efforts "relatively little is known about students' developmental processes and conceptual understanding in curricula that involve learning programming and/or computational modeling in conjunction with scientific concepts and representational practices" (Basu et al. 2017, p. 2). One challenge that remains to be addressed is how to successfully integrate computational thinking into STEM education (Rubinstein and Chor 2014). More, STEM education faculty has made attempts to define what science learning and teaching would be like in the area of NGSS and in a fashion integrated with CT. Weintrop et al. (2016) defined computational thinking in the context of mathematics and science. Weintrop et al. (2016) argue that "science and mathematics are meaningful contexts in which we can successfully "situate the concepts and practices of computational thinking." They argue this is the case because this is the "way mathematicians and scientists are using computational thinking to advance their disciplines" (Weintrop et al. 2016, p. 128). They propose a taxonomy that consists of four main categories: "data practices, modeling and simulation practices, computational problem solving practices, and systems thinking practices" (p. 128). The authors note, "Although we present our taxonomy as a set of distinct categories, the practices are highly interrelated and dependent on one another" (p. 134). The authors developed this framework through the analysis of various sources related to computational thinking. These include reviewing existing computational thinking literature, interviewing computational scientists, and reviewing computational thinking-based lesson plans. After the research team developed a tentative taxonomy, they consulted an expert panel (consisting of STEM teachers and researchers) to finalize their taxonomy (Fig. 13.1).

A similar framework has been proposed by Sengupta et al. (2013). The components of their framework include "1) Relationship between CT and Scientific Expertise, 2) Selection of a Programming Paradigm, 3) Selection of Curricular Science Topics, 4). Principles for System Design" (p. 353). In reference to the first aspect of this framework, the authors point out the role of abstractions and computational models in the sciences with a particular attention to how both scientists and software engineers use abstractions to design systems and solve problems. The authors make an explicit and intentional reference to the importance of modeling in science and how computational thinking is an essential practice in developing scientific models. They also highlight the pedagogical affordances of computational models for science learning in this aspect of their framework. The second aspect of the proposed framework focuses on the importance of selecting a *programing paradigm* that is conducive for the targeted domain and expected learning outcomes. This is important because each programming paradigm provides different affordances for the types of inquiry activity and scientific and computational

Data Practices	Modeling and Simulation Practices	Computational Problem Solving Practices	System Thinking Practices
Collecting Data	Using Computational Models to Understand a Concept	Preparing Problems for Computational Solutions	Investigating a Complex System as a Whole
Creating Data	Using Computational Models to Find and Test Solutions	Programming	Understanding the Relationships within a System
Manipulating Data	Assessing Computational Models	Choosing Effective Computational Tools	Thinking in Levels
Analyzing Data	Designing Computational Models	Assessing Different Approaches/Solutions to a Problem	Communicating Information about a System
Visualizing Data	Constructing Computational Models	Developing Modular Computational Solutions	Defining Systems and Managing Complexity.
		Creating Computational Abstractions	
		Trouble Shooting and Debugging	

Fig. 13.1 Computational thinking in mathematics and science taxonomy adaopted from Weintrop et al. 2016. https://link.springer.com/content/pdf/10.1007%2Fs10956-015-9581-5.pdf

practices students are expected to engage in and the artifacts they are expected to produce. The third component of the framework proposed by Sengupta et al. deals with selection of science topic covered by the curriculum. The final component of the proposed framework focuses on the "principles for system design." By system the authors refer to the learning modules that integrate both science and computational thinking. The principle guide includes "a) supporting low-threshold as well as high ceiling learning activities; b) design of programming primitives, c) supporting algorithm visualization; and d) sequencing learning activities in a constructivist fashion" (p. 353). Sengupta et al.'s (2013) model is based on a learning progression for students to jointly develop computational thinking proficiency, conceptual understanding, and engagement with epistemic practices of target domain.

Each of the frameworks/taxonomies presented provides a different perspective on the integration of CT into science. These taxonomies are critical not only for curriculum development efforts but also for accurately assessing the type of knowledge and skills that are being promoted by CT-based STEM curricula. Moreover, such taxonomies could help teacher education faculty to develop more relevant content for professional development programs for teachers. Finally, these efforts can initiate a meaningful conversation among STEM education colleagues about the strengths and weaknesses of proposed taxonomies/ frameworks regarding integration of CT into STEM curricula. Such conversation in turn can inform and lead to more relevant and effective programming related to curriculum development, professional development, and assessment efforts.

13.5 CT-Based Learning Environments

Some scholars have also investigated the process of learning in CT-based learning environments such as NIELS (NetLogo Investigations in Electromagnetism, CTSiM (Computational Thinking in Simulation and Modeling) and *Scalable Game Design*. NetLogo is a multi-agent-based modeling environment in which the user can create and/or interact with thousands of "agents," whose behavior is controlled by simple rules, and it is through the interaction of these agents that complex, emergent phenomena are generated (Sengupta et al., p. 30). CTSiM is a learning environment for K-12 science that is based on a computational thinking approach (Basu et al. 2015; Sengupta et al. 2013). "The system consists of an agent-based, visual programming and modeling platform where students can model, simulate, and study science processes to simultaneously learn about domain-general computational concepts and practices and relevant science phenomena" (Basu et al. 2017, p. 2).

Scalable Game Design (SGD) is a program designed to "motivate, engage and educate secondary students in designing complex games" (Reppening, et al. 2010, p. 1). The program has evolved from an after-school program to a full-blown IT curriculum that is now being implemented in K-12 classrooms. One critical aspect of this curriculum is that even teachers without computer science backgrounds can teach concepts targeted by SGD curriculum (Reppening et al. 2014). The project is currently being adopted by more than 25 school districts in the United States. The SGD program aims to increase the cognitive complexity of learning tasks that middle school students would typically engage. More precisely, they wanted to shift the focus of instruction from Web browsing, use of Web application, and keyboarding to engaging students in computational thinking defined by problem-solving and creativity. The program first targeted middle school as students start to develop their identities at the middle school age (Gootman 2007) and now is offered for different grade levels ranging from elementary to college. The program aims to expose students to computational thinking through games and science visualization and thus encourage and motivate students to pursue advanced learning opportunities and possibly careers in computer science (Reppening et al. 2010). Reppening and his colleagues argue that complex educational activities such as programming "must be heavily scaffolded" (p. 2). Therefore, their approach has two stages: first they motivate students to engage in computational thinking activities through game design and then leverage the skills acquired in this process to engage students in development of science visualizations. They also argue that in order for programming to be interesting, programming tasks must be "grounded in students' interests," promote problem-solving and critical thinking, and allow for and foster "student creativity" (p. 2). However, achieving these goals has not been easy. Therefore, STEM educators have been investing time and resources into researching different motivational and pedagogical models to integrate computational thinking into K-12 education in an effective manner (Barr and Stephenson 2011; Reppening et al. 2010; Sengupta et al. 2013; Weintrop et al. 2016; Wilson et al. 2010).

Basu et al. (2017) designed a study to investigate both "specific issues with integrating CT with middle school science instruction to support students' science and CT learning, and the types of difficulties students face when learning in CTSiM learning environment and "the kinds of support they require to overcome these challenges" (p. 2). After providing an extensive review of literature on how students learn and in what ways they struggle with both learning about and doing science in simulation-based learning environments and programming, they elaborate on the characteristics of the CTSiM learning environment, the curriculum they have designed, their intervention, and the findings of their research study.

Basu et al. worked with 15 6th grade students as they engaged in 7 learning activities in 2 different units. Each set of activity focused on one science domain: one set of activities (n = 7) focused on kinematics and the other set (n = 7) on ecology. Students spent 3 h each day working with a member of the research team learning the content through CTSiM for 3 days for each learning activity. The researchers worked directly with the students to make observations and provide guidance through leading questions when they struggled. They collected both observational and interview data. The researchers "characterize the types of challenges the students faced while working with CTSiM" through their analyses.

The results showed the CTSiM learning environment made significant contributions to students' learning gains. Moreover, scaffolding made a big difference on students' learning gains in science. The authors report that students who received one-on-one scaffolding showed improved learning gains between pre- and posttests: F for kinematics ($F(1,21) = 4.101$, $p < 0.06$), and for ecology ($F(1,21) = 37.012$, $p < 0.001$). The authors reported several challenges that the students faced while learning in the CTSiM environment. The authors reported that students experienced several challenges while learning science through a CT-based STEM curriculum. The first challenges were associated with content knowledge due to lack of prior knowledge or gaps in content knowledge. The second reported challenge was associated with modeling practices. Students specifically experienced challenge in "representing scientific concepts and processes as computational models and refining constructed models (partial or full) based on observed simulations" (Basu et al. 2017, p. 19). Another reported challenge experienced by the students was "agent-based thinking challenges"—"expressing agent behaviors as computational models", "understanding how individual agent interactions lead to aggregate-level behaviors, and the consequences of agent behavior changes on the aggregate behavior" (p. 19). The final type of challenges experienced by the students included programming-related challenges. The reported programming challenges include (1) challenges in understanding the semantics of domain-specific primitives; (2) challenges in using computational primitives like variables, conditionals, nesting, and loops to build programs (i.e., behaviors); (3) procedurality challenges; (4) modularity challenges; (5) code reuse challenges; and (6) debugging challenges)" (p. 20).

Collectively, these results show us that integrating CT into STEM curricula is a novel goal and as shown in this pullout study and some other studies can make significant contributions to student learning. However, the challenges students face in learning STEM concepts and practices in different CT-based learning environments and in different domains and contexts must be observed, studied, and characterized. Unless we study and understand how students learn in these learning environments and what specific challenges they experience and types of support they need to overcome the observed challenges, we will not be able to develop effective CT-based STEM curricula. Such knowledge is also necessary for designing effective teacher professional development programs. Teachers who are knowledgeable of these challenges can be guided by experts to adopt pedagogical strategies to address these challenges.

13.6 Computer Science in K-12 Education

Computer education at the K-12 level typically has been limited to after-school programs and summer camps specifically designed to increase female and racial minority students' interest in computing (Basu et al. 2017; Margolis and Goode 2016; Yadav et al. 2017). However, there has been an increasing push to make computer science education as part of official K-12 curriculum (Margolis et al. 2014; Royal Society 2012). As a result, several curriculum development and teacher education programs have been developed to address the challenge. While it is impractical to present all of these efforts, we will mention few efforts that have been widely received by the CS community in the United States. Three curriculum programs that have received wide acceptance from the CS community are Exploring Computer Science, Scalable Game Design, and AP CS Principles.

Exploring Computer Science is a yearlong course consisting of 6-week long units. The purpose of the course is to engage students both in computer science content and computational practices. ECS curriculum creates opportunities for students to utilize a variety of computational tools and engage in problem-solving using computing practices in a culturally relevant fashion. One of the strengths of this curriculum program is that both "assignments and instruction are contextualized to be socially relevant and meaningful for diverse students" (n.p).

The ECS curriculum consists of human computer interaction, problem-solving, programming, Web design, computing and data analytics, and problem-solving units. The *human-computer interaction* unit gives the students diverse opportunities to explore a variety of websites and Web applications and discuss issues of privacy and security. Students learn the characteristics that make certain tasks easy or difficult for computers and how these differ from those that humans characteristically find easy or difficult. The *problem-solving unit* provides students with opportunities to become "computational thinkers" by applying a variety of problem-solving techniques as they create solutions to problems that are situated in a variety of contexts. In the *programming unit*, students design algorithms and create programming solutions to a variety of computational problems using an iterative development process in scratch. The *Web design* unit challenges students to develop a Web page by applying their knowledge of algorithms, abstraction, and Web page design. In the *computing and data analysis* unit, students use computers to translate, process, and visualize data in order to find patterns and test hypotheses around community-related issues. The *robotics unit* introduces students to robotics as an advanced application of computer science that can be used to solve problems in a variety of settings from business to healthcare. Students explore how to integrate hardware and software in order to solve problems. More detailed description of each unit can be found at http://www.exploringcs.org/curriculum.

AP Computer Science (CS) Principles teaches core computational concepts and practices in a multidisciplinary approach. The course introduces students to programming, abstractions, algorithms, large data sets, the Internet, cybersecurity concerns, and computing impacts from a creativity and problem-solving perspective. Students' knowledge and skills are assessed through multiple choice test, two performance-based tasks, and analysis of students' digital portfolio.

13.7 Computer Science and Professional Development of Teachers

Computer science education has been part of K-12 education in several parts of the world for a long time. The United States has started an aggressive campaign to elevate the status of computer science in K-12 classrooms in the early 2010s. NSF supported the program CS 10K's goal to prepare and place 10,000 CS-certified teachers in K-12 classrooms across the country (Cuny 2012). India is another country that has placed an increasing emphasis on CS education in its curriculum. For instance, India has increased the hours spent on computer science from 160 to 180 for the 9th and 10th grades and from 140 to 180 for post-secondary school (Kim et al. 2015b). Korean Ministry of Education initiated a plan for computing education with software emphasis in K-12 curricula (Kim et al. 2015b). The program integrates computing concepts into curriculum starting with the elementary school and all the way to 12th grade. Israel has incorporated CS as part of school curriculum since 1970 and more strategically since the 1990s (Gal-Ezer and Stephenson 2014). High school students interested in computer science usually take three 90-hour course works during 10th through 12th grades. Israeli curriculum has made it optional for highly motivated and capable students to take additional CS courses that engage them in more rigorous course work. However, this option is not widely accessible for every student (Gal-Ezer and Stephenson 2014). Russia has been teaching "The Foundations of Informatics" since 1985 in its high school curriculum, whose goal has been to promote "algorithmic thinking and computer literacy among students" (Khenner and Semakin 2014, p. 1). This course has been updated several times in the last three decades but a main focus is to foster students' "algorithmic thinking, systems thinking skills, skills of formalization and systematization, and skill of accessing ICT tools to solve problems" (Khenner and Semakin 2014, p. 5). The Russian government has simultaneously prepared thousands of mathematics and physics teachers to teach the informatics course.

In the United States, integration of computing in school curriculum has been a bit challenging. States have not been proactive in designing policies around CS education or teacher education around CS. Similarly, schools are not yet ready to fully embrace the CS education initiative. However, despite institutional resistance from schools, computer science education continues to become a game player in primary and secondary schools across the globe (Basu et al. 2017; Yadav et al. 2017). For instance, in the United States alone, more than 2000 high schools started to offer new computer science courses (Cuny 2016). The United Kingdom (The Royal Society 2012) and Australia have developed rigorous policies and programs around computing. Mexico has started its own initiatives to make computing an integral part of school curriculum (Escherle et al. 2016). Turkey is considering to make coding as a mandatory course starting from elementary school, all the way to high school. Efforts in Europe and Asia follow the same pattern. South Korea has started a new initiative in computer science with a focus on software development (Kim et al. 2015b).

While there is an increasing synergy in making CS a core curriculum course, and computational thinking a core learning outcome in STEM courses, certifying sufficient number of teachers who can teach CS and training enough STEM teachers who can integrate computational thinking into their curriculum have been a challenge for the STEM community (CSTA 2013; Ericson et al. 2008; Gal-Ezer and Stephenson 2010; Lye and Koh 2014; Menekse 2015; Ni and Guzdial 2012; Yadav et al. 2017). The complexity of teacher preparation stems from challenges around recruitment, the confusion about what types of knowledge and skills need to be taught, and how teacher quality should be measured in CS as well as the place of CS in teacher preparation programs (Gal-Ezer and Stephenson 2010; Goode et al. 2014; CSTA 2013). Gal-Ezer and Stephenson (2010) state:

> Because so few countries or states/provinces require or allow for teachers to be certified specifically as computer science teachers, very few teacher preparation institutions provide programs with rigorous and relevant computer science training. In the absence of clear and specific requirements for computer science, these institutions have little or no incentive to address the needs of computer science teachers. (p. 63)

This is the case partly because interest in teaching CS in schools is relatively new; policy makers, schools of education, and computer science departments are not prepared to deal with this suddenly emerging need and the challenge it has posed to the community. Nevertheless, the issue of preparing future CS teachers has received attention from several scholars (Ericson et al. 2008; Gal-Ezer and Stephenson 2010; Goode et al. 2014; Lapidot and Hazzan 2003; Yadav and Korb 2012; Yadav et al., 2014). The CS education community has embraced two popular models, one in which students are recruited from computer science major and are supported through pedagogical courses and clinical experiences to become certified.

In the other model, the focus is on recruiting and preparing teachers from other content areas through professional development to become CS teachers (CSTA 2013; Yadav et al. 2017).

13.7.1 Preservice Teacher Education Models

Since the synergy for teaching computer science in the US schools started, CS educators have come up with various models for preparing future teachers. Goode developed a course for teacher education programs in California around the *Exploring Computer Science* curriculum, which has now been adopted in several other states and serves as a model for professional development activities in the United States. The course creates learning opportunities for students to "explore computer science as a discipline that encourages inquiry, creativity, and collaboration" and "models the investigative nature of computing" through hands-on activities which "allow students to gain insights about teaching and learning core computing concepts and practices in classroom settings. The course focuses on "methods of eliciting, understanding and assessing" students' conceptual understanding of core computing concepts and practices. The course also equips prospective teachers with knowledge and skills to engage student in inquiry-based learning activities. Students explore a variety of instructional resources, tools, and virtual environments to support their students' engagement with computing concepts in an effective manner (Goode 2011).

Some other systematic models have also been developed to infuse computer science in secondary STEM teacher preparation program. For instance, in Idaho, colleagues have developed a program called IDoCode. The IDoCode is a 35-credit graduate certificate program designed for teachers to get certified to teach computer science. Preservice teachers take a series of computer science courses and a mix of courses around pedagogy of computing to get certified to teach computer science (more information about the program can be accesses at https://coen.boisestate.edu/idocode/students/cs-masters-stem-education/).

The UTeach Institute has also developed a certification route for computer science majors to get certified to teach computer science in secondary schools. The program is consistent with their model for math and science certification in which students get certified to teach math and science by taking a series of courses as part of a minor and some clinical experiences in local classrooms under the mentorship of master teachers. Students who choose the CS path are required to take 24 h in CS and UTeach's pedagogy courses. While a viable option, the program has not been as successful as their science and mathematics certification pathway programs as computing jobs are abundant, and teacher salaries cannot compete with the salaries offered to the graduate of computer science majors. However, this may change in the future with more rigorous recruitment methods and more competitive incentives for teachers of computer science.

13.7.2 In-Service Professional Development Programs

Goode and her colleagues have developed a professional development program for CS teachers building on their successful preservice teacher education model. The ECS professional development program focuses on content, pedagogy, and participants' belief systems, provides in-class coaches who can help the teachers implement the content learned in the summer workshops, and provides the opportunity to be part of a professional learning community (Margolis et al. 2012). The key features of their PD program are that the program (1) immerses teachers into inquiry-based learning and teaching practices; (2) focuses on discussion of equity-related issues and culturally relevant pedagogy; (3) focuses on analysis of instructional practices, through a teacher-learner-observer model in which teachers take turns planning and delivering CT lessons in teams, giving and receiving feedback through structured debriefing sessions with fellow teachers; and (4) focuses on development and sustainability of an ongoing professional learning community (Goode et al. 2014). Collectively, these targeted educational experiences create rich opportunities for deep and reflective learning and, therefore, increases teachers' knowledge of both content and pedagogy.

The *UTeach Institute* has developed a 5-day summer professional development program where they expose teachers to computing content and pedagogy with ongoing support during school year. *The Scalable Game* design group has also developed an effective professional development program for teachers to implement their curriculum in their classrooms.

TEALS (Technology Education and Literacy in Schools) is another successful program that has received significant attention from CS community and school districts in the United States. This program establishes connections between professionals from industry and teachers to bring rigorous computer science to the nation's classrooms. Collaborating with industry professionals has several benefits. First, industry professionals have robust knowledge of computing concepts and practices. Therefore, they can help teachers to develop and strengthen their knowledge of the subject matter. Second, industry professional can help teachers to present content in a way that is relevant to students' personal lives thus increasing students' motivation to more effectively engage with learning core computing concepts and practices. Finally, they can serve as role models and inform and excite students about career opportunities in CS-related fields. One strength of the TEALS is that it offers three different support models for teachers to choose from. These models include (1) classroom enrichment model, (2) co-teaching model, and (3) the lab support model. The classroom enrichment model is designed for teachers who already possess significant content knowledge and confident in pedagogy of computing. The industry professional's role is to enrich teachers' curriculum by demonstrations or other motivating activities. In the co-teaching model, industry professionals are responsible for and provide instruction with teachers in the classroom for the entire school year. They co-plan and teach with teachers so the teacher can strengthen their content knowledge and improve their pedagogical knowledge related to teaching core computing concepts and practices. In this model, the industry professionals gradually hand the responsibilities over to the teacher over the course of 2 years. Once confident in content and pedagogy of computing, the teacher starts to assume the primary planning, teaching, and assessment responsibilities. In the *lab support* model, the teacher who may not be confident in their content knowledge assumes the primary responsibility for instruction; however, they are aided by teaching assistants who have strong commend of content and can answer difficult questions and assist students with challenging assignments and projects.

13.8 Discussion

There has been a growing synergy around integrating computer science in K-16 school curriculum and CT in STEM curricula particularly. Although considerable effort has been put into developing curricula and after-school programs to broaden participation of underrepresented populations, there are still significant challenges that remain to be addressed both by the computer science education and STEM education community. These challenges include but not limited to "defining a learning progression and curriculum, assessing student achievement, preparing teachers, and ensuring equitable access" (Weintrop et al. 2016, p. 130). To address these challenges, STEM education community has started to develop curriculum integration models (Joyner et al. 2014; Perkovic et al. 2010; Roschelle et al. 2000; Rubinstein and Chor 2014; Schanzer et al. 2015; Sengupta and Wilensky 2009; Tan and Biswas 2007; Wilensky and Reisman 2006), study how students learn STEM concepts, and engage in STEM practices in an integrated learning environment (Basu et al. 2017; Sengupta et al. 2013; Sengupta and Farris 2013; Sengupta et al. 2015). For instance, Northwestern group has successfully developed computational thinking-based lessons ranging from population biology, DNA sequencing, ohm laws, kinetic molecular theory, to chemical reactions, and these lessons have been successfully used in schools to engage student in computational thinking (Weintrop et al. 2016), and Vanderbilt group is working on both the developments of new curricula and studying how students learn core science concepts in this new learning environment.

While there are different models for infusing CT into school curriculum, one thing is very clear: we no longer can separate computer science or computational thinking from STEM education. Whether taught in stand-alone CS courses, or as part of STEM courses, computational thinking and the pedagogy around computational thinking are going to drive the future of teacher education, research, and curriculum development efforts in STEM education. Therefore, it is imperative that we start engaging in conversations around teacher preparation and pedagogy of teaching CT through STEM. This integration perspective may challenge our current assumptions about ways to prepare future mathematics and science teachers and promote and measure teachers' pedagogical content knowledge.

While our assumptions about what is worth to measure in STEM has already been challenged with the NGSS, with more emphasis on CT skills in STEM, both the content and methods of our assessments will also need to change. While curriculum development and assessments are relatively easier to develop and implement, recruitment and preparation of teachers and professional development of STEM teachers to teach CT in their curriculum will continue to present unique challenges to our community. Convincing practicing teachers to learn about new concepts and practices of an unfamiliar territory and the pedagogy associated with this new domain knowledge can be quite challenging. However, we have experience and resources that can help us to successfully respond to these emerging challenges. Despite our motivation to tackle these emerging issues around integration of CT in school curriculum and teacher education, the biggest challenge will be getting support from policy makers and preparing school leaders to understand the importance of CT and dedication to invest in CT-based efforts and the complexity of how schools function as a system and as a community.

Apart from teacher education, there are new perspectives on how students will learn in a computer-based learning environment in which they are no longer the receivers of knowledge but the authors of knowledge. In this data-rich, model-based, and computational-driven learning environment, students take ownership, and this model challenges the traditional roles that the teachers and students have assumed. Instead of transmitting the expert knowledge to their students, teachers now must assume the primary role of a facilitator in this learning environment, scaffold student learning, and monitor their progress in real time through formative assessment. In order for teachers to successfully achieve this new responsibility, they must develop new ways of thinking, acquire new content knowledge, and develop pedagogical content knowledge necessary for this new learning environment and changing goals.

References

Ashcraft, C., & Blithe, S. (2009). *Women in IT: The facts*. Washington, DC: National Center for Women and Information Technology. Retrieved from http://www.ncwit.org/sites/default/files/legacy/pdf/NCWIT_TheFacts_rev2010.pdf

Association for Computing Machinery. (2016). *K–12 Computer Science framework*. Retrieved from http://www.k12cs.org

Astrachan, O., Cuny, J., Stephenson, C., & Wilson, C. (2011). The cs10k project: Mobilizing the community to transform high school computing. In *Proceedings of the 42nd ACM Technical symposium on Computer Science Education (SIGCSE)* (pp. 85–86). American Computing Machinery.

Augustine, N. R. (2005). *Rising above the gathering storm: Energizing and employing America for a brighter economic future*. Washington, DC: National Academies Press.

Bailey, D., & Borwein, J. M. (2011). Exploratory experimentation and computation. *Notices of the American Mathematical Society, 58*(10), 1410–1419.

Barr, V., & Stephenson, C. (2011). Bringing computational thinking to K-12: What is involved and what is the role of the computer science education community? *ACM Inroads, 2*(1), 48–54.

Basu, S., Sengupta, P., & Biswas, G. (2015). A scaffolding framework to support learning of emergent phenomena using multiagent based simulation environments. *Research in Science Education, 45*(2), 293–324. https://doi.org/10.1007/s11165-014-9424-z.
Basu, S., Biswas, G., & Kinnebrew, J. S. (2017). Learner modeling for adaptive scaffolding in a Computational Thinking-based science learning environment. *User Modeling and User-Adapted Interaction, 27*(1), 5–53.
Ben-Ari, M. (2004). Situated learning in computer science education. *Computer Science Education, 14*(2), 85–100. https://doi.org/10.1080/08993400412331363823.
Blikstein, P., & Wilensky, U. (2009). An atom is known by the company it keeps: A constructionist learning environment for materials science using agent-based modeling. *International Journal of Computers for Mathematical Learning, 14*(2), 81–119.
Brennan, K., Valverde, A., Prempeh, J., Roque, R., & Chung, M. (2011). More than code: The significance of social interactions in young people's development as interactive media creators. In T. Bastiaens & M. Ebner (Eds.), *Proceedings of world conference on educational multimedia, hypermedia and telecommunications* (pp. 2147–2156). Chesapeake: AACE.
Buckingham, D. (2015). Do we really need media education 2.0? Teaching media in the age of participatory culture. In T. Lin, D. Chen, & V. Chai (Eds.), *New media and learning in the 21st century* (pp. 9–21). Singapore: Springer.
Burgstahler, S., Ladner, R., & Bellman, S. (2012). Strategies for increasing the participation in computing of students with disabilities. *ACM Inroads, 3*(4), 42–48.
College Board. (2016). AP computer science principles curriculum framework 2016–2017. https://securemedia.collegeboard.org/digitalServices/pdf/ap/ap-computer-science-principles-curriculum-framework.pdf
College Board. (2017). AP computer science principles. Retrieved from https://apcentral.collegeboard.org/pdf/apcomputer-science-principles-course-and-exam-description.pdf
Computer Science Teacher Association Standards [CSTA] Task Force. (2011). *K-12 Computer Science standards*. Revised 2011. http://csta.acm.org/Curriculum/sub/K12Standards.html
Computer Science Teachers Association [CSTA]. (2013). *Bugs in the system: Computer science teacher certification in the U.S.* New York: Author. Retrieved from http://csta.acm.org/ComputerScienceTeacherCertification/sub/CSTA_BugsInTheSystem.pdf
Cuny, J., Snyder, L., & Wing, J. M. (2010). Demystifying computational thinking for non-computer scientists. Unpublished manuscript in progress, referenced in http://www.cs.cmu.edu/~CompThink/resources/TheLinkWing.pdf
Cuny, J. (2012). Transforming high school computing: A call to action. *ACM Inroads, 3*, 32–36.
Cuny, J. (2016, February). *CS Education: Catching the Wave*. Proceedings of the 47th ACM Technical Symposium on Computing Science Education. Association for Computing Machinery
Denning, P. J. (2017). Computational thinking in science. *American Scientist, 105*(1), 13–17. https://doi.org/10.1511/2017.124.
Dickes, A., Sengupta, P., Farris, A. V., & Basu, S. (2016). Development of mechanistic reasoning and multi-level explanations in 3rd grade biology using multi-agent based models. *Science Education*. https://doi.org/10.1002/sce.21217.
Emmott, S. (Ed.). (2006). *Towards 2020 Science*. Cambridge, UK: Microsoft Research.
Ericson, B., Armoni, M., Gal-Ezer, J., Seehorn, D., Stephenson, C., & Trees, F. (2008). *Ensuring exemplary teaching in an essential discipline: Addressing the crisis in computer science teacher certification*. Final report of the CSTA Teacher Certification Task Force. ACM.
Escherle, N. A., Ramirez-ramirez, S. I., Basawapatna, A. R., Maiello, C., & Nolazco-florez, J. A. (2016). *Piloting computer science education week in Mexico*. In Proceedings of the 47th ACM Technical Symposium on Computer Science Education (SIGCSE '16) (pp. 431–436). Memphis, TN. doi:https://doi.org/10.1145/2839509.2844598.
Feurzeig, W., Papert, S., & Lawler, B. (2011). Programming-languages as a conceptual framework for teaching mathematics. *Interactive Learning Environments, 19*(5), 487–501.

Furber, S. (2012). *Shut down or restart? The way forward for computing in UK schools.* Technical report. London: The Royal Society.

Gal-Ezer, J., & Stephenson, C. (2010). Computer science teacher preparation is critical. *ACM Inroads, 1*(1), 61–66.

Gal-Ezer, J., & Stephenson, C. (2014). A tale of two countries: Successes and challenges in k-12 computer science education in Israel and the United States. *ACM Transactions on Computing Education (TOCE), 14*(2), 8.

Gholipour, B. (2017). Discovery of 18 new autism-linked genes may point to new treatments. Scientific American, March.

Goode, J. (2011). Exploring computer science: An equity-based reform program for 21st century computing education. Journal for Computing Teachers. Retrieved from http://www.iste.org/store/magazines-and-journals/downloads/jct-downloads.aspx

Goode, J., Margolis, J., & Chapman, G. (2014). Curriculum is not enough: The educational theory and research foundation of the exploring Computer Science professional development model. *SIGCSE, 2014*, 493–498.

Gootman, E. (2007, March 17). The critical years: For teachers, middle school is test of wills. *New York Times.*

Guzdial, M. (1995). Software-realized scaffolding to facilitate programming for science learning. *Interactive Learning Environments, 4*(1), 1–44.

Guzdial, M. (2008). Education: Paving the way for computational thinking. *Communications of the ACM, 51*(8), 25–27. https://doi.org/10.1145/1378704.1378713.

Hambrusch, S., Hoffmann, C., Korb, J. T., Haugan, M., & Hosking, A. L. (2009). A multidisciplinary approach towards computational thinking for science majors. *ACM SIGCSE Bulletin, 41*, 183–187.

Jona, K., Wilensky, U., Trouille, L., Horn, M.S, Orton, K., Weintrop, D., & Beheshti, E. (2014). *Embedding computational thinking in science, technology, engineering, and math (CT-STEM).* Presented at the Future Directions in Computer Science Education Summit Meeting, Orlando.

Joyner, D. A., Goel, A. K., & Papin, N. (2014). *MILA-S: Generation of agent-based simulations from conceptual models.* In Proceedings of the 19th International conference on intelligent user interfaces (pp. 289–298). Haifa, Israel.

Kemp, P., Wong, B., & Berry, M. (2016). *The Roehampton annual computing education report 2015 data from England.* Retrieved from https://drive.google.com/file/d/0B1xf_L-jCIzYZmZDbFAzb3BPUEk/view

Khenner, E., & Semakin, I. (2014). School subject informatics (computer science) in Russia: Educational relevant areas. *ACM Transactions on Computer Education, 14*(2), 1–10. https://doi.org/10.1145/2602489.

Kim, B., Pathak, S. A., Jacobson, M. J., Zhang, B., & Gobert, J. D. (2015a). Cycles of exploration, reflection, and consolidation in model-based learning of genetics. *Journal of Science Education and Technology, 24*(6), 789–802. https://doi.org/10.1007/s10956-015-9564-6.

Kim, D. K., Jeong, D., Lu, L., Debnath, D., & Ming, H. (2015b). Opinions on computing education in Korean K-12 system: Higher education perspective. *Journal of Computer Science Education, 25*(4), 371–389.

Kolodner, J. L., Camp, P. J., Crismond, D., Fasse, B., Gray, J., Holbrook, J., et al. (2003). Problem-based learning meets case-based reasoning in the middle-school science classroom: Putting learning by design (tm) into practice. *The Journal of the Learning Sciences, 12*(4), 495–547.

Lapidot, T., & Hazzan, O. (2003). Methods of teaching a computer science course for prospective teachers. *Inroads – The Sigcse Bulletin, 35*(4), 29–34.

Lehrer, R., & Schauble, L. (2006). Cultivating model-based reasoning in science education. In R. K. Sawyer (Ed.), *The Cambridge handbook of the learning sciences* (pp. 371–388). New York: Cambridge University Press.

Lin, C.C., Zhang, M., Beck, B., & Olsen, G. (2009). *Embedding computer science concepts in K-12 science curricula.* In Proceedings of the 40th ACM technical symposium on computer science education (pp 539–543). New York: ACM.

Lye, S. Y., & Koh, J. H. L. (2014). Review on teaching and learning of computational thinking through programming: What is next for K-12? *Computers in Human Behavior, 41*, 51–61.
Margolis, J., & Goode, J. (2016). Ten lessons for CS for all. *ACM Inroads Magazine, 7*(4), 52–56.
Margolis, J., Goode, J., & Ryoo, J. (2014). Democratizing computer science knowledge. In educational leadership, STEM for all December 2014/January 2015 | Volume 72 | Number 4. p. 48–53.
Margolis, J., Ryoo, J. J., Sandoval, C. D. M., Lee, C., Goode, J., & Chapman, G. (2012). Beyond access: Broadening participation in high school computer science. *ACM Inroads, 3*(4), 72–78.
Marling, C., Juedes, D. (2016). *CS0 for computer science majors at Ohio University* (pp. 138–143). New York: ACM. http://329Stocker Center.
Mehalik, M. M., Doppelt, Y., & Schunn, C. D. (2008). Middle-school science through design-based learning versus scripted inquiry: Better overall science concept learning and equity gap reduction. *Journal of Engineering Education, 97*(1), 71–85.
Menekse, M. (2015). Computer science teacher professional development in the United States: A review of studies published between 2004 and 2014. *Computer Science Education, 25*(4), 325–350.
National Research Council. (2011a). *Committee for the workshops on computational thinking: Report of a workshop of pedagogical aspects of computational thinking*. Washington, DC: National Academies Press.
National Research Council. (2011b). *Learning science through computer games and simulations*. Washington, DC: The National Academies Press.
National Research Council. (2013). *A framework for K-12 science education: Practices, crosscutting concepts, and core ideas*. Washington, DC: The National Academies Press.
National Science Foundation [NSF]. (2013). *Broadening Participation in Computing Alliance Program (BPC-A)*. Retrieved on April 18, 2017, from https://www.nsf.gov/funding/pgm_summ.jsp?pims_id=503593
National Science Foundation. (2016). Broadening Participation in Computing Alliance Program (BPC-A) Retrieved from https://nsf.gov/funding/pgm_summ.jsp?pims_id=503593&org=DIS&from_org=DIS
Nersessian, N. J. (1992). How do scientists think? Capturing the dynamics of conceptual change in science. In R. N. Giere (Ed.), *Cognitive models of science* (pp. 3–45). Minneapolis: University of Minnesota Press.
Ni, L., & Guzdial, M. (2012). *Who am I? Understanding high school computer science teachers' professional identity*. In Proceedings of the 43rd ACM technical symposium on computer science education (pp. 499–504). Raleigh, NC.
Papert, S. (1980). *Mindstorms: Children, computers, and powerful ideas*. New York: Basic Books.
Papert, S. (1991). Situating constructionism. In I. Harel & S. Papert (Eds.), *Constructionism* (pp. 1–11). Norwood: Ablex.
Penner, D. E. (2000). Cognition, computers, and synthetic science: Building knowledge and meaning through modeling. *Review of Research in Education, 25*, 1–36.
Perković, L., Settle, A., Hwang, S., & Jones, J. (2010). A framework for computational thinking across the curriculum. In *Proceedings of the fifteenth annual conference on Innovation and technology in computer science education* (pp. 123–127). ACM.
Repenning, A., Webb, D., & Ioannidou, A. (2010). *Scalable game design and the development of a checklist for getting computational thinking into public schools*. In Proceedings of the 41st ACM Technical Symposium on Computer Science Education (SIGCSE '10) (pp. 265–269). New York: ACM Press.
Repenning, A., Webb, D. C., Koh, K. H., Nickerson, H., Miller, S., Brand, C., Horses, I. H. M., Basawapatna, A., Gluck, F., Grover, R., Gutierrez, K., & Repenning, N. (2014). Scalable game design: A strategy to bring systemic computer science education to schools through game design and simulation creation. *ACM Transactions on Computing Education (TOCE), 15*(2), 11. https://doi.org/10.1145/2700517.

Roschelle, J., Kaput, J., & Stroup, W. (2000) SimCalc: Accelerating student engagement with the mathematics of change. In *Learning the sciences of the 21st century: Research, design, and implementing advanced technology learning environments* (pp 47–75).

Rubinstein, A., & Chor, B. (2014). Computational thinking in life science education. *PLoS Computational Biology, 10*(11). https://doi.org/10.1371/journal.pcbi.1003897.

Ryoo, J., Goode, J., & Margolis, J. (2016). It takes a village: Supporting inquiry- and equity-oriented computer science pedagogy through a professional learning community. *Computer Science Education*. https://doi.org/10.1080/08993408.2015.1130952.

Schanzer, E., Fisler, K., Krishnamurthi, S., & Felleisen, M. (2015). *Transferring skills at solving word problems from computing to algebra through Bootstrap*. In Proceedings of the 46th ACM Technical symposium on computer science education (pp. 616–621). New York: ACM.

Selby, C. C. (2015). *Relationships: Computational thinking, pedagogy of programming, and Bloom's Taxonomy*. In Proceedings of the Workshop in primary and secondary computing education (pp. 80–87). New York: ACM.

Sengupta, P., & Farris, A. V. (2013). *Learning kinematics in elementary grades using agent-based computational Modeling: A visual programming based approach*. In Proceedings of the 11th International conference on interaction design & children (pp 78–87).

Sengupta, P., & Wilensky, U. (2009). Learning electricity with NIELS: Thinking with electrons and thinking in levels. *International Journal of Computers for Mathematical Learning, 14*(1), 21–50.

Sengupta, P., Kinnebrew, J. S., Basu, S., Biswas, G., & Clark, D. (2013). Integrating computational thinking with K-12 science education using agent-based computation: A theoretical framework. *Education and Information Technologies, 18*(2), 351–380.

Sengupta, P., Krishnan, G., Wright, M., & Ghassoul, C. (2015). Mathematical machines and integrated stem: An Intersubjective constructionist approach. In S. Zvacek, M. Restivo, J. Uhomoibhi, & M. Helfert (Eds.), *Computer supported education, Communications in Computer and Information Science* (Vol. 510, pp. 272–288). Cham, Switzerland: Springer.

Sherin, B. L. (2001). A comparison of programming languages and algebraic notation as expressive languages for physics. *International Journal of Computers for Mathematics Learning, 6*(1), 1–61.

Smith, M. (2016). *Computer science for all*. Washington, DC: Office of Science and Technology Policy, Executive Office of the President Retrieved from https://www.whitehouse.gov/blog/2016/01/30/computer-scienceall

Simard, C., Stephenson, C., & Kosaraju, D. (2010). *Addressing Core Equity Issues in K–12 Computer Science Education: Identifying barrier and sharing strategies*. Palo Alto, CA: The Anita Borg Institute and the Computer Science Teachers Association.

Tan, J., & Biswas, G. (2007). *Simulation-based game learning environments: Building and sustaining a fish tank*. In Proceedings of the First IEEE International Workshop on Digital game and intelligent toy enhanced learning (pp. 73–80). Jhongli, Taiwan.

Taub, R., Armoni, M., Bagno, E., & Ben-Ari, M. (2015). The effect of computer science on physics learning in a computational science environment. *Computer Education, 87*, 10–23.

Teodte, R., & Aydeniz, M. (2015). Computational thinking and impacts on K-12 science education. Published in *Proceedings of the 2015 IEEE Frontiers in Education Conference* (FIE), DOI: https://doi.org/10.1109/FIE.2015.7344239

The National Center for Women & Information Technology (NCWIT). (2012). *NSF PI Meeting*. http://www.ncwit.org/sites/default/files/legacy/pdf/Pre-Service%20Curriculum%20-%20Goode.pdf

The Royal Society. (2012). *Shut down or restart: The way forward for computing in UK schools*. Retrieved from http://royalsociety.org/education/policy/computing-in-schools/report/

The UK Department for Education. (2013). *National curriculum in England: Computing programmes of study*. Available [Online] https://www.gov.uk/government/publications/national-curriculum-in-england-computing-programmes-of-study/national-curriculum-in-england-computing-programmes-of-study

The White House. (2016). *Computer science for all*. Retrieved from https://www.whitehouse.gov/blog/2016/01/30/computer-science-all

Vee, A. (2013). Understanding computer programming as a literacy. *Literacy in Composition Studies, 1*(2), 42–64.

Weintrop, D., Beheshti, E., Horn, M., Orton, K., Jona, K., Trouille, L., & Wilensky, U. (2016). Defining computational thinking for mathematics and science classrooms. *Journal of Science Education and Technology, 25*(1), 127–147. https://link.springer.com/article/10.1007%2Fs10956-015-9581-5

Wilensky, U. (1995). Learning probability through building computational models. *Proceedings of the 19th International Conference on the Psychology of Mathematics Education*. Recife, Brazil, July 1995.

Wilensky, U., & Reisman, K. (2006). Thinking like a wolf, a sheep or a firefly: Learning biology through constructing and testing computational theories—An embodied modeling approach. *Cognition & Instruction, 24*(2), 171–209.

Wilensky, U., Brady, C., & Horn, M. (2014). Fostering computational literacy in science classrooms. *Communications of the ACM, 57*(8), 17–21.

Wilkerson-Jerde, M. H. (2014). Construction, categorization, and consensus: Student generated computational artifacts as a context for disciplinary reflection. *Educational Technology Research and Development, 62*(1), 99–121.

Wilson, C., Sudol, L. A., Stephenson, C., & Stehlik, M. (2010). *Running on empty: The failure to teach K-12 computer science in the digital age*. New York: The Association for Computing Machinery and the Computer Science Teachers Association.

Wing, J. (2010). Computational thinking: What and why? Unpublished manuscript in progress, Available [Online] http://www.cs.cmu.edu/~CompThink/resources/TheLinkWing.pdf

World Bank. (2016). *World Development Report 2016: Digital dividends*. Washington, DC: World Bank. doi:10.1596/978-1-4648-0671-1. License: Creative commons attribution CC BY 3.0 IGO http://documents.worldbank.org/curated/en/896971468194972881/pdf/102725-PUB-Replacement-PUBLIC.pdf

Yadav, A., & Korb, J. T. (2012). Learning to teach computer science: The need for a methods course. *Communications of the Association for Computing Machinery, 55*, 31–33.

Yadav, A., Gretter, S., Hambrusch, S. S., & Sands, P. (2017). Expanding computer science education in schools: understanding teacher experiences and challenges. *Computer Science Education, 26*(4), 235–254. https://doi.org/10.1080/08993408.2016.1257418.

Yadav, A., Mayfield, C., Zhou, N., Hambrusch, S., & Korb, J. T. (2014). Computational thinking in elementary and secondary teacher education. *ACM Transactions on Computing Education, 14*(1), 1–16.

Chapter 14
Susceptibility to Learn Computational Thinking Against STEM Attitudes and Aptitudes

Ana Calderon

14.1 Introduction

Computational thinking (henceforth abbreviated to CT) (Papert 1996; Guzdial 2008; Wing 2008) has received increasing attention amongst researchers and practitioners in the education field. Of particular importance is the recent trend to incorporate CT and related concepts at earlier stages of a pupil's education (Calderon et al. 2015; Bers et al. 2014). This initiative is based on the belief that by doing so, pupils' problem-solving skills and ability for logical thinking will be enhanced and that pupils will benefit greatly throughout their learning journey. The importance CT takes in compulsory education is best evidenced by changes in curriculum, policy and practice. For instance, the European Commission's science and knowledge service has recently released a report on "Developing Computational thinking in Compulsory Education". Not only does it provide comprehensive text on CT skills for children in compulsory education, but it also presents recent findings, based on research to support the policy-making process within Europe (Bocconi et al. 2016). Other examples include CT skills being incorporated into K–12 (NRC 2011) and recent significant educational reforms in the UK, with a focus on the need for digitally confident citizens and on enabling digital innovation (UK Digital Skills Taskforce 2014; House of Lords 2015). Moreover, in 2014, a new subject, computing, was introduced in the English National Curriculum, aiming to address computational thinking skills at varying key stages with similar changes introduced in Wales (Arthur et al. 2013).

This shift in perspective presents great opportunities but also challenges for those working in computer science education in higher education, particularly as notions

A. Calderon (✉)
Department of Computing and Information Systems, Cardiff Metropolitan University, Wales, UK
e-mail: acalderon@cardiffmet.ac.uk

M. S. Khine (ed.), *Computational Thinking in the STEM Disciplines*,
https://doi.org/10.1007/978-3-319-93566-9_14

that used to be reserved for post-compulsory and higher educational stages are now being pushed further and further down the educational ladder. It is imperative that, as computer science becomes one of the core educational components of primary and secondary school, educators at higher education level adopt common themes and present further challenges to students. For instance, there is an opportunity to get students more involved with complex societal problems, and aiding in achieving their solution, there is an opportunity to move towards teaching students skills in complex problem-solving, in particular, by employing advanced computational thinking techniques. Hence, although most of the effort has thus far been concentrated on compulsory education, our work focuses on HE and post-compulsory and higher education, as those working in this area will face increasing challenges and also will be best equipped to advise those introducing CT in post-compulsory and higher education. Hence, these considerations are timely with recent development.

14.2 Computation Thinking: Main Components

There are many views of computational thinking (NRC 2010); some researchers adopt the original notions of procedural thinking, as developed by Papert (1980). This work was the first to introduce computational thinking and views it as a step-by-step list of detailed and unambiguous instructions such that it can be interpreted and executed by an automated agent. Other researchers argue that computational thinking is a way of enabling humans to solve complex problems, by generating powerful tools to do so. Whatever viewpoint adopted, most researchers seem to agree that computational thinking is an integral part of computer science and that skills obtained from it can be transferred to problems in other subjects.

To date there is no consensus regarding the definitive or necessary components of computational thinking, although we note that recent efforts (Kalelioglu et al. 2016) have begun to provide a unifying framework for CT. Also, Barr et al. (2011) provide an operational definition of computational thinking, aimed at giving educators a definition they can incorporate in their teachings. At present, it is commonly agreed that computational thinking involves a combination of, at least, the following five concepts:

- Decomposition

  The process of breaking a complex problem into smaller sub-problems, in a manner that allows for understanding how their solutions can be placed together to form a solution to the original complex problem.
- Data representation

  As suggested by the title, studying data representation enables pupils to use CT methods for representing information.
- Algorithmic thinking

Algorithmic thinking allows students to solve problems by generating a step-by-step set of unambiguous instructions that can be followed by autonomous agents. This is similar to the original definition of computational thinking (Papert 1980).

- Abstraction

  Once patterns have been recognized, one can abstract away from the particular the details and create a general framework for solving several similar problems.

These concepts are linked together to solve a complex problem and are typically taught as such, as oppose to completely separate notions. Moreover, as mentioned in the introduction, several researchers and practitioners agree that CT elements can be readily applied to enhancing problem-solving skills in several subjects (Wing 2011). Validating this claim is perhaps ecologically impossible. However, if one was to test it, the most readily place one would expect this conjecture to hold is within the reach of computer science, as it is the most closely related to computational thinking; once sufficient evidence has been gathered, one may then investigate more distant STEM-based subjects. This conjecture is the motivation for our work; we do not claim to have attempted a solution to it, and we choose to investigate the opposite direction, whether students who are more able and more interested in STEM subjects are those who more readily learn CT elements or whether CT really is sufficiently generalized that all academic backgrounds, aptitudes and attitudes learn it equally readily. Essentially the question of interest to this work is whether computational thinking is unbiased towards varying preferences and abilities (humanities vs STEM), and it thus indeed seems to be the case. We stress that we do not claim to prove the conjecture but instead offer evidence that seems to strengthen it. Of particular interest to us is whether students with higher aptitudes (during secondary school education) in STEM or humanities are more susceptible to the learning of CT components and related concepts.

14.2.1 Relevant Literature

Following its introduction (Papert 1980) and then popularization (see, for instance, Wing (2008)), computational thinking has been applied in a wide range of educational domains and investigations. We now highlight the main contributions to the area. As highlighted in the previous section, most work has focused on primary and secondary school education, and we aim at filling the gap regarding how to then challenge pupils at higher education; hence, the gap in literature forces our literature review to focus mainly on compulsory education.

Visual languages (Selker Koved 1988; Chang et al. 1994) and serious games have recently been used to teach pupils, particularly of a young age, computing concepts mainly related to programming (see, for instance, Kazimoglu et al. [2012]). This is often linked to computational thinking. For instance, Koh et al. (2010) created a visual tool to measure transfer of computational thinking skills, from educational games to science simulations. The particular games where the measurements were taken were created by students, implemented using computational thinking tools. On the other hand, Repenning and Ioannidou (2010) reports on a research project on developing and testing strategies for incorporating computational thinking in curricula, teacher training and authoring schools. Of particular interest was the achievement of a balanced curriculum that incorporates game designs and computational thinking skills.

So far we have surveyed the literature for computational concepts taught with electronic means, but there are growing research and practice in the teaching of computational thinking without the use of a computer, with the most prominent examples found in Adams et al. (2005). Other important examples and resources include Curzon (2013), Boyle et al. (2012) and Ball et al. (2012). Also, although most of our literature survey has focused on computational thinking and its applicability to computer science, it can influence (and be influenced by) research in varied subjects from STEM-based knowledge to humanities. For instance, Bundy (2007) describes workshops held with professionals and researchers across a wide range of subjects reporting on the impact computational thinking research has had on their fields. The importance of computational thinking in varying subjects is perhaps best evidenced by its applicability to digital humanities (Gold 2012).

14.3 Investigation

As mentioned, our focus was on understanding how a preference for STEM or for humanities, as well as aptitude, is impacted by the ability to comprehend CT and CT-related concepts. Our investigations spanned 3 years, with the first 2 years focusing on gathering data linking particular concepts of CT to particular aptitudes. Short sessions were designed to introduce CT concepts at the start of term, which were then tested throughout the term. Once data seemed to confirm that there is no difference in CT, whereas more traditionally taught modules did appear to suffer bias, we moved on to creating a stand-alone CT module, to be taught alongside other computer science modules. We then (on the third and final year of the investigation) analysed how students performed in the subjects "Mathematics for Computing" and "Introduction to Programming" and the cohort outperformed previous years. The stand-alone module included aspects of the introductory sessions held for the benefit of this investigation; we will describe the course in detail once the data has been presented.

14.3.1 Methodology for First Investigation

The first part of our investigation focused on particular concepts of CT, namely decomposition, algorithmic thinking and abstraction. The focus was on learner

Table 14.1 Grouping of participating students

Group A	Mathematics, computing and physics
Group B	History, literature and drama

performance of those components, compared to programming susceptibility in students. The results were analysed against preferences in STEM and humanities. To decide which students' data would be used in the investigation, we looked for their highest achievements in their last 2 years of schooling and classified them according to the most commonly occurring subjects; for humanities, these were history, literature and drama, and for STEM, they were physics, mathematics and computing (see Table 14.1 above). To decide on aptitude, we analysed their grades, and if their highest was on group A of the table below, then we rated them as aptitude in STEM and, if their highest grades were in group B, as aptitude in humanities. The reason for not setting a universal boundary (a particular grade or percentage) is that doing it individually allows for measurement of students' aptitudes with regards to their own abilities. Also not all subjects were included; we choose to restrict to students with a particular choice of A-levels and found we could get the most participants by opting to focus on the subjects specified on Table 14.1.

To measure attitude (in our setting, this reads as personal preference), we asked all students to complete an attitude survey. We built the questionnaire by modifying a version of the Mathematics Attitude Test (Alken 1974) to cater for varying subjects. For the stem questionnaire (focused on computing and mathematics), we used five subscales:

- Programming confidence (PC)
- Computer science confidence (CC)
- Affective engagement (AE)
- Behavioural engagement (BE)
- Confidence in using technology (TC)
- Attitude to the use of technology to learn mathematics (MT)

In each of these categories, students were asked to answer questions according to the scale below. To score, similar to variations of MAS (Mathematics Atitude Scale) (Tapia and Marsh 2004), we used the sum of chosen ratings:

(1) hardly ever, (2) occasionally, (3) about half the time, (4) usually, (5) nearly always

In the interest of reproducibility, the test is specified below:

(PC) I am confident with programming.
(PC) I know I can handle difficulties in programming.
(PC) I can get good results in programming.
(PC) I think like a programmer.

(CC) I am confident with computer science.
(CC) I know I can handle difficulties in computer science.

(CC) I can get good results in computer science.
(CC) I think like a computer scientist.

(MT) Maths is more interesting when using graphics calculators and computers.
(MT) Graphics calculators and computers help me learn maths better.
(MT) Using graphics calculators and computers in maths is worth the extra effort.
(MT) I like using graphics calculators and computers in maths.

(TC) I am good at using computers.
(TC) I can fix a lot of computer problems.
(TC) I am good at using and adapting to new technologies.
(TC) I can master any computer programs needed for school.

(BE) If I make mistakes, I work until I have corrected them.
(BE) If I can't do a problem, I keep trying different ideas.
(BE) I try to answer questions the teacher asks.
(BE) I concentrate hard in maths.

(AE) Learning computing is enjoyable.
(AE) I am interested to learn new things in computing.
(AE) I get a sense of satisfaction when I solve computing problems.
(AE) In computing you get rewards for your efforts.

As previously mentioned, the score was taken as the sum of individually chosen ratings. Hence, the maximum score is 120, and we defined an individual as having a positive attitude in STEM if they scored anything strictly above 72, with no scores less than or equal to 2 for any given rating.

14.4 Results Following the First Stage of the Investigation

We now present some statistical conclusions derived from our study. We compared performance in programming against performance in particular elements of CT while looking for bias in the categories of STEM and humanities. The particular elements of CT we focused on are those considered essential components of the pedagogy, namely, abstraction, decomposition and algorithmic thinking. The results for each of these aspects will be presented separately, followed by a general observation of performance in computational thinking and programming, taken as average of individual student's performance variation between overall CT and programming. We also analysed individual variations in each of the CT components.

We first discuss some significant findings in the performance of the particular cohort in programming. The programming module's main focus was on:

- Syntax: Formal languages, regular expressions, lexical analysis, context-free grammars, parsing and context sensitivity.
- Semantics: Both operational and denotational were introduced, along with their main concepts.
- Programming paradigms.
- Complexity and measuring complexity of algorithms.

Performance was measured as the grade obtained in a coursework (that did not count for their final degree classifications) aimed at assessing their ability to:

- Interpret software requirements from a given scenario.
- Structure programs using key programming constructs.
- Identify and implement modular elements of programs.
- Demonstrate an appreciation of the key principles of user-centric design to design and develop appropriate user interfaces.

The particular task was to produce a piece of software to allow the user to catalogue various forms of media, according to specified classifications and information described to pupils in a coursework brief. This is a fairly standard type of coursework for a programming class. This was an intentional choice, aimed at minimizing influences of particular work choices during result analysis.

The programming distribution of grades (for the first year, we conducted this investigation) can be found below (Fig. 14.1).

The STEM group's performance was 67.3%, with a standard deviation of 57.6, and those with an aptitude in humanities was 57.6% with a standard deviation of 11.74; this suggests a potential bias for stem-based abilities.

We will now explain the data for computational thinking performance, broken into the varied elements, together with descriptions of the sessions.

14.4.1 Decomposition

Teaching of decomposition was done by first explaining it to students as a way to solve a complex problem. This was firstly achieved by asking students to think about the problem in terms of its parts, mentioning that the same process can apply to algorithms, systems, etc. The first concern was to ensure students understood that the division of parts had to be so that the overall problem could be solved by, instead of attempting it, solving and evaluating the individual parts (i.e. it had to be "compositional"). This is important as it enables making complex problems more

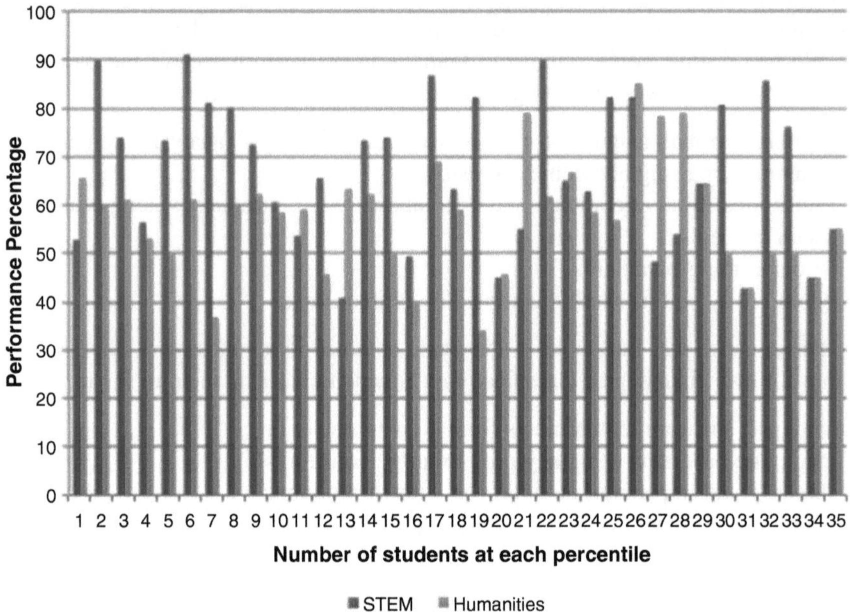

Fig. 14.1 Distribution of grades for programming against humanities and STEM preferences at A-levels

manageable and less difficult to solve. The whole class then engaged in a thought experiment where students were asked to think how they would assign sub-problems in developing a particular piece of software. This included role assignments, e.g. perhaps some would be responsible for the interface. It also included questions about how a breakdown into smaller components would be achieved. Finally students had to ask and answer questions regarding the specifics behind the breakdown of the problem.

To assess students' ability to solve problems by first applying decomposition, students were given Lego Mindstorms and requested to complete particular shape themes (these included a companionship robot, an animal robot, a robot to aid in exploratory scientific missions, e.g., exploring a dangerous unneutered terrain). The students were told the robots would also need to be programmed to perform particular actions. They were requested to solve this using their computational skills, in particular by analysing the problem and decomposing into smaller problems that had to be pieced together; this activity was done in groups of four. Performance was measured by focusing on the following elements:

- Variety of moves the robot was capable of performing
- Ensuring the robot had been programmed for independent action
- Creativity and complexity of shape and mechanics
- Programming, including explanation from students about its inner workings
- Ability of the student to explain concisely what they had done

Mean performance was 68.3% for humanities (with a variance of 15.1) and 70.8% (with a variance of 20.1) for STEM (full figures can be found in Table 14.2 below) (Fig. 14.2).

14.4.2 Abstraction

Each week, for 12 weeks, students were given 2 h of lectures on abstraction, followed by a 2-hour workshop, broken into two stages. We now give a brief overview of the sessions. They began with a gentle introduction to abstraction

Table 14.2 Grades from cohorts before and after introduction of CT module

	Before CT introduction		After CT introduction	
	STEM	Humanities	STEM	Humanities
Mean performance	52.9	48.9	61.6	61.0
Variance in grades	23.1	34.9	23.9	24.2

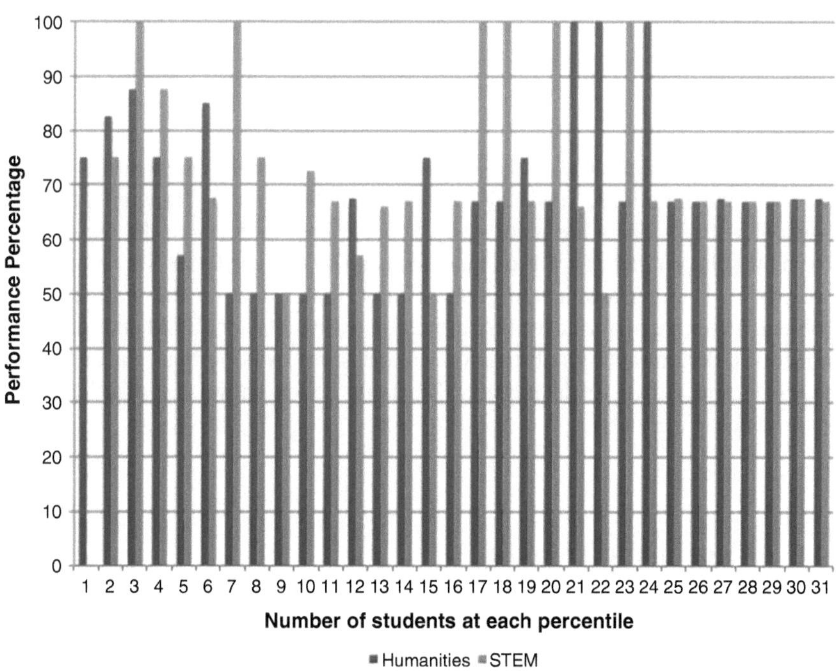

Fig. 14.2 Distribution of grades for decomposition against humanities and STEM preferences at A-levels

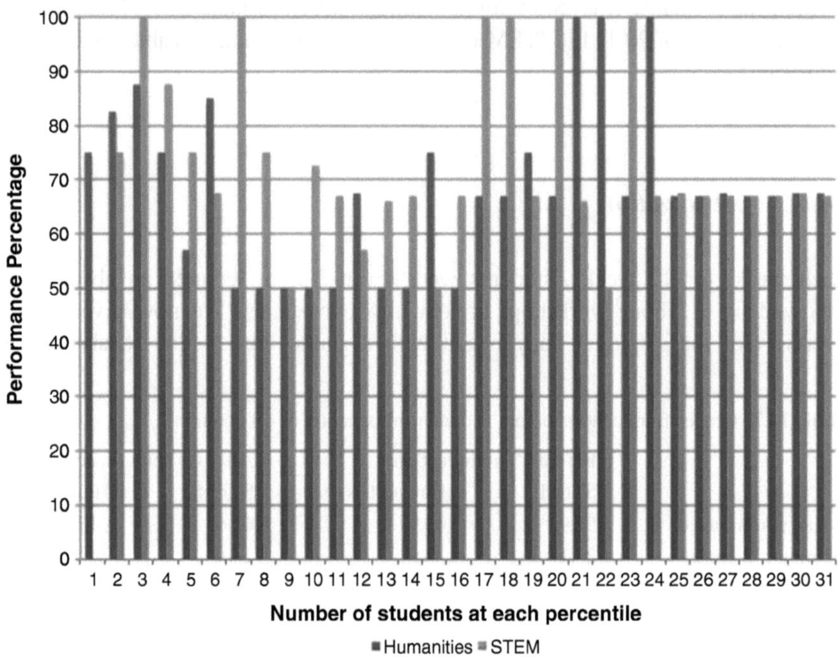

Fig. 14.3 Distribution of grades for abstraction against humanities and STEM preferences at A-levels

using ideas from Bell et al. (1998). Once students were confident in the simple exercises, we introduced a visual challenge, in the form of Bongard problems (Foundalis 2013). These are notoriously difficult, and the focus of the session was not on the solution, rather it was on them focusing on finding details not relevant and focusing on learning abstraction geometrically, rather than on the more standard problems. Average performance for abstraction was 61.5% for STEM and 61.0% for humanities. Performance for abstraction was measured by performance in an activity modified from Adams et al. (2005). The table below shows the distribution of performance (Fig. 14.3).

14.5 Algorithmic Thinking

Likewise for nearly all components (decomposition excluded), algorithmic thinking was first presented during two sessions of 2-hour lecture sessions and then followed by a 2-hour workshop.

During the workshop, students were reminded of what algorithmic thinking consists of and then given practical exercises, adapted from Curzon (2014), such as an adaptation of the "Knight's Tour" and "Kakuro, Sudoku and Computer Science". Students were then requested to think of daily activities and write pseudocode for them, and finally given an activity similar to "Marching Orders Activities" (Bell et al. 1998).

Performance was measured by asking students to consider a person with locked-in syndrome, only able to blink, and write pseudocode for a communication method, assuming the person would have a helper (human) to aid in reading. For instance, a solution could be to blink once for "yes" and twice for "no", while the helper chooses each alphabet letter. Of course this is quite inefficient, and grouping the letters or other methods leads to more efficient solutions and algorithms. The students were told to carefully consider the efficiency and complexity of their algorithms, prior to writing their pseudocode.

Pupils' performance was judged based on:

- Appropriateness of suggestions regarding how the patient could communicate with the (human) helper. For instance, blinking once for yes and twice for no (henceforth referred to as the base case)
- The particulars of the method with regard to its efficiency, as compared to the base case
- Comparative advantages and disadvantages against the base case
- Careful discussion of what other problems need solving, for instance, how to deal with:
  - Extra characters: punctuation, digits, etc.
  - Accidental blinking

Average performance for algorithmic thinking was 68.8% for STEM and 67.5% for humanities. The table below shows the individual distributions of grades for the activity just described (Fig. 14.4).

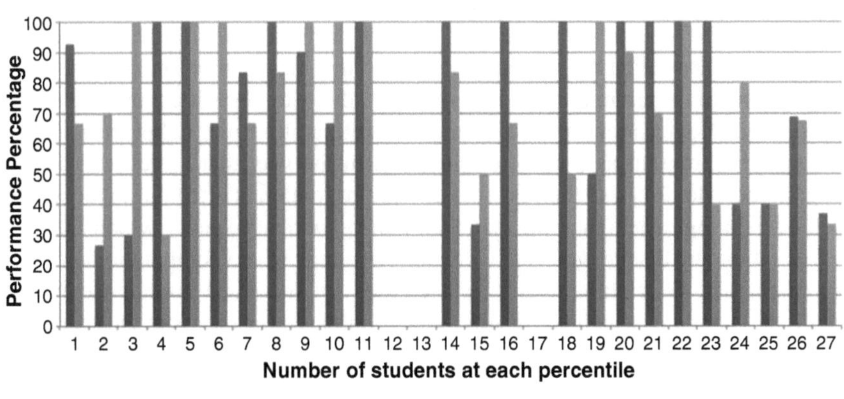

Fig. 14.4 Distribution of grades for algorithmic thinking against humanities and STEM preferences at A-levels

14.6 Discussion of Results

Discrepancies in programming performance, for STEM against humanities, were higher than those for computational thinking concepts, and that was the focus of this investigation. The aim was to discover whether there would be a discrepancy change in one group over the other, and there is some statistical evidence to suggest that the answer is negative; hence, CT might be less biased to varied academic backgrounds than programming. More illustrative content is given by observing differences in performance between programming and each of the CT components we tested in each student. In order to understand individual differences, we analysed individual student performances and measured it against each computational thinking element. The result was that, as compared to introductory programming, overall individual students performed 5.5% better in decomposition, 6.4% in algorithmic thinking and 4.8% better in abstraction. Hence, there is some further indication for the need of a module entirely dedicated to computational thinking. To complete this project, we designed such a module, targeting computer science and software engineering students (first year, first semester). The course was intensive and ran for 12 weeks with 4 hours each week; we then compared the performance (in a mathematics module) of students who took the course against those who did not (all other modules were kept constant); for ethical reasons, this had to be done with students from different cohorts, and we acknowledge this may cloud the conclusion but still provides enough content to merit mention and further investigation suggestions. The design of the course and results of the comparison are discussed in the next section.

Teaching programming at higher education is notoriously difficult, and designing a computational thinking course to go alongside it is of relevance to educators worried about the impact of introductory programming courses. In particular, we conjecture that a computational thinking module might aid in reducing disengagement in other courses, in particular programming courses. One of the main difficulties with regards to programming is how difficult it is to find a challenging level during classes that fits all pupils. Although this may be true for most subjects, in the case of programming, what varies is the way in which problems are approached. Novices tend to focus on a single line at a time, rather than understanding the program as a complex problem, in the way that more experienced programmers do. It is also the case that novice programmers often fail to recognize their own failings at a higher rate than their more experienced peers, which can lead to frustration and disengagement. These are two areas where computational thinking might aid; frustration might be decreased by increased confidence in the parallel CT module (which our results show is unbiased regardless of the academic background of pupils). Moreover, disengagement because of varying approaches to the problem can also be fixed by a parallel CT module, since one of its main foci is on viewing computing and social problems and generating methods for systematically solving them through a particular set of skills. Hence, there is some indication that novices would benefit by transferring this view to their introductory programming courses. There are often other reasons for dropping out of a programming course or disengaging with it; these may be reasons beyond computational thinking's reach in terms of minimizing negative effects. However, we have identified a few main reasons accepted by educators as significant with regard to student numbers and shown some evidence that computational thinking might help reduce these.

14.6.1 Designing a CT Module

To test feasibility of a separate CT module, we designed a new course for software engineering and computer science students. To test its effectiveness, in particular on the learning of mathematics, the students were taught a 12-week course on CT skills (4 h each week). The particulars of the sessions are as follows: 2 h of theoretical understanding of CT components and 2 h of practical sessions with complex problems challenging pupils' understanding in each of the three computational thinking elements (abstraction, abstract thinking and decomposition). Each component was also complemented by concepts linked to computer science, for instance, algorithmic thinking included lessons in complexity of algorithms, but these were presented in a sufficiently general manner that the module could still be stand-alone and, importantly, taught to students with a wide range of backgrounds.

The main focus of this pilot study was to design a course so that, upon completion, students would be able to demonstrate understanding of fundamental concepts in computer science and software engineering, in particular being able to apply a range of approaches to solving data-driven and computational problems. During delivery of the course, the focus was on giving pupils the skills needed so they were able to demonstrate an appreciation of the role of computer science and computational thinking in the modern world. The first session of the course consisted of an explanation of what computational thinking is, including a brief introduction into each of its components. The remaining sessions were then responsible for their detailed explanations. A summary of introductory material into each component is given now.

Abstraction was first presented as a formal definition with examples. We stated it as the removal of unnecessary data, considered as noise, to make the problem simpler; it was recapped during the introduction to decomposition, due to how coupled the concepts are. We highlighted that, as humans, we abstract to problem-solve regularly; this was followed by examples of human conversation, where detail is unimportant, cases in which the wrong amount of detail is abstracted (both over and under the expectancy of the other conversationalist). Samples of questions posed to students included:

> If a friend asks you about their holidays, would you give them an hour-by-hour account, or abstract some detail so only the highlights would be communicated?

> On the other hand, if a waiter asks of your lunch choice, would the general category "food" suffice?

Students were also taught how to abstract by parameterization (describing named computations) and by specification (ignoring how computations are performed).

This was then followed by highlights of the important role abstraction serves to computer science, as well as some considerations of how it is applied in exemplary problems. For instance, with regard to programming, we essentially explained it as hiding some complexity, for example, by saying that in a language such as Python to simply say "Hello World", it takes a few lines; however, it contains thousands of machine code, which the programmer is usually blind to. We also used the introduction of abstraction to discuss a layered view of computer science, in terms of abstraction layers (turning into a class discussion). Suggested layers included:

- Theory/mathematics
- Applications
- Operating system, kernel
- Assembler, computer architecture
- Computer organization
- Digital logic
- Electronics
- VLSI design
- Silicon wafer design
- Physics

Algorithmic thinking was presented with the usual definition of an algorithm, that is, *an unambiguous list of sequential steps to be followed, as a solution to a problem.* Also included was a brief introduction to pseudocode (thus allowing students to write solution algorithms to problems from the first session). Amongst the topics more directly linked to computer science were:

- Equipping students with skills necessary to describe the steps in the program development process
- Equipping students with skills necessary to introduce algorithms formally
- Equipping students with skills necessary to describe program data in a general setting
- Ensuring students were able to design algorithmic solutions to complex problems and knew how to verify and test their solutions
- Analysing algorithms, showing correctness of algorithms. For instance, proving recursive algorithms are correct, using mathematical induction
- Using loop invariants in proofs

Decomposition, similarly to the previous concepts, was first presented via a formal definition. We stated it as the ability to break down complex problems into manageable smaller ones, in a manner such that the sub-problems can be composed together to form a larger one. This was followed by motivational examples of complex problems and finally comparatively less complex problems that the students could work on forming compositional sub-problems. With regard to linking it to computer science, we focused on how functionality is organized within programming. The core idea was to explain how programming languages allow for the breaking up of code into pieces such that, if the subdividing is done sensibly, then editing and debugging are more easily achieved. Other topics relating decomposition to programming included links with structured programming, object-oriented decomposition (breaking into smaller classes and objects). In addition, students also understood that decomposition is not universally applicable, that there are some problems too complex, with too many unknowns that we cannot use decomposition to devise a solution. Decomposition was naturally introduced in relation to abstraction. The particular lesson plan included:

- Introduction to decomposition with motivational examples, followed by the main steps involved, namely:
- Identifying components, aiming at minimizing component dependency, and reducing coupling
- Deciding which information needs to be private and encapsulating it in particular modules (information hiding)
- Modelling decomposition by designing the modules, including pseudocode
- Using abstraction to remove unnecessary detail and aid in simplifying the problem

14.7 Performance in Mathematics Subjects Post-Computational Thinking Course

We found that the discrepancy in grades for a particular mathematics module was smaller than in previous years (as described in Table 14.2), observing a decrease in discrepancy by 19.2% for STEM and by 30.7% for humanities. Moreover, the actual performance of students in the specific mathematics module improved by 8.7% for STEM and humanities by 12.1%. We acknowledge that this was done with two different cohorts, and there might be some effects on the performance due to varied participants; however, the change in variance is quite significant and merits further investigation.

Performance in the mathematics module was measured according to standard conditions for such a module; we will describe it in details soon. First, a description of the core content of the module is given. It was concerned with facilitating the understanding and implementations of various mathematical concepts underpinning computer science, with a focus on:

- *The axiomatic method:* basic concepts, axioms, definitions, theorems, finite and infinite sets, natural numbers and induction
- *Logic:* statements, truth values, Boolean operators, laws of propositional logic, predicates, quantifiers and laws of predicate logic
- *Sets:* connection between sets and predicates, operations on sets and laws of set operations. Functions, sequences and relations
- *Algorithms:* basic complexity theory, rates of growth, time and space constraints, common data structures and sorting algorithms. Interpreting graphs. Iteration and recursion
- *Number theory:* representations, permutations and combinations and foundations of cryptography
- *Automata, grammars and languages:* finite-state machines, finite-state automata and Turing machines
- *Statistics and probability:* definitions, conditional probability, Bayes' theorem, expectation, variance, standard deviation and analysis of simple datasets

For further information on how the concepts were introduced to students, we refer the reader to the supplemental reading on that course, namely, Lehman et al. (2010), Boolos et al. (2007) and Johnsonbaugh (2013).

The performance in both cases (before and after introduction of a focused computational thinking module) was measured with a similar exam. Students were requested to sit an in-class test and answer questions on a variety of topics (as above). Common questions included completing the truth table for a Boolean formula (different formulae were given to different cohorts) and using logical analysis of the definitions of union and intersection of sets (and therefore rules for conjunction, disjunction and De Morgan's), to prove a particular statement (different statements were given to different cohorts). Also, the grading scheme was similar in both cases.

In order for a student to achieve a first class (higher than 70 percentage), the student's work had to demonstrate excellent understanding of the main mathematical concepts in computing science, including key mathematical terminology, notation and the formal definitions and proofs. Also, the work had to display an excellent awareness of how discrete mathematics applies to computation and its application to real-world problems.

In order to achieve a second class, upper division (between 60 and 69 percentile), the submission had to demonstrate a good understanding of the key mathematical concepts in computer science, including key mathematical terminology, notation and formal definitions and proofs. Also, the work had to display an awareness of how discrete mathematics applies to computations and its application to real-world problems.

In order to achieve a second class, lower division (between 50 and 59 percentile), the submission had to demonstrate some understanding of the main mathematical concepts in computer science, including some of the key mathematical terminology, notation and formal definitions and proofs. Also it was required that the work displayed an awareness of how discrete mathematics applies to computation and its application to real-world problems.

In order for work to be classed as third division (between 40 and 49 percentile), the submission had to demonstrate satisfactory understanding of the main mathematical concepts in computer science, including some of the key mathematical terminology, notation and formal definitions and proofs. Also it needed to display some awareness of how discrete mathematics applies to computation and its application to real-world problems.

Finally, a "fail" entailed a weak submission that demonstrated poor understanding of the main mathematical concepts in computing science, including with very little understanding of some of the key mathematical terminology, notation and formal definitions and proofs and, also, with little awareness of how discrete mathematics applies to computation and its application to real-world problems.

14.8 Performance in Introductory Programming Courses Post-Introduction of Computational Thinking Course

The results found were similar to those in the previous section (as per Table 14.2 above). There is some statistical analysis that indicates that there was an increased performance in programming for both groups. It is important to acknowledge that these had to be done with different cohorts for ethical reasons (it would not be possible to offer a group of students the course and have a "control group" who did not receive a beneficial pedagogical intervention, as they would be disadvantaged); hence, the particular findings need to be taken in context, and it indicates an increase, but we cannot conclude for certain that there was an increase in performance.

We detailed how performance was measured in mathematics, and we now give, in similar detail, the performance in programming. The particular course measured performance in programming by identifying and grading the following components:

- *Production and organization of code*

  More specifically, the focus was on usage of key programming constructors, as well as effective approach to specific paradigms (for instance, object-oriented programming).
- *Graphical user interface*

  More specifically, the focus was on intuitive designs with consideration of basic concepts of usability engineering, as well as how professional and aesthetically pleasing the interface was.
- *Interpretation of user requirements*

  More specifically, the focus was on how the assignment was interpreted and the production of software that met the requirements, together with evidence of testing and debugging of code on several development stages.

14.9 Conclusion and Directions for Future Work

Computational thinking's efficacy and transferability onto varying subjects and skills is often taken as a tautology. We set out the beginnings of an investigation into whether aptitudes and attitude to varying subjects has an impact into the ability to learn CT. Our results indicate that CT learning has no discriminatory effect on the large groups of STEM and humanities. We have focused only on these two general categories, but since our results have thus far have been positive, we argue that there is sufficient motivation to analyse whether there is a difference when dealing with smaller subclasses of subjects (both within and across the broader categories we investigated).

For our studies, we paired an aptitude and attitude in mathematics and computer science, but could not similarly pair an attitude in humanities as, to the best our knowledge, no such tests exist. Given that humanities are outside the author's expertise, further work will involve a multi-disciplinary team investigating how attitude in learning humanities, or lack thereof, might be impacted differently in humanity subjects against computational thinking learning. The results found could then lead to a bigger investigation on how computational thinking skills can be transferred to subjects more cognitively distant from computer science, particularly outside the sciences.

We categorized students who classified high on attitude and aptitude in specific subjects of STEM and humanities (chosen to maximize the available number of participants in our sample); we then analysed the data against performance in programming and in varying components of computational thinking. To measure attitude, we modified the MAS scale to cater for our investigations; and, in order to measure aptitude, we took the highest grades individuals obtained in each of the classifying groups for STEM and humanities.

Our results indicate the variance in performance is smaller in computational thinking concepts than in programming. An interesting direction for further work would be to ask whether the same holds true for other traditional computer science courses (against the same components of computational thinking).

In addition to our initial investigation, in order to test feasibility of designing a stand-alone CT module and understanding how it might affect mathematical concepts, we conducted a pilot study (described in the previous session), by encompassing such a course with a new cohort of students. Following the pilot course, analysis in student performance was indicative of a positive correlation with the course and increased performance in a particular mathematical module. This merits further investigation; in particular, it would be interesting to see whether subjects a bit "further" from computational thinking would be similarly impacted, for instance, by taking a sample of students enrolled in both humanities and STEM subjects (at higher education level) and seeing whether a short CT course can aid in their problem-solving skills. Importantly, a positive correlation between performance and introduction of the CT course would indicate that CT problem-solving abilities can indeed be transferred to varied subjects. Also, investigating how specific these findings are to higher education, and whether they scale further down pupils' learning journeys, would have significant consequences regarding an understanding of when to introduce CT in pupils' learning. Likewise, future work will also contain a similar experimental setup with adults who have been away from educational settings and in a work environment for a significant amount of time. Again, if positive correlations were found, this would have significant consequences and consist of further evidence into how efficient (and transferable to a variety of skills and subjects) computational thinking truly is.

Finally, it is worth emphasizing that teaching introductory programming at higher education is notoriously difficult. The cohort of students enrolled in a programming introductory class (at the institution where the experiments took place) typically contains students with varying academic backgrounds. Moreover, students will have had varying levels of exposure to (both formal and informal) programming concepts prior to the beginning of their higher education journeys. The discrepancy in ability and prior exposure between students makes delivery of material and, in particular, its difficulty level challenging to deliver. Disengagement from students is commonly observed due to material being too difficult or not sufficiently challenging. Recently, we have witnessed the increased awareness on the importance on "software carpentry", and teaching such skills requires academic teaching staff to move beyond demonstrating technical skills, to aid students in their ability to explain the rational thoughts behind decisions taken in solutions to complex problems, typically associated with developing software. It has been argued that computational thinking is amongst the pillars underpinning software development and that it is advantageous to expose pupils to such skills in conjunction with teaching programming. Our work indicates that there is indeed a positive correlation between such exposure and increased performance in students' programming skills. However, we can also argue that such an emphasis on CT in conjunction with a more formal delivery of programming could aid in minimizing disengagement risks, and we leave such an investigation to further work. Our conjecture is based on disengagement due to the wrong level of difficulty causing demotivation amongst students. It is evident that other disengagement causes may not be impacted by the introduction of a parallel computational thinking course, and we would like to further analyse the potential of such links. Therefore, we also leave to future work an attempt at establishing links between disengagement and pupils' achievements in programming and links between disengagement and pupils' achievements in computational thinking subjects, with the aim of establishing a statistically significant correlation. We are, however, aware that it would be difficult to quantifiably establish the introduction of a computational thinking module as the positive cause in lowering drop-out rates and disengagement amongst a particular cohort, in particular because we could not test students who did and did not complete the computational thinking module and whether they left the course or not, as it would be unethical to have a negative and irreversible effect on pupils' learning.

We will also conduct a broader investigation to the one reported here, namely, understanding whether humanities against STEM preferences suffer bias in the ability to engage and learn computational thinking skills, in non-student populations.

This investigation will be an attempt to establish if similar results can be reproduced with a broader sample of the population rather than university students, and varying ages, educational backgrounds and careers will be taken into account.

References

Adams, R., Bell, T., McKenzie, J., Witten, I. H., & Fellows, M. (2005). *Computer science unplugged: An enrichment and extension programme for primary-aged children.*

Alken, L. R. (1974). Two scales of attitude toward mathematics. *Journal for Research in Mathematics Education, 5*, 67–71.

Arthur, S., Crick, T., & Hayward, J. (2013). *The ICT Steering Group's Report to the Welsh Government.* http://learning.wales.gov.uk/docs/learningwales/publications/131003-ict-steering-group-report-en.pdf (retrieved: 1 May, 2015).

Ball, C., Moller, F., & Pau, R. (2012, November). *The mindstorm effect: a gender analysis on the influence of LEGO mindstorms in computer science education.* In Proceedings of the 7th workshop in primary and secondary computing education (pp. 141–142). ACM.

Barr, D., Harrison, J., & Conery, L. (2011). Computational thinking: A digital age skill for everyone. *Learning & Leading with Technology, 38*(6), 20–23.

Bell, T., Witten, I., & Fellows, M. (1998). *Computer science unplugged.* www.csunplugged.org.

Bers, M. U., Flannery, L., Kazakoff, E. R., & Sullivan, A. (2014). Computational thinking and tinkering: Exploration of an early childhood robotics curriculum. *Computers & Education, 72*, 145–157.

Bocconi, S., Chioccariello, A., Dettori, G., Ferrari, A., & Engelhardt, K. (2016). *Developing computational thinking in compulsory education-Implications for policy and practice* (No. JRC104188). Joint Research Centre (Seville site).

Boolos, G. S., Burgess, J. P., & Jeffrey, R. C. (2007). *Computability and logic* (5th ed.). Cambridge: Cambridge University Press.

Boyle, R. D., Dee, H. M., & Labrosse, F. (2012, November). *Technocamps: bringing computer science to the far west.* In Proceedings of the 7th workshop in primary and secondary computing education (pp. 147–148). ACM.

Bundy, A. (2007). Computational thinking is pervasive. *Journal of Scientific and Practical Computing, 1*(2), 67–69.

Calderon, A. C., Crick, T., & Tryfona, C. (2015, July). *Developing computational thinking through pattern recognition in early years education.* In Proceedings of the 2015 British HCI conference (pp. 259–260). ACM.

Chang, S., Korfhage, R.R., Levialdi, S., & Ichikawa, T. (1994). Ten years of visual languages research, IEEE, pp. 196–205.

Curzon, P. (2013, November). *cs4fn and computational thinking unplugged.* In Proceedings of the 8th workshop in primary and secondary computing education (pp. 47–50). ACM.

Curzon, P. (2014). *Computer science activities with a sense of fun: Tour guide V1.1 14 Dec 2014).* Created by Paul Curzon, Queen Mary University of London for Teaching London Computing: http://teachinglondoncomputing.org

Foundalis. (2013). Research on the bongard problem, retrieved June 5th 2018. http://www.foundalis.com/res/diss_research.html

Gold, M. K. (2012). *Debates in the digital humanities.* Minneapolis: University of Minnesota Press.

Guzdial, M. (2008). Education: Paving the way for computational thinking. *Communications of the ACM, 51*(8), 25.

House of Lords Select Committee on Digital Skills. (2015). *Make or Break: The UK's digital future.* http://www.publications.parliament.uk/pa/ld201415/ldselect/lddigital/111/11102.html (retrieved: 28 April, 2015).

Johnsonbaugh, R. (2013). *Discrete mathematics* (7th ed., pp. 257–280). Pearson. ISBN: 0131593188.

Kalelioglu, F., Gülbahar, Y., & Kukul, V. (2016). A framework for computational thinking based on a systematic research review. *Baltic Journal of Modern Computing, 4*(3), 583.

Kazimoglu, C., Kiernan, M., Bacon, L., & Mackinnon, L. (2012). A serious game for developing computational thinking and learning introductory computer programming. *Procedia-Social and Behavioral Sciences, 47*, 1991–1999.

Koh, K. H., Basawapatna, A., Bennett, V., & Repenning, A. (2010, September). *Towards the automatic recognition of computational thinking for adaptive visual language learning*. In Visual Languages and Human-Centric Computing (VL/HCC), 2010 I.E. Symposium on (pp. 59–66). IEEE.

Lehman, E., Leighton, T., & Meyer, A. R. (2010). *Mathematics for computer science*. Technical report, 2006. Lecture notes.

National Research Council. (2010) *Report of a workshop on the scope and nature of computational thinking*(Chapter 2, p. 4). Washington, DC: The National Academies Press. doi:https://doi.org/10.17226/12840.

NRC. (2011). *A framework for K-12 science education: Practices, crosscutting concepts, and core ideas*. Washington, DC: The National Academies Press.

Papert, S. (1980). *Mindstorms: Children, computers, and powerful ideas*. New York: Basic Books, Inc.

Papert, S. (1996). An exploration in the space of mathematics educations. *International Journal of Computers for Mathematical Learning, 1*, 95–123.

Repenning, A., Webb, D., & Ioannidou, A. (2010, March). *Scalable game design and the development of a checklist for getting computational thinking into public schools*. In Proceedings of the 41st ACM technical symposium on Computer science education (pp. 265–269). ACM.

Selker, T., & Koved, L. (1988). *Elements of visual language*. IEEE, pp. 38–44.

Tapia, M., & Marsh, G. E. (2004). An instrument to measure mathematics attitudes. *Academic Exchange Quarterly, 8*(2), 16–22.

UK Digital Skills Taskforce (2014). *Digital skills for tomorrow's world*. http://www.ukdigitalskills.com/wp-content/uploads/2014/07/Binder-9-reduced.pdf (retrieved: 28 April, 2015).

Wing, J. (2008). Computational thinking and thinking about computing. *Philosophical Transactions of the Royal Society A, 366*(1881), 3717–3725.

Wing, J. M. (2011, March). Computational thinking. In VL/HCC (p. 3).

Chapter 15
Mapping Computational Thinking for a Transformative Pedagogy

Michael Vallance and Phillip A. Towndrow

15.1 Introduction

Computer science practitioners have called upon education ministries to adopt Computational Thinking as a foundation for twenty-first-century learning and skills development at all age levels. Computational Thinking is aimed at ways of understanding human behavior through the analytical thinking processes associated with solving problems, mostly in the fields of science, technology, engineering, and math (STEM) education. Accordingly, a problem-solving process, informed by the characteristics of Computational Thinking, requires active and experiential behaviors of both learners and instructors within a multidisciplinary education environment. In this chapter it is proposed that educational robot activities offer such a context to discover more about the Computational Thinking of learners and that this can be achieved by considered pedagogy and informed task design.

The chapter is a descriptive, case study account to identify Computational Thinking in beginner educational robot activities, reminiscent of the late Seymour Papert's development of LOGO and Mindstorms. Challenging tasks are provided to university undergraduate students in Japan who then design, program, and implement unique robot solutions. The interpretation of data from ten tasks undertaken by four undergraduate university students over two semesters will demonstrate how they mapped their mental models of problem-solving (i.e., their Computational Thinking) to the resulting robot program solutions. The case study is then used to explain how pedagogy and task design inform a new learning paradigm.

The chapter is organized as follows. First, the resurgent interest in Computational Thinking is discussed. After that a method for capturing students' problem-solving

M. Vallance (✉)
Department of Media Architecture, Future University Hakodate, Hokkaido, Japan

P. A. Towndrow
National Institute of Education, Nanyang Technological University, Singapore, Singapore

M. S. Khine (ed.), *Computational Thinking in the STEM Disciplines*,
https://doi.org/10.1007/978-3-319-93566-9_15

processes as illustrative flowcharts is presented. The context of using robots in an imaginary disaster situation is then discussed, along with the project tasks, participants, and instrument. This is followed by a detailed account of the results. The chapter concludes with an informed discussion of pedagogy and task considerations that support Computational Thinking.

15.2 Background

Forty years ago the late Seymour Papert wrote about a schizophrenic split between the disciplines of science and humanities in his seminal book, *Mindstorms: Children, Computers, and Powerful Ideas* (Papert 1980). He advocated that computers could "break down the line between the two cultures" (*ibid*; p. 38) and subsequently promote a "less dissociated cultural epistemology" ((*ibid*; p. 39). He reasoned that computers could be used to "challenge current beliefs about who can understand what and at what age" (*ibid*; p. 4) and "can help people form new relationships with knowledge that cut across the traditional lines separating humanities from sciences" (*ibid*; p. 4). As we near the end of the second decade of the twenty-first century, Papert's philosophical beliefs once again resonate in education research and practice in the form of Computational Thinking. The resurgence of Papert's "powerful ideas" can be attributed to an influential article by Jeanette Wing (2006) who posited that "computational thinking involves solving problems, designing systems, and understanding human behavior, by drawing on the concepts fundamental to computer science..." (p. 33).

Crucially, Computational Thinking is something people do, not computers. It involves logical thinking, the ability to recognize patterns, and the capacity to structure problems so that computers can be used, if necessary, to help develop solutions (cf., Liukas 2015). It uses a set of concepts such as abstraction, decomposition, algorithmic logic, and pattern recognition to process and analyze data and create real and virtual artifacts in solving problems. Influential MIT professor Mitch Resnick advocates Computational Thinking be a device for conceptualizing learning for students to:

- Formulate problems in a way that enables us to use a computer and other tools to help solve them
- Logically organize and analyze data
- Represent data through abstractions such as models and simulations
- Automate solutions through algorithmic thinking (a series of ordered steps)
- Identify, analyze, and implement possible solutions with the goal of achieving the most efficient and effective combination of steps and resources
- Generalize and transfer this problem-solving process to a wide variety of problems (Brennan and Resnick 2012)

The Computer Science Teachers Association (CSTA) and the International Society for Technology in Education (ISTE) in the USA are actively promoting the inclusion of Computational Thinking in school and higher education curricula. Barr and Stephenson (2011) proposed that educational policies need to present a single message at federal, state, and local levels about the importance of Computational Thinking in K-12 education. They recommended that Computational Thinking be incorporated throughout the entire K-12 experience and all preservice teachers be provided preparation classes on Computational Thinking across all disciplines.

In the UK in 2012, the Royal Society (2012) in a report entitled "Shut Down or Restart?" recommended a transformation of the ICT (information communication technology) subject to adopt Computational Thinking in both the primary and secondary school curricula. Consequently, in 2014, the UK's National Curriculum introduced a new subject called computing to replace ICT. The compulsory subject was developed on the supposition that, "Computational thinking is a skill that all pupils must learn if they are to be ready for the workplace and able to participate effectively in the digital world" (NAACE 2014).

Like many other accounts of education excellence, Finland stands out as a role model. Toikkanan (2015) states, "No other country has the same approach as Finland. The Finnish curriculum includes coding as a mandatory, cross curricular theme starting from first grade." Of the 20 other European countries surveyed by Toikkanan, only the UK and Belgium have computing as a mandatory requirement. The remaining countries offer the subject only as an option (*ibid*).

It took nearly 10 years for Wing's influential article on Computational Thinking to be translated into Japanese (cf., Nakashima 2015), but even in Japan, where a philosophical gulf exists between the sciences and humanities (Berlin 1974), Computational Thinking is gaining traction (Computational Thinking for All 2017).

At this juncture, it may appear that academics in the computer science and information technology disciplines are asserting their beliefs upon all of education and that Computational Thinking may mistakenly be considered the act of working on a digital device. However, using a word processor obviously involves the use of a computer, but it would be a stretch of the imagination to associate cognitive actions while typing be considered Computational Thinking. It has been argued that Computational Thinking does not even need computers, as it is an approach to problem-solving that uses strategies such as logic (algorithms), ideas (abstraction), and the removal of errors (debugging) (Yadav et al. 2011). However, Dede et al. (2013) suggest that computers, machines, and digital devices have already altered our interactions with technology so that "... new forms of expression that are emerging today have significant implications for how we engage and interact with machines" (p. 4). In other words, we are learning "in" technology (Schrader 2008; Vallance and Towndrow 2016), and our creativity is augmented by Computational Thinking. Consequently, it is reasoned that students who are engaged in Computational Thinking processes across the curriculum can begin to see relationships between the subjects they study at school or higher education and life outside the classroom. Therefore, Computational Thinking is as relevant to the arts, humanities, and social science disciplines as it is to the disciplines of science, technology, engineering, and math.

In this chapter we will attempt to illustrate Computational Thinking. We want students to make sense of their processes through reflection and articulation and to reconsider and refine processes so they can articulate their understanding as they proceed through the problem-solving procedure. We want them to hypothesize, experiment (*what if*), design, interpret, and reflect. Supported by a "learning by design" pedagogy, students will engage in the design and subsequent use of the physical, digital, or virtual representations for modeling and reasoning (Papert 1980). For instance, the "learning by design" pedagogy allows students to design and use representations of something such as a 3D model, an infographic, a podcast, or a robot to understand the "what" and "why" of their learning processes and acquired knowledge. As a result, it has been posited that the development of students' Computational Thinking skills can be closely aligned with the students' design-based computational representational practices (Basu et al. 2016), facilitated by a transformative pedagogy and contextualized by tasks incorporating educational robot activities. Our goal is to show how practitioners can view those representations.

15.3 Method

The method used in our project has been informed by Atmatzidou and Demetriadis' (2014) paper "How To Support Students' Computational Thinking Skills In Educational Robotics Activities." Their students worked in small groups, guided by worksheets, to solve problems. They used a model of Computational Thinking concepts consisting of abstraction, generalization, algorithmic logic, modularity, and decomposition. Data was collected from pre- and post-implementation quizzes, questionnaires, and interviews from 35 participants over a course of 11 tasks each lasting 2 h. The interpretation of the data revealed that their students became familiar with the concepts of algorithm logic, modularity, and decomposition, but abstraction and generalization proved more challenging. The researchers suggested that more engagement with complicated problems was required so that students could reflect further on their solutions. In our project we were limited to four students but took the opportunity to develop tasks and reflection data that would, it was anticipated, provide that deeper understanding of the students' awareness of the concepts of Computational Thinking. To that end we provided a post-task worksheet for students to discuss and illustrate their knowledge via algorithmic flowcharts and Computational Thinking concept linkages with their robot program solutions.

15.3.1 Educational Robot Activities

Informed by Atmatzidou and Demetriadis (2014), our project adopted the programming of LEGO Mindstorms EV3 robots to navigate mazes of measurable

Fig. 15.1 LEGO Mindstorms EV3 robot example

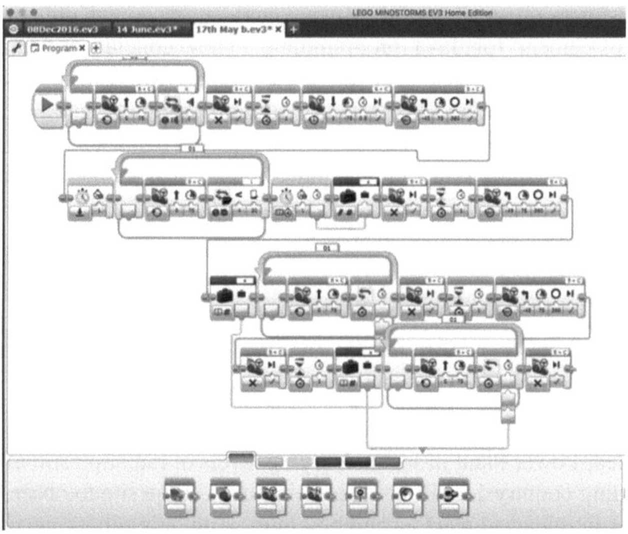

Fig. 15.2 Mindstorms EV3 program example with blocks, loops, and variables

complexity. LEGO Mindstorms is often thought of as simply a toy for children interested in building robots. LEGO markets Mindstorms for children aged 10 and over. But this downplays the enormous versatility LEGO Mindstorms has to offer learners of all ages. The programming of a LEGO Mindstorms robot (see Fig. 15.1) begins with the drag and drop graphical user interface (see Fig. 15.2) that enables commands to be downloaded from a computer to the "brick" (a programmable microcomputer with a processor, flash memory, and Linux operation system). At the beginning, this allows new users to appreciate the concept of procedures. Adding sensors to a robot requires learners to then consider sensor values, variables, arrays, and logic. Learners become engaged in testing and adjusting their programs in attempts to succeed in their desired outcomes.

LEGO Mindstorms is not only for school children but for any aged learner starting out in programming in any STEM discipline. Research by Lui et al. (2010) used LEGO Mindstorms with university computer science students to promote self-directed learning. Popelka and Nožička (2014) utilized LEGO Mindstorms as a simulation of robotic systems using the programming language C#, a PID (proportional-integral-derivative) controller, Visual Studio software, and MonoDevelop framework, tools far beyond the level of a 10-year-old and certainly more appropriate for university undergraduates. They concluded that, "Based on discovered facts LEGO Mindstorms can be considered a low-cost and capable kit to simulate real robotics systems" (p. 1128). Catlin and Blamires (2010) adopted LEGO Mindstorms to promote Computational Thinking which involves mathematical modeling, inductive thinking, and experimentation. Turner and Hill (2008) used LEGO Mindstorms as a prerequisite for teaching Java programming at a university. They focused on problem-solving and robot maze emulation. Lew et al. (2010) used LEGO Mindstorms on an advanced software engineering course utilizing the leJOS firmware replacement for Java programming. They concluded that there were sufficient technical challenges with the use of LEGO Mindstorms that they will continue to be used in future semesters. To sum up this small sample of literature, due to its versatility, LEGO Mindstorms is misrepresented as a simple toy and can be as powerful a robotic tool as any child, adult, or computer science student wishes to imagine.

15.3.2 Tasks

Ten tasks were implemented over two semesters: 15 weeks per semester. The narrative for the tasks was informed by the continuing disaster at the Fukushima Daiichi Nuclear Power Plant in Japan. Varying levels of radiation still afflict the site and surrounding countryside. Robots are being used at the site to observe, monitor, clean, and repair damaged reactors and buildings. And only robots can enter parts of the plant off-limits to humans due to the radiation levels. The task narratives are imaginary but are aimed at creating an awareness and a curiosity of the daily challenges faced by the plant workers.

In our project tasks, robots have to be customized to maneuver obstacles (Tasks T1, T2, T3), turn on auxiliary machinery (Tasks T4, T5, T6), climb steps (Task T8), and synchronize with other robots (Tasks T7, T9). All robots use a combination of motors and sensors. The tasks were not solely a trial-and-error, discovery implementation. Of course, giving students opportunities to tinker was essential in the sessions, but occasionally the instructor taught new concepts and EV3 capabilities. For instance, programming structures for accelerating and decelerating a robot proved useful. Without knowing this, the students' robots simply moved at one particular speed, and when stopping, inertia would move the robot off its trajectory. The concept of self-regulated motors and the physical use of gears were also practiced during task solution considerations. Moreover, additional tools such as the Pixy camera from the Carnegie Mellon Robotics Institute (http://www.cmucam.org/projects/cmucam5) and associated programming were introduced. The Pixy camera could be programmed with a PID (proportional-integral-derivative) controller to recognize and track objects. The Pixy camera was considered in Task 9. Also, the students were instructed how to create EV3 MyBlocks that could then be recycled in later tasks. This proved extremely useful as the MyBlocks saved much time and enhanced the aesthetic elegance of the EV3 program.

15.3.3 Participants

Four second year, male undergraduate intelligent systems students were purposively selected. The students studied Java programming in their regular course but were not instructed in the development of flowcharts or their associated semiotic symbols in any programming classes. The students also had prior experience at a beginner level with LEGO Mindstorms EV3 robot construction kits and the associated software in a prior project with the same researcher (Vallance and Goto 2015). They were therefore all familiar with the physical LEGO Mindstorms components such as the range of sensors and possible robot constructions. They were also familiar with the semiotic representations of programming within the EV3 software such as motors, loops, sensors, and read-write variables and how to wire and download the programs. However, they were not considered to have advanced knowledge of EV3 software, and so tasks could be designed that would challenge the students.

Over two semesters of 15 weeks per semester, the students met once a week for 2 hours. There was no time limit on completing each task, and the more challenging tasks carried over from 1 week to the next. The emphasis was on successful task completion and reflection on the procedures needed to undertake the task (i.e., solve a problem). To promote constructive conversations among the four students, it was decided to request one collaboratively developed solution, both in construction of a robot and its associated EV3 program solution, and one post-hoc flowchart drawn together for each task immediately after the task was successfully completed.

15.3.4 Instrument

A post-task A3-sized paper worksheet was used to collect students' reflections (see Figs. 15.3, 15.4, 15.5, 15.6, 15.7, and 15.8, for examples). The worksheet aimed to determine evidence of the Computational Thinking concepts: generalization, modularity, abstraction, decomposition, and algorithmic logic.

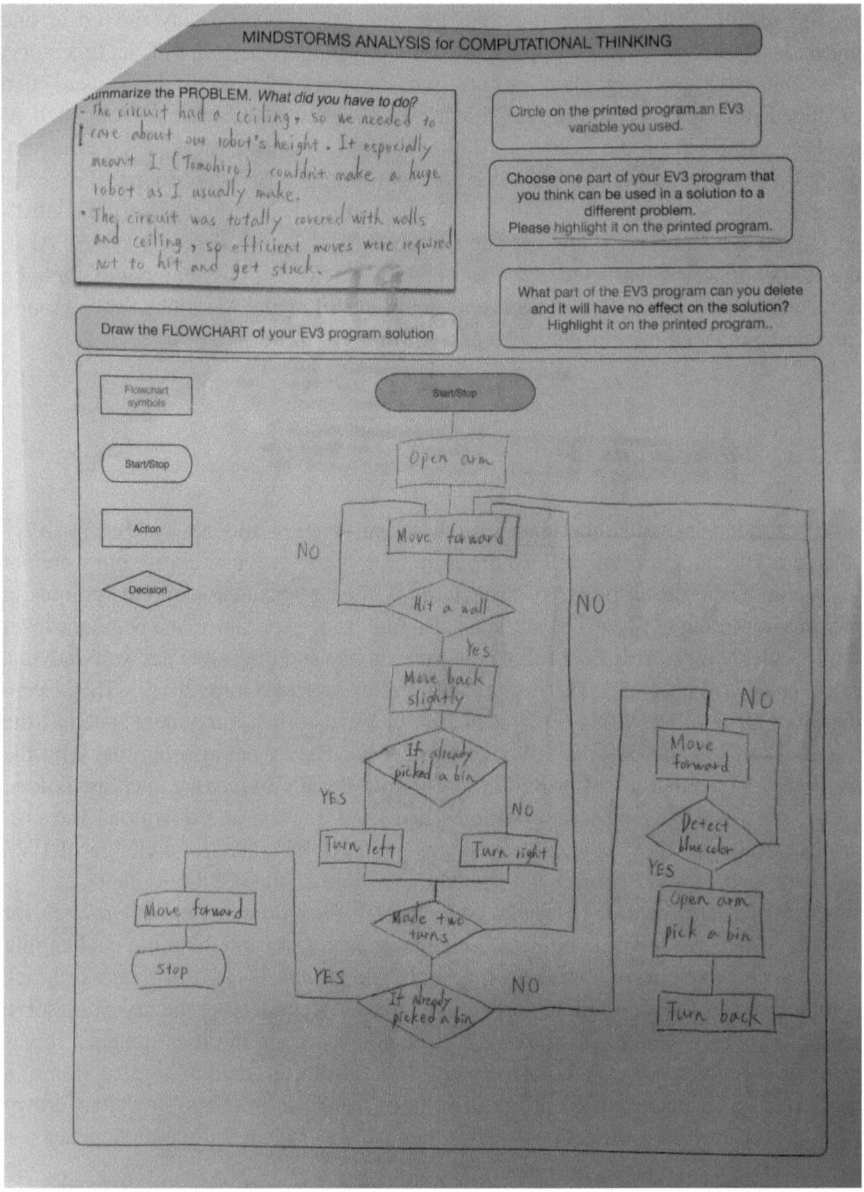

Fig. 15.3 Flowchart

15.3.4.1 Generalization

Generalization is being able to transfer one problem-solving process to another problem or a variety of problems. For the development of Computational Thinking

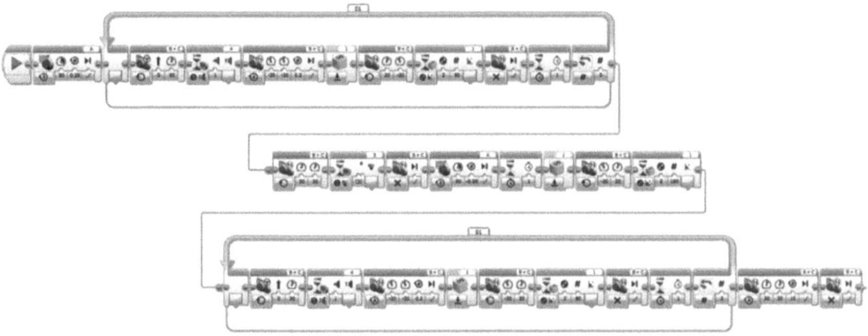

Fig. 15.4 EV3 program

in our context, there needed to be a recognition that an existing EV3 program solution, or parts of it, can be adopted in another, different problem. The most common transfer of EV3 program blocks from one solution to another solution would be the read-write variable block. Therefore, if students used a variable in their solution, then it is anticipated they could transfer that variable to another solution at a later task. To determine such recognition, we simply asked students to highlight an EV3 program block or series of blocks that they would consider in a future task. In the post-task discussion with the researcher, the students indicated whether they were able to either highlight the block/s (O) or not (X) on their post-task worksheet.

15.3.4.2 Modularity

Modularity is the development of autonomous processes that encapsulate a set of often used commands to perform a specific function and might be used in the same or a different problem. For the development of Computational Thinking, there needed to be a recognition of autonomous sections of code that could be copied and later used for the same or a future task problem. To determine such recognition, we simply asked students to highlight an EV3 series of blocks that they could be considered operable in a future task. In the post-task discussion with the researcher, the students indicated whether they were able to either highlight the block/s (O) or not (X) on their post-task worksheet.

15.3.4.3 Abstraction

Abstraction is the process of creating something simple from something complicated by leaving out the irrelevant details and also finding relevant patterns. For the development of Computational Thinking, students needed to be able to separate important information from redundant information. To determine whether students could recognize redundancy in the EV3 solutions, we asked them what part of the

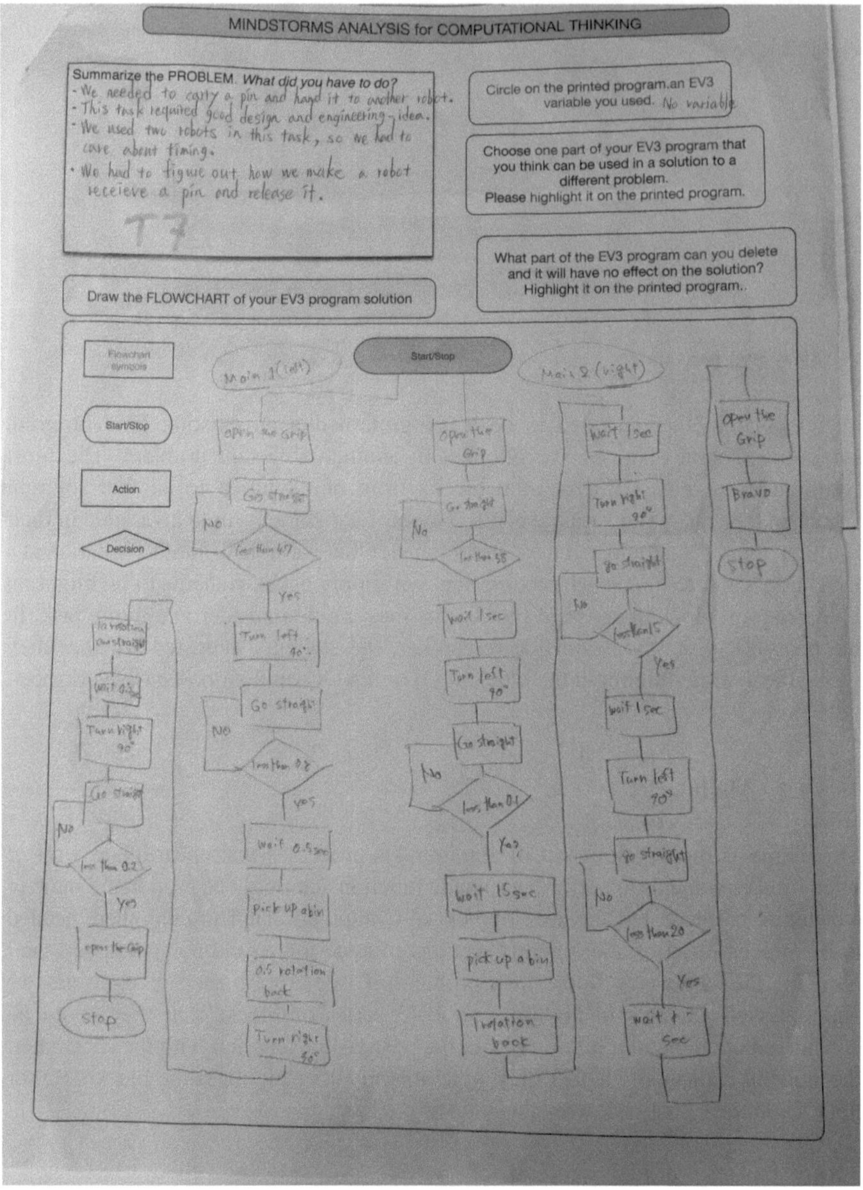

Fig. 15.5 Flowchart

EV3 program (e.g., a block, a loop, a variable) could be deleted that would have no effect on the current solution. In the post-task discussion with the researcher, the students indicated whether they were able to either highlight the block/s (O) or not (X) on their post-task worksheet.

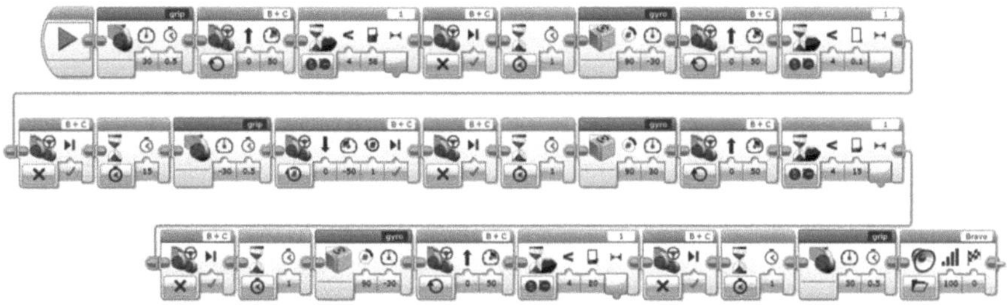

Fig. 15.6 EV3 program

15.3.4.4 Decomposition

Decomposition is the process of breaking problems down into smaller parts so that the problem may be more easily solved. For the development of Computational Thinking, students needed to show a problem being broken into smaller parts (i.e., micro problems). To determine students' ability to break up the overall task problem into smaller components, we asked them to write about the problem in detail on the post-task worksheet. In the post-task discussion with the researcher, the students indicated whether they were able to either break up the overall problem into smaller parts (O) or not (X) on their post-task worksheet.

15.3.4.5 Algorithmic Logic

Algorithmic logic is the practice of writing step-by-step, specific, and unambiguous instructions for carrying out a process. For the development of Computational Thinking, there needed to be explicit wording of the steps in the algorithm and, where applicable, an effort to find the most effective algorithm. The algorithm in this project was the development of a semiotic flowchart of the student's procedure in solving each task problem. A metric for determining the extent of algorithmic logic was initially considered. Vallance et al. (2015) quantified Mindstorms EV3 program solutions as a metric entitled robot task complexity (RTC). This was a numerical value derived from summating the EV3 program's semiotic blocks and associated variables used in each solution. In their research, this metric was then compared with a circuit task complexity (CTC) metric derived from the navigation, maneuvers, and obstacles represented by each problem (cf., Barker and Ansorge 2007; Vallance et al. 2015).

However, the flowcharts developed in this project take the concept of robot task complexity further by charting representations of students' reflective thinking processes of the procedures they utilized to complete each task problem. Therefore, assigning numerical values to a flowchart would not make sense. It is more prudent to seek an interpretation of the flowcharts. To do this the researcher can compare the flowchart to the EV3 program solution. As the researcher "walks through" the

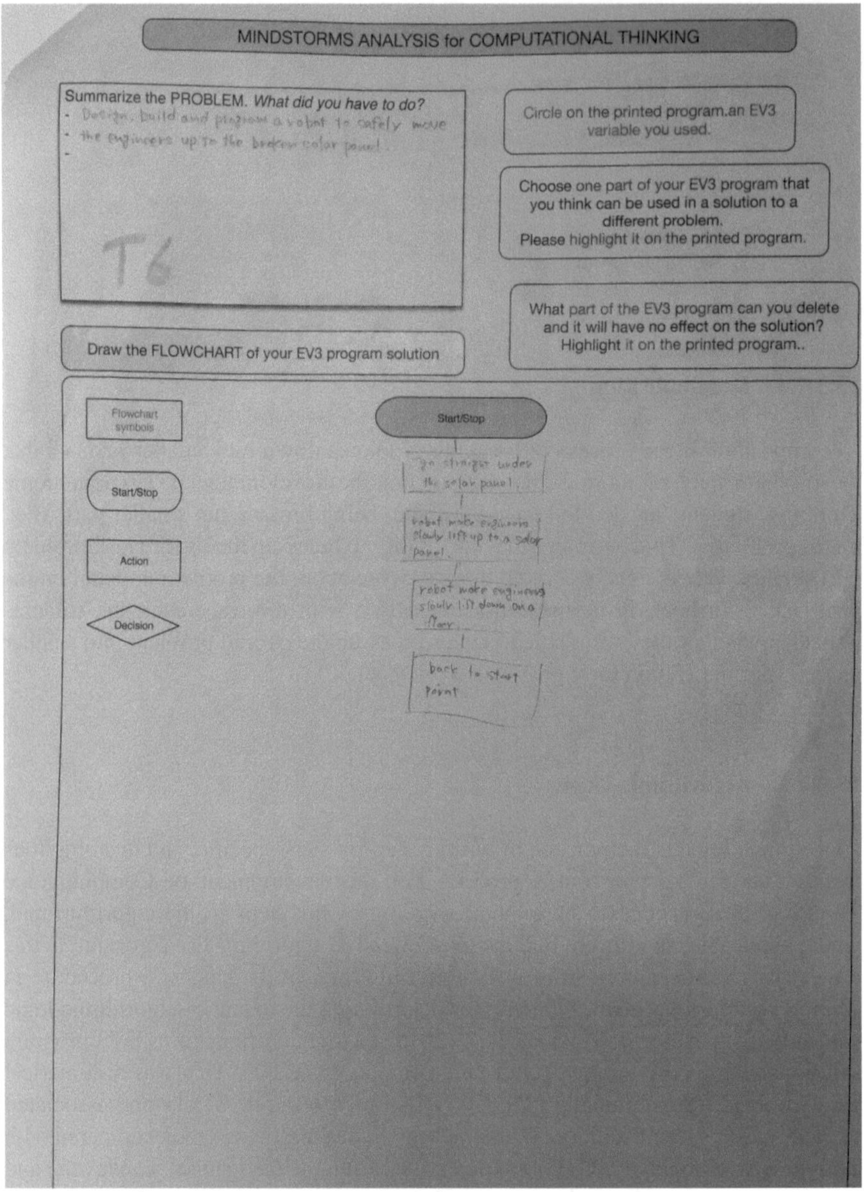

Fig. 15.7 Flowchart

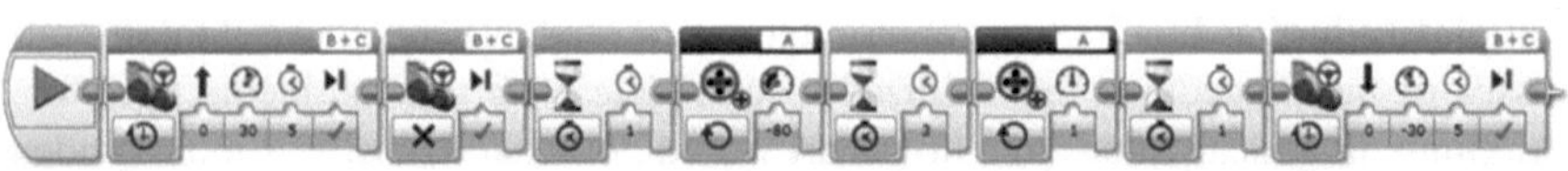

Fig. 15.8 EV3 program

Table 15.1 Estimation of flowchart and robot solution

O	Recognizable. The flowchart is a clear indication of the EV3 program solution	See example 1
Δ	Partially recognizable. The flowchart partially illustrates the EV3 program solution, but some additional work is required	See example 2
X	Not recognizable. The flowchart does not clearly demonstrate the EV3 program solution	See example 3

student's algorithmic thinking process, the researcher can attempt to simultaneously follow the semiotic symbols of the EV3 program solution. The aim is to decide how closely the flowchart can be used to determine the programmed robot solution (see Table 15.1). To explain this, three examples of algorithmic logic from our data are shown below.

Example 1 In this task the student's flowchart is recognizable when compared to the EV3 program solution.

Summary The flowchart clearly illustrates the students' procedures. By reading the flowchart, one can simultaneously follow the process within the EV3 program.

Example 2 In this task the student's flowchart is only partially recognizable when compared to the EV3 program solution.

Summary The students' flowchart covers two robots which worked in sequence. The EV3 program contained customized MyBlock processes. This made it difficult to coordinate the flowchart with the EV3 program. To understand the EV3 program, one has to open the software and view the MyBlock constructions.

Example 3 In this task the student's flowchart is unrecognizable when compared to the EV3 program solution.

Summary The flowchart was minimal and lacked the detail expected (and previously drawn). The EV3 program appears straightforward but contains a customized MyBlock. The combination of flowchart and EV3 program was confusing.

15.4 Results

This section will explain the data collected from the four participants. Table 15.2 shows the ten task problems to solve; task T10 is text only so is added below. Table 15.3 illustrates the combined Computational Thinking data from the ten tasks. Our results cannot be generalized due to the limited number of students involved. However, it is an attempt to inform practitioners how Computational Thinking can be recognized, illustrated, and analyzed.

Task 10: Build two basic robots to demonstrate the use of all the EV3 sensors. Then run the two robots together to demonstrate how the sensors operate.

Table 15.2 Tasks

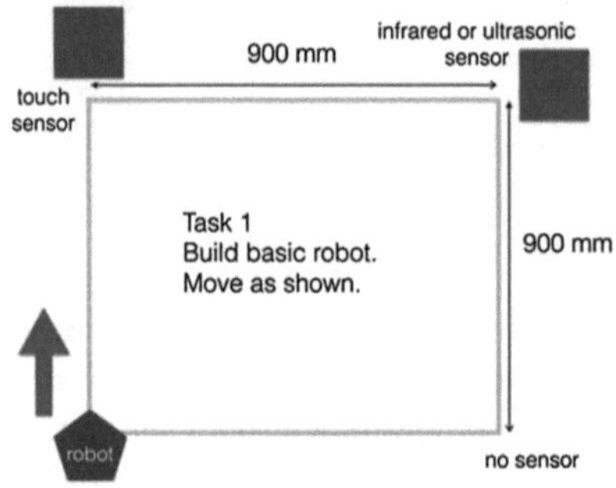 Task T1	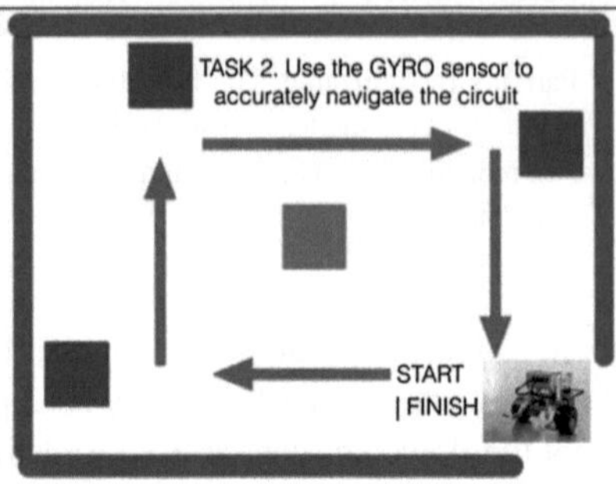 Task T2
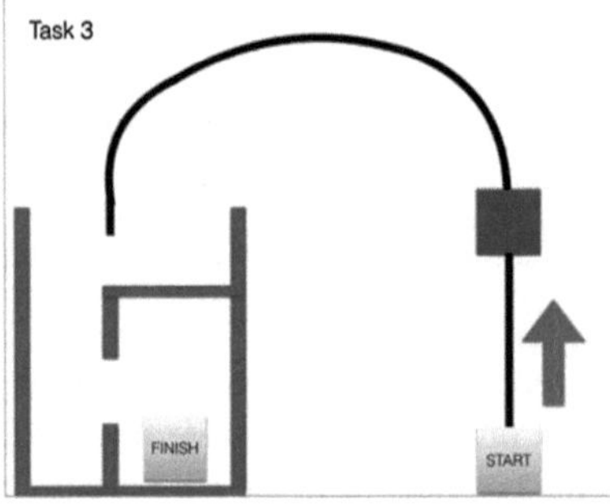 Task T3	 Task T4
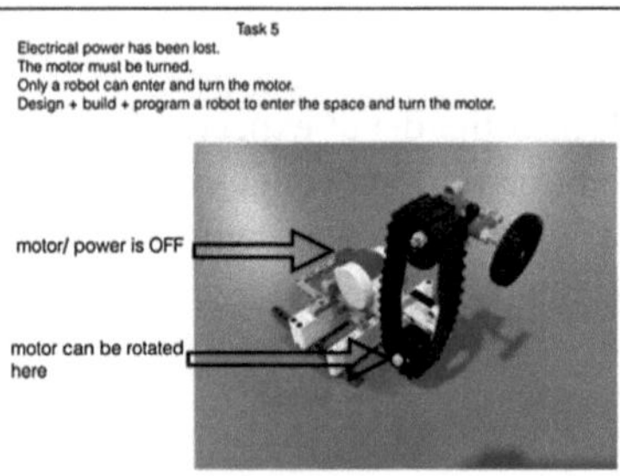 Task T5	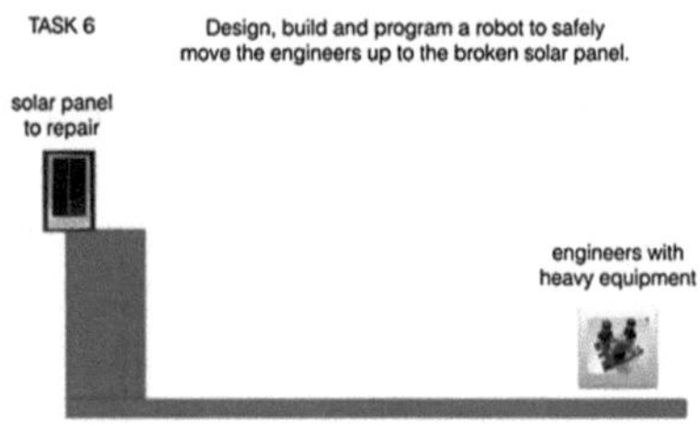 Task T6

(continued)

Table 15.2 (continued)

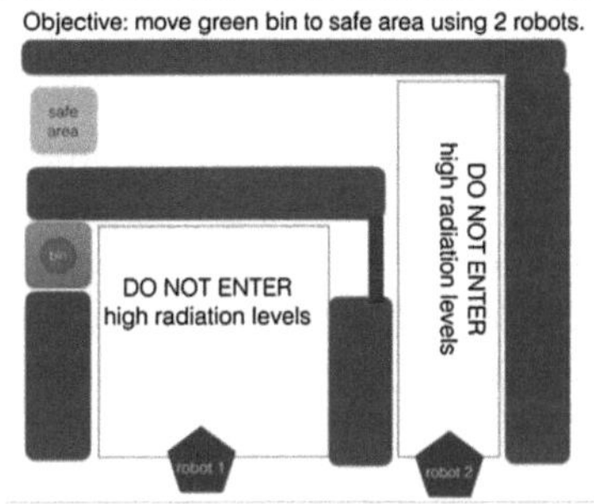

Task T7

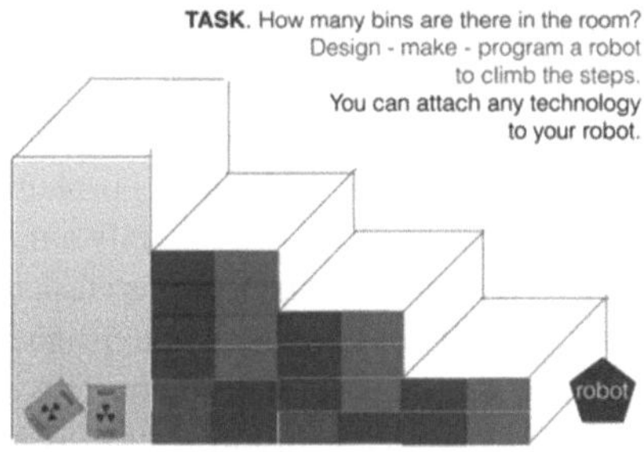

Task T8

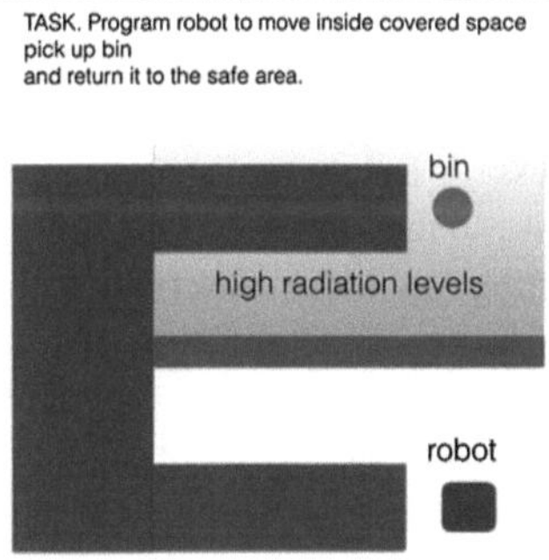

Task T9

Table 15.3 Computational thinking data

Task	Generalization	Modularity	Abstraction	Decomposition	Algorithmic logic
T1	O	O	O	△	O
T2	O	O	O	O	△
T3	X	O	O	O	O
T4	X	O	X	△	△
T5	X	X	X	△	△
T6	X	△	△	△	△
T7	X	O	O	O	O
T8	△	X	X	O	O
T9	O	O	X	O	O
T10	X	O	O	O	O

15.4.1 Generalization

Generalization was the most difficult concept for students to grasp and recognize in their solutions. In the initial tasks T1 and T2, the students were able to consider their solution (the designed robot and EV3 program components) and solution process in another context such as the next problem. This was due to the tasks being quite easy for the students as they were mainly a reminder of a previous EV3 robot project (cf., Vallance & Goto 2015). However, tasks T3 to T8 proved more challenging, and the students could not specifically envisage how their problem-solving process could be applied to a more general problem. This may be the fault of the instructor as he failed to use pedagogical techniques such as elicitation to draw out any ideas of generalization directly from the students during the post-task group interviews. Over time though, as students gained experience in more effective problem-solving processes, the concept of generalization was partially understood (as revealed in task T9). However, it was a surprise that students did not indicate positively in Task 10 as this task involved use of all sensors: surely students would be able to generalize the use of sensors to another problem-solving process. A follow-up interview revealed that students knew that task T10 was the final task, so they believed that they did not need to consider any future problems. Reflecting on these observations, it has become apparent that more time is required to discuss generalization of problem-solving to other contexts, to elicit explicit additional examples from the students, and then to further reflect by checking understanding either via instructor-student question and answers or student-to-student peer reviewing.

15.4.2 Modularity

Modularity is quite similar to generalization but operates at a more specific, deeper level in the problem-solving process. Modularity in computing relates to functions and is the recognition of sections of code that can be used in different parts of the same, or other, solution. Students in this project were asked to indicate sections of their EV3 program solutions that could be used later. Modularity recognition proved problematic for tasks T5, T6, and T8. However, these tasks did not involve robot navigation but activating external auxiliary machinery (T5, robot to manually turn a motor; T6, robot to lift engineers up to a solar panel; T8, robot to climb steps and photograph damaged radioactive bins). It is posited that the EV3 program solutions for tasks T5, T6, and T8 were very specific to their designed robot and the associated problem.

15.4.3 Abstraction

The challenge was to see if students could ignore unimportant details. One way to do this was for students to identify a part of their problem-solving process and/or an EV3 program block that could, after reflection, be deleted and have no detrimental impact on their solution. Students were able to look at their solutions and take out parts of the solutions in five of the ten tasks. Common to all five recognized tasks was that they involved navigation and maneuvering. Tasks T1, T2, T3, and T7 challenged students to navigate obstacles in predetermined circuits. Task T10 involved two EV3 robots synchronizing with each other but still required the robots to navigate a circuit. The other five tasks included the use of additional physical components such as a solar panel and a fan. Students considered the programs of these five tasks to be specific to the given challenge and concluded that there was no opportunity within these solutions to delete any section of their problem-solving process or EV3 program blocks.

15.4.4 Decomposition

Decomposition involves breaking the problem-solving process into smaller parts. In a computer program, smaller parts can be easier to understand and later maintain (such as debugging a problematic program). Tasks T4, T5, and T6 involving the mechanisms proved partially problematic to itemize. On the whole though, students could adequately decompose their problem-solving processes. This was due to prior programming experiences acquainting the students with the process of breaking code into smaller chunks.

15.4.5 Algorithmic Logic

Immediately after each task was completed, the students illustrated their workflow as a flowchart (see Figs. 15.3, 15.4, 15.5, 15.6, 15.7, and 15.8 above). Rectangular boxes, for instance, represent processing activities, while diamond boxes illustrate decisions. It was posited that the flowchart would correspond to the resultant EV3 program solutions. Flowcharts for tasks T1, T3, T7, T8, T9, and T10 were indeed recognizable when compared to the EV3 programs. However, Table 15.3 shows that the flowcharts that attempted to explain solutions involving mechanisms (T4, T5, and T6) were partially recognizable; i.e., it was difficult to distinguish the students' flowchart thinking process with the EV3 program procedure. In these tasks the students had to consider the actions of sensors and data and therefore found it challenging to integrate into their logical workflow "thinking" process.

15.5 Discussion

If we accept that the previously mentioned tasks and flowcharts have the potential to promote and develop students' problem-solving processes within a particular context, then two central questions arise concerning the aptness and capability of Computational Thinking as an instructional approach within and, most certainly, beyond STEM. These are: (1) *How* can Computational Thinking transform teaching and learning in classrooms? (2) *What* are the implications of adopting Computational Thinking for teachers, learners, and learning resources as a guiding pedagogy? In this discussion, we address these broad issues in two main, yet complementary, directions: task design and pedagogy.

15.5.1 Task Design

While we fully accept that instructional interactions between teachers, learners, and learning resources are multiple, simultaneous, and continuous, it is useful (both descriptively and analytically) to distinguish between two types of "work" in classrooms: activities and tasks. In our viewpoint and understanding, the activities that students do in classrooms usually involve low cognitive demand, such as quiet seatwork, for example, filling in a worksheet or practicing test questions (Towndrow 2007). In contrast, tasks are larger schemes designed to lead learners to the achievement of specified outcomes (Seedhouse 2005). According to Blumenfeld et al. (1987), instructional tasks are the organizational backbone of classroom interactions in the crucial sense that teachers combine various task elements together to shape or frame how students learn.

When seen academically, tasks are the key ways in which students experience a particular curriculum (Doyle 1983). From this perspective, tasks determine not only what content students learn but also how they think about, develop, make sense of, and apply the content knowledge they encounter. Consequently, whereas some tasks involve students at a surface level, others engage them more deeply by requiring the interpretation, flexibility, shepherding of resources, and construction of meanings. These are areas where Computational Thinking has a major part to play so long as the tasks (as we have characterized them so far) involved are designed aptly and well by teachers (and students).

For us, digital tasks (whatever their specific content and objectives) require teachers' design (Towndrow 2005). This is similar but not identical to the way in which an architect might devise a blueprint for a building or structure (cf., Wiggins and McTighe 2005). By way of illustration, here are four types of task structure for a teacher's consideration (adapted from Towndrow and Vallance 2004):

- *Single output/outcome, single strategy***.** This is where there is a single known or acceptable answer to a question or problem and there is only one way to reach it. For example, find the length of the unknown side of a right-angled triangle.

- *Single output/outcome, multiple strategies***.** Given a single objective, there is more than one known or acceptable way of achieving it, for example, losing weight or saving money.
- *Multiple outputs/outcomes, single strategy***.** There is only way to achieve more than one outcome, for example, using a data set derived from standard formulae and accepted procedures, to produce tables, graphs, and charts to address more than one research question.
- *Multiple outputs/outcomes, multiple strategies***.** Tasks of this kind are highly flexible and usually involve solving ill-structured (*c.f.,* Jonassen 2000) or "wicked" problems (Rith and Dubberly 2007).

This learning task classification has several uses in planning and/or interpreting tasks. In our case examples above, some of the work required the students to enact tasks where there was a single or fixed aim requiring an EV3 programming solution (e.g., navigate a closed space, negotiate hurdles, or move an object to a predetermined destination). Additionally, in some cases, there was potentially more than one acceptable way or method of reaching an end goal (e.g., T4, T5, T6) although this may not have been immediately obvious to the students at the time of enactment. However, beyond the confines of task completion in a specific instance, the scope of Computational Thinking is potentially *transcendent*. This is particularly the case with generalization and abstraction where learners may want or need to think beyond their immediate circumstances toward other scenarios and outcomes.

With this information, we are now able to make two points with relevance to the *hows* of Computational Thinking. First, we maintain that it requires a vision and understanding of classroom work where tasks are not simply agendas or checklists of things to get done. Rather, tasks—when designed appropriately—can be the carriers or vehicles of serious and demanding intellectual context and a basis for engagement in multiple ways and directions.

Second, tasks can be the amplifiers of student agency. Take again the issue of trying to find ways to express generalization and abstraction explicitly. One possible method could be for students to engage with more complicated or complex problems and then discuss them alongside deeper task enactment reflections. Another angle might be to open or ease the scope of the robot programming tasks to allow for more than one output. This would allow students to generate multiple ideas, make more decisions, and chart further courses depending on their *own* intentions or purposes. Of course, ill-structured problem-solving is harder to evaluate and assess (because there is no set or accepted answer to the issues involved), but many of the situations we have to deal with in life outside of school are just like this.

Does Computational Thinking have a role to play? Perhaps, but much depends on how we view the purposes of teaching and learning more generally, and this leads us to a short consideration of our second discussion question concerning pedagogy.

15.5.2 Toward a Pedagogy of Computational Thinking

For those researchers and educators who tend to think of teaching and learning in mostly pragmatic terms (what works, curriculum coverage, and strict accountability), "pedagogy" is synonymous with what teachers do—teaching. Indeed, when compliance to certain structures, policy directives, or ideologies is valued by the members of educational communities, then pedagogy—in terms now of methods or strategies in a teacher's toolkit—is often listed as effective and efficient means toward an end. However, we believe this view of pedagogy is limited and limiting given the scope and intentions of Computational Thinking.

According to Alexander (2008), pedagogy is not something that can be prescribed or even described in official educational documents such as curricula or policy briefs. In his view:

> … pedagogy is the act of teaching together with its attendant discourse of educational theories, values, evidence and justifications. It is what one needs to know, and the skills one needs to command, in order to make and justify the many different kinds of decision of which teaching is constituted. (*ibid.*, p.47)

In this definition, Alexander is careful to note that pedagogy "is" the act of teaching but there is more to it than that. In fact, we can understand pedagogy as a case or argument for itself in terms of the values, evidence, and justifications we (as teachers, curriculum developers, educational policymakers, etc.) create and use that necessarily go beyond the classroom, beyond the school, and beyond governments or boards of education toward society itself. Pedagogy, in a crucial sense, characterizes who we are as people and what it is we aspire to in life. It also has much to do with how we think we can best achieve the "tasks" we set ourselves and which tasks we think, know, or believe are important and meaningful in our lives (e.g., free speech or self-determination). These are matters to be discussed, debated, and deliberated on constantly. In other words, pedagogy is dynamic, shifting, and necessarily contestable. What counts today pedagogically may not hold tomorrow but at least those who are informed would know why and how things have changed over time.

It is also right to view Computational Thinking pedagogically in order to assess its full potential in the classroom and learning more generally. We do not wish to argue that all problems or issues in life or the school curriculum can be solved computationally. That would be untenable and perhaps even undesirable. Rather, as a system of thought and disposition, Computational Thinking offers a set of principles, practices, and procedures that potentially foster clarity, purposefulness, empowerment, reflection, understanding, critique, and creativity. And when combined with other ways of viewing the world, those who can think and act computationally, we believe, bring a unique and valuable perspective to the table.

What then would be some of the implications of adopting Computational Thinking for teachers, learners, and learning resources as a guiding pedagogy in schools and beyond? We begin with teachers.

While it is undeniable that factual and procedural knowledge are both important and necessary in school, and life, more broadly, there are certain types of task design that are more suited to the achievement of open-ended, flexible, and alternative outcomes. Teachers, then, who want to design tasks involving or requiring Computational Thinking, would need to acknowledge that in certain circumstances there might be more than one way of achieving a desired academic outcome as clearly illustrated in our case study tasks. This is not to advocate an "anything goes" mentality. Rather, tasks when viewed pedagogically require explanation and justification. This means that teachers not only need to know what to do and how to do things but also be prepared to "explain" why they have chosen their particular modes of operation in specific circumstances. Such explanations, we argue, would be part of a pedagogy of Computational Thinking in the making, the circumstance where teachers think computationally about their work and its importance as a mode of instruction and set of principles for life.

Importantly, for teachers, Computational Thinking also has the potential for the design of a series of tasks that build on each other conceptually and practically, as demonstrated in our case study. Alexander (2008) stresses the importance of *cumulation*, that is, "knowing what has gone before, learning from it, evaluating it, and building on it" (p. 68). Pedagogically, it is vital for both teachers and students to know where they started in a task, where they are now, and what it is that they are trying to do. The robot tasks and semiotic representations of developing solutions in our case study clearly illustrated these conditions. It is only under the circumstances of working toward task completion that teachers' feedback and scaffolding make sense.

When viewed operationally and pragmatically, learners are often cast at the passive receiving end of teaching and instruction. Their obligation is to complete work that has been set for purposes they may not always be aware of. Our case examples above illustrate a different experiential perspective. The robot programming tasks and the reflection activities that followed were designed to make explicit for students the thinking they used to complete the work set. If this is a worthy aim (and we believe it is), then students have an essential part to play in enacting (not just doing) learning that involves Computational Thinking. As a result, they should be proactive and prepared to map, for example, their lines of thinking *even when not asked to do so by the teacher.* They should also be willing (and able, after a while) to suggest different ways in which tasks can be designed and redesigned in their enactment. For example, if a student saw or realized that a task specification was faulty or that there might be a different (and more elegant) way of achieving a previously set goal, then they should consider it perfectly reasonable to suggest a task modification or solution that fits their own learning aims and ambitions in a better way. By this, teachers and learners could enter into pedagogic partnerships that are mutually informative and collaborative (cf., Prensky 2010).

Finally, while learning or instructional resources (e.g., books, LEGO blocks, sensors, and motors) are inanimate and therefore lack inherent purposes of their own (we might argue), the ways in which teachers and learners approach, identify, and recruit them might be open to review with Computational Thinking or any kind of subject-based instruction. We subscribe to the view that teaching and learning can be supported in different ways for varying purposes, at many levels and with different consequences. For example, in an edited book chapter, Cohen et al. (2002) mention three kinds of interdependent instructional resources that shape and influence classroom interactions. These are:

- *Conventional*—Teachers' formal qualifications, books, facilities, expenditures, class size, time, libraries, laboratories
- *Personal*—Teachers' and students' knowledge and skills (these influence perceptions and uses of conventional resources)
- *Environmental*—Leadership, academic norms, and institutional structures (these also affect whether and how teachers and students use conventional resources)

In our opinion, this conception of instructional resources is expansive and demanding as it includes tangible and immaterial items. However, in the interests of supporting a wide range of computational skills and dispositions, personal and environmental factors and considerations cannot be ignored. Rather, they should be cultivated, promoted, and used as widely as possible to ensure comprehensiveness and flexibility.

15.6 Conclusion

The chapter has been a descriptive account of undergraduate university students designing, building, and programming LEGO Mindstorms EV3 robots to solve problems set as tasks in an imaginary disaster scenario. It was found that students were able to express recognition of the Computational Thinking concepts of modularity, decomposition, and algorithmic logic but had difficulty expressing explicit recognition of generalization and abstraction. This finding, as a result of our qualitative interpretations of the students' flowcharts, supports the quantitative data outcomes of Atmatzidou and Demetriadis' (2014) research.

Demonstrating that our tasks had merit, we then critiqued our Computational Thinking observations in terms of implications for teaching and learning. To do so, we addressed two main issues, namely, task design and pedagogy.

We proposed that a task is considered to be a structure that intends to lead learners to the achievement of a specific objective. Tasks determine not only what content students learn but also how they think about, develop, make sense of, and apply the knowledge they encounter. Four types of task structure were highlighted: single output/outcome, single strategy; single output/outcome, multiple strategies; multiple outputs/outcomes, single strategy; and multiple outputs/outcomes, multiple strategies. Consequently, explicitly incorporating Computational Thinking enables learners and instructors to go beyond specified task completion criteria and consider other scenarios and outcomes. However, such "generalization" and "abstraction" were considered most challenging to the students in our case study.

This led us to consider a pedagogy of Computational Thinking. It was argued that although pedagogy is often considered solely an act of teaching, its performance embeds values, evidence, justifications, and characteristics of who we are. For instance, as mentioned above, teachers designing tasks requiring Computational Thinking strategies would need to acknowledge that in certain circumstances there might be more than one way of achieving a desired academic outcome. This means that teachers not only need to know what to do and how to do things but also be prepared to "explain" why they have chosen their particular modes of operation in specific circumstances. This becomes a part of the pedagogy of Computational Thinking in the making, where teachers think computationally about their work and its importance as a mode of instruction and set of principles for life. Learners too have an important role in reciprocating explanations as they too need to explain their thinking, as illustrated in the reflection maps in our case study. As teachers and learners enter this constructive partnership, improved task specifications may be suggested, existing solutions may be modified, and alternate interpretations may be designed.

Finally, through our case study implementations, interactions, and reflections, we have proposed that a transformed pedagogy can be supported by tasks that engage learners in Computational Thinking. Our students not only illustrated their "thinking" with semiotic flowcharts but began to "think about thinking" as they reflected not only on their task solutions but also alternative ideas and the transfer of their ideas to other problems. We believe this is what Seymour Papert (1980) had in mind when he talked about powerful ideas:

> And for me what is the most important in this is that through these experiences these children would be serving their apprenticeships as epistemologists, that is to say learning to think articulately about thinking. (p.27)

Acknowledgments The research is supported by JAIST *kakenhi* [grant number 15K01080]. Many thanks to students in the iVERG lab at Future University Hakodate, Japan.

References

Alexander, R. (2008). *Essays on pedagogy*. London: Routledge.

Atmatzidou, S., & Demetriadis, S. (2014, July 18). *How to support students' computational thinking skills in educational robotics activities*. In Proceedings of 4th International workshop teaching robotics, teaching with robotics & 5th International conference robotics in education, Padova, Italy.

Barker, S. B., & Ansorge, J. (2007). Robotics as means to increase achievement scores in an informal learning environment. *Journal of Research in Technology and Education, 39*(3), 229–243.

Barr, V., & Stephenson, C. (2011). Bringing computational thinking to K-12: What is involved and what is the role of the computer science education community? *ACM Inroads, 2*(1), 48–54.

Basu, S., Biswas, G., Sengupta, P., Dickes, A., Kinnebrew, J. S., & Clark, D. (2016). Identifying middle school students' challenges in computational thinking-based science learning. *Research and Practice in Technology Enhanced Learning, 11*, 13.

Berlin, I. (1974). *The divorce between the sciences and the humanities*. The second Tykociner memorial lecture, University of Illinois. Available from http://berlin.wolf.ox.ac.uk/published_works/ac/divorce.pdf. Accessed 28 January, 2016.

Blumenfeld, P. C., Mergendollar, J., & Swarthout, D. (1987). Task as a heuristic for understanding student learning and motivation. *Journal of Curriculum Studies, 19*, 135–148.

Brennan, K., & Resnick, M. (2012). *Using artifact-based interviews to study the development of computational thinking in interactive media design*. Paper presented at annual American Educational Research Association meeting, Vancouver, Canada.

Catlin, D., & Blamires, M. (2010) *The principles of educational robotic applications (ERA)*. In Constructionism 2010: Constructionist approaches to creative learning, thinking and education: Lessons for the 21st century, proceedings for constructionism 2010: The 12th EuroLogo conference, Paris.

Cohen, D. K., Raudenbush, S. W., & Ball, D. L. (2002). Resources, instruction, and research. In F. Mosteller & R. Boruch (Eds.), *Evidence matters: Randomized trials in education research* (pp. 80–119). Washington, DC: Brookings Institution Press.

Computational Thinking for All. (2017, February 27). *Symposium for information education*. The University of Tokyo Graduate School of Information Science and Technology.

Dede, C., Mishra, P. and Voogt, J. (2013). *Working group 6: Advancing computational thinking in 21st century learning*. Available from http://www.curtin.edu.au/edusummit/local/docs/Advancing_computational_thinking_in_21st_century_learning.pdf. Accessed 28 February, 2015.

Doyle, W. (1983). Academic work. *Educational Researcher, 53*(2), 159–199.

Jonassen, D. H. (2000). Toward a design theory of problem solving. *Educational Technology Research & Development, 48*(4), 63–83.

Lew, M. W., Horton, T. B., & Sherriff, M. S. (2010). *Using LEGO MINDSTORMS NXT and LEJOS in an Advanced Software Engineering Course*. In 23rd IEEE Conference on Software Engineering Education and Training (CSEE&T), Pittsburgh, pp. 121–128.

Liukas, L. (2015). *Hello ruby. Adventures in coding*. Indiana: Macmillan.

Lui, A. K., Ng, S. C., Cheung, H. Y., & Gurung, P. (2010). Facilitating independent leaning with LEGO Mindstorms robots. *ACM Inroads, 1*(4), 49–53.

NAACE. (2014). *Computing in the national curriculum*. Available from http://www.computingatschool.org.uk/data/uploads/cas_secondary.pdf Accessed October 20, 2014.

Nakashima, H. (2015, June). Computational thinking. *Japanese Translation*, *56*(6), Available from https://www.cs.cmu.edu/afs/cs/usr/wing/www/ct-japanese.pdf. Accessed December 29, 2015.

Papert, S. (1980). *Mindstorms: Children, computers, and powerful ideas*. New York: Basic Books.

Popelka, M., & Nožička, J. (2014). Lego Mindstorms as a simulation of robotic systems. *International Journal of Computer, Control, Quantum and Information Engineering, 8*(7).

Prensky, M. (2010). *Teaching digital natives: Partnering for real learning*. Thousand Oaks: Corwin.

Rith, C., & Dubberly, H. (2007). Why Horst W. J. Rittel matters. *Design Issues, 23*(1), 72–74.

Schrader, P. G. (2008). Learning in technology: Re-conceptualizing immersive environments. *AACE Journal, 16*(4), 457–475.

Seedhouse, P. (2005). "Task" as research construct. *Language Learning, 55*(3), 533–570.

The Royal Society. (2012). *Shut down or restart? The way forward for computing in UK schools*. London: The Royal Society.

Toikkanan, T. (2015). *Coding in school: Finland takes lead in Europe*. Available from http://legroup.aalto.fi/2015/11/coding-in-school-finland-takes-lead-in-europe/ Accessed January 31, 2017.

Towndrow, P. A. (2005). Teachers as digital task designers: An agenda for research and professional development. *Journal of Curriculum Studies, 37*(5), 507–524.

Towndrow, P. A. (2007). *Task design, implementation and assessment: Integrating information and communication technology in English language teaching and learning*. Singapore: McGraw-Hill.

Towndrow, P. A., & Vallance, M. (2004). *Using IT in the language classroom: A guide for teachers and students in Asia* (3rd ed.). Singapore: Longman.

Turner, S. J., & Hill, G. (2008). Robotics within the teaching of problem-solving. *ITALICS Innovations in Teaching and Learning in Information and Computer Sciences, 7*(108).

Vallance, M., & Goto, Y. (2015, September 20–24). *Learning by TKF to promote computational participation in Japanese education*. In Proceedings of the 43rd International conference on engineering pedagogy. World Engineering Education Forum. Florence, Italy.

Vallance, M., & Towndrow, P. A. (2016). Pedagogic transformation, student-directed design and computational thinking. *Pedagogies: An International Journal, 11*(3).

Vallance, M., Martin, S., & Naamani, C. (2015). A situation that we had never imagined: Post-Fukushima virtual collaborations for determining robot task metrics. *International Journal of Learning Technology, 10*(1), 30–49.

Wiggins, G., & McTighe, J. (2005). *Understanding by design* (2nd ed.). Alexandria: Association for Supervision and Curriculum Development.

Wing, J. M. (2006, March). Computational thinking. *Communications of the ACM, 49*(3).

Yadav, A., Zhou, N., Mayfield, C., Hambrusch, S., & Korb, J. T. (2011). *Introducing computational thinking in education courses*. In Proceedings of ACM Special Interest Group on Computer Science Education, Dallas, TX.

MIX
Papier aus verantwortungsvollen Quellen
Paper from responsible sources
FSC® C105338
FSC
www.fsc.org

If you have any concerns about our products,
you can contact us on
ProductSafety@springernature.com

In case Publisher is established outside the EU,
the EU authorized representative is:
Springer Nature Customer Service Center GmbH
Europaplatz 3, 69115 Heidelberg, Germany

Printed by Libri Plureos GmbH
in Hamburg, Germany